Remote Sensing Assessment of Sustainable Development in the Hindu Kush-Himalayan (HKH) Region

Shanlong Lu · Harrison Odion Ikhumhen · Xin Du ·
Tingting Lv · Jia Song · Xidong Chen

Remote Sensing Assessment of Sustainable Development in the Hindu Kush-Himalayan (HKH) Region

Springer

Shanlong Lu
International Research Center of Big Data
for Sustainable Development Goals (CBAS)
Aerospace Information Research Institute
Chinese Academy of Sciences
Beijing, China

Harrison Odion Ikhumhen
International Research Center of Big Data
for Sustainable Development Goals (CBAS)
Aerospace Information Research Institute
Chinese Academy of Sciences
Beijing, China

Xin Du
Aerospace Information Research Institute
Chinese Academy of Sciences
Beijing, China

Tingting Lv
Aerospace Information Research Institute
Chinese Academy of Sciences
Beijing, China

Jia Song
Institute of Geographic Sciences
and Natural Resources Research
Chinese Academy of Sciences
Beijing, China

Xidong Chen
The Division of Landscape Architecture
Future Urbanity and Sustainable
Environment (FUSE) Lab
Hong Kong, China

ISBN 978-981-96-8674-2 ISBN 978-981-96-8675-9 (eBook)
https://doi.org/10.1007/978-981-96-8675-9

This Springer imprint is published by the registered company Springer Nature Singapore Pte Ltd.
The registered company address is: 152 Beach Road, #21-01/04 Gateway East, Singapore 189721, Singapore

Foreword

The Hindu Kush-Himalayan (HKH) region, which spans eight countries—Afghanistan, Bangladesh, Bhutan, China, India, Myanmar, Nepal, and Pakistan—is a globally significant area known for its ecological diversity, freshwater resources, and rich world heritage. This region supplies water to billions of people and supports vital ecosystems. However, it faces growing challenges, including climate change, glacial retreat, unsustainable development, and resource scarcity, all of which threaten the livelihoods and resilience of its inhabitants.

This book, *Remote Sensing Assessment of Sustainable Development in the Hindu Kush-Himalayan (HKH) Region*, addresses critical issues through innovative research that integrates long-term satellite remote sensing data, advanced modeling techniques, and geospatial analyses. By examining key elements such as the impacts of climate change on agricultural sustainability, land use transitions, water resources, and ecosystem health, this work provides actionable insights into the dynamic interactions between human activities and environmental systems. The findings presented in this comprehensive assessment not only enhance our understanding of the region's vulnerabilities but also offer forward-looking strategies to mitigate risks, adapt to changing conditions, and promote sustainable development.

At its core, this research addresses critical scientific questions: How does climate change impact hydrological cycles in glacier-fed river basins? What are the implications for food and water security in downstream communities? How can remote sensing technologies be effectively utilized to monitor transboundary water resources? The study's innovative methodological framework employs high-resolution satellite imagery, machine learning algorithms, and model results to assess spatiotemporal changes across various Sustainable Development Goals (SDGs) indicators. These advancements are essential for bridging data gaps in the HKH region, enhancing evidence-based policymaking, and promoting regional collaboration.

Importantly, this work aligns with the SDGs, highlighting the essential need for comprehensive approaches to sustainability in mountain regions. The research outcomes are intended to assist decision-makers, scientists, and practitioners in formulating policies and strategies that balance economic growth, environmental conservation, and social equity.

As climate change continues to reshape our planet, the experiences of the HKH region provide valuable insights for global sustainability efforts. I hope that this work will inspire ongoing innovation and collaboration, driving progress toward a sustainable and equitable future for the HKH region and beyond.

Huadong Guo
Director-General
International Research Center of Big
Data for Sustainable Development
Goals
Member of the UN 10-Member Group
of TFM for SDG (2018–2021)
Beijing, China

Preface

The Hindu Kush-Himalayan (HKH) region holds immense ecological, hydrological, and cultural significance. However, this fragile ecosystem is confronted with unprecedented challenges due to climate change, population pressures, and unsustainable development practices. Acknowledging the urgent need to tackle these complex sustainability issues, this book presents a comprehensive study based on advanced remote sensing technologies and innovative analytical frameworks.

The primary objective of this work is to provide a robust scientific foundation for understanding and addressing the complex impacts of climate change on agricultural systems, water resources, and ecosystems in the HKH region. By utilizing satellite remote sensing data, machine learning models, and dynamic spatial-temporal analyses, we have developed innovative methodologies for monitoring and assessing key Sustainable Development Goals (SDGs) indicators. Our research spans land use patterns, changes in glacial and snow cover, hydrological variability, and vegetation dynamics, offering critical insights into the region's vulnerabilities and adaptive capacities.

One of the key technological advancements highlighted in this study is the integration of high-resolution satellite imagery with artificial intelligence models to map surface water distribution and crop productivity. Furthermore, the utilization of cloud-based platforms, such as Google Earth Engine, has enabled large-scale, real-time data processing, significantly improving the accuracy and applicability of our findings for policy development and regional planning.

This study presents several key findings. First, it demonstrates the accelerating retreat of glaciers and snowpacks, which has profound implications for downstream water availability and hydropower generation. Second, our analyses of agricultural land use and crop yield trends reveal increasing pressure on food production systems due to shifting precipitation and temperature patterns. Third, we highlight the cascading effects of ecosystem degradation on biodiversity and human livelihoods, emphasizing the necessity for integrated management strategies.

We believe these findings hold significant value for advancing both regional and global sustainable development. Policymakers, researchers, and practitioners will discover actionable recommendations to promote resilient water management, sustainable agricultural practices, and ecosystem conservation.

This book is the result of dedicated efforts by a multidisciplinary team of experts. We extend our deepest gratitude to all the contributors whose hard work and insights have made this publication possible. Special thanks go to the editorial team at Springer for their meticulous guidance and support throughout the editing and publication process. Finally, we sincerely hope that this book will serve as a valuable resource for all readers.

Dr. Shanlong Lu
Professor, International Coordinator of SDG 6
International Research Center of Big Data for Sustainable Development Goals (CBAS)
Aerospace Information Research Institute
Chinese Academy of Sciences
Beijing, China

Acknowledgements We gratefully acknowledge the financial support provided by the National Key R&D Program of China for the project titled "Remote Sensing Assessment and Capacity Building of Sustainable Development in the Hindu Kush-Himalayan Region" (2021YFE0117800).

We would like to express our sincere gratitude to the International Cooperation Center of the Ministry of Science and Technology of China, the Aerospace Information Research Institute of the Chinese Academy of Sciences (AIRCAS), the Institute of Geographic Sciences and Natural Resources Research of the Chinese Academy of Sciences (IGSNRR), the International Centre for Integrated Mountain Development (ICIMOD), and North China University of Water Resources and Electric Power (NCWU) for their invaluable management, logistical, and technical support throughout the project.

We extend our heartfelt appreciation to Mr. Birendra Bajracharya, Dr. Rajesh Bahadur Thapa, and Mr. Sudip Pradhan from ICIMOD; Prof. A. K. M. Saiful Islam and Prof. Tarekul Islam from the Bangladesh University of Engineering and Technology (BUET); and Asst. Prof. Binod Baniya from Tribhuvan University for their insightful contributions to the chapters and their thorough manuscript reviews. We are also grateful to Prof. Qiangzi Li and Prof. Zui Tao from AIRCAS, Prof. Xiang Zhou from the Institute of Electrical Engineering at the Chinese Academy of Sciences, and Prof. Yunqiang Zhu from IGSNRR for their expert guidance on project research and modifications to the book draft.

We would like to acknowledge the dedicated efforts of all project team members: Cong Du, Xinya Kuang, Hailong Tang, Mingyang Li, Chun Fang, Yong Wang, Xinru Li, Yuan Guo, Zijuan Zheng, Hongyan Wang, Yuan Zhang, Yunqi Shen, Jingbo Hu, Yong Dong, Haoxuan Hu, Shuguang Gong, Sifeng Yan, Yuxuan Huang, Houyu Liang, Mingjian Zhai, Yinhang Li, and Ruoxi Li from AIRCAS; Xiangbin Yan, Jiazheng Zhu, Luosheng Xia, and Chenyang He from IGSNRR; and Xin Wei, Shangfei Guo, and Xiaoquan Yang from NCWU. We are deeply grateful to them for their time, expertise, and enthusiasm in the data collection, processing, and validation that underpin this book. Their commitment and hard work have been the foundation of our achievements.

Without the generous support and collaboration of these institutions and individuals, this book would not have been possible. We extend our heartfelt gratitude to everyone involved.

Competing Interests The authors have no competing interests to declare that are relevant to the content of this manuscript.

Contents

List of Figures

List of Tables

Chapter 1
Introduction

The Hindu Kush Himalayan (HKH) region, often referred to as the "Third Pole," holds immense significance as a crucial ecological and hydrological system supporting the livelihoods of nearly two billion people. Encompassing eight countries—Afghanistan, Bangladesh, Bhutan, China, India, Myanmar, Nepal, and Pakistan—this region provides vital freshwater resources through its glacier-fed river systems. However, it faces increasing threats from climate change, glacial retreat, and human activities, necessitating a comprehensive understanding of its dynamic environmental and socio-economic systems.

1.1 Global Climate Change Trends

High alpine areas are recognized for sharing common features such as steep topography, cold climates, and geographical isolation. These regions, characterized by cryospheric elements like glaciers, snow cover, and permafrost, have significant global impacts, particularly on downstream water availability. In 2010, approximately 671 million people lived in high mountainous regions within 100 km of cryospheric zones (Jones & O'Neill, 2016) (Fig. 1.1). By 2050, this population is expected to reach between 736 and 844 million (Gao, 2020). Understanding climate change's influence on cryosphere services and related socio-economic processes is critical.

S. Lu et al., *Remote Sensing Assessment of Sustainable Development in the Hindu Kush-Himalayan (HKH) Region*, https://doi.org/10.1007/978-981-96-8675-9_1

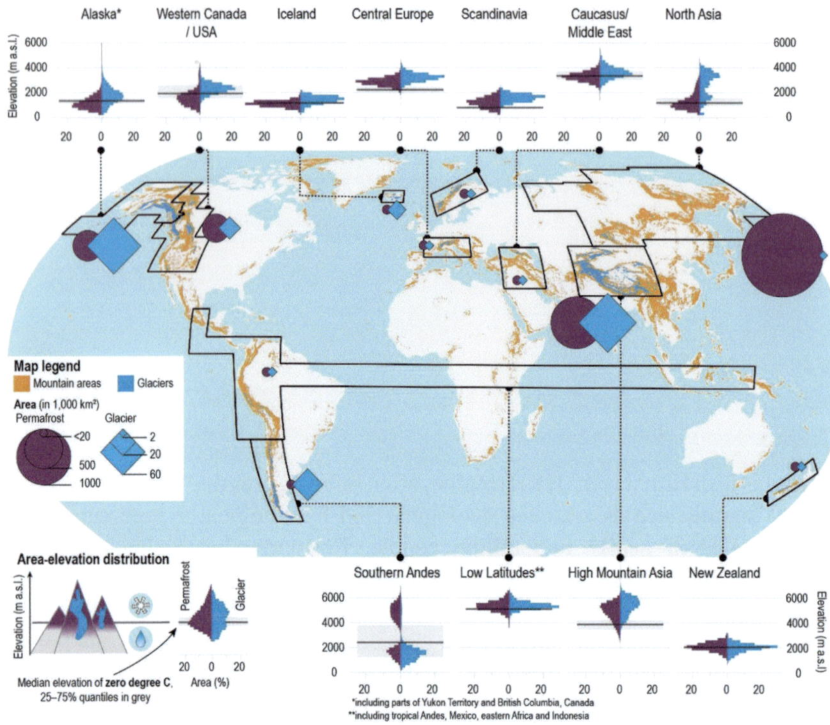

Fig. 1.1 Distribution of mountain areas (orange shading) and glaciers (blue), as well as regional summary statistics for glaciers and permafrost in the mountains (IPCC, 2022a, b)

1.1.1 Variations in Surface Air Temperature (SAT) Across Various Height Bands

Recent analyses highlight that mountainous region are warming at rates exceeding the global average. The IPCC (2018a, b) reports an SAT increase of 0.3 °C per decade in areas such as Western North America, the European Alps, and High Mountain Asia, surpassing the global warming rate of 0.2 °C per decade (IPCC, 2018a, b). Seasonal warming patterns vary: the European Alps experience amplified summer warming, while winter warming dominates the Tibetan Plateau (Zheng et al., 2022). Elevation-dependent warming (EDW) further amplifies temperature increases, with observed warming often exceeding 4000 m above sea level on the Tibetan Plateau (Gao et al., 2018). However, contradictory findings in other regions, such as the Antarctic Ice Sheet and Southern Himalaya, show variable warming across elevations (Xie et al., 2023).

Anthropogenic greenhouse gas emissions are the primary drivers of global warming, with SAT in 2011–2020 rising 1.1 °C above the pre-industrial levels

(Calvin et al., 2023). Increased emissions, land use changes, and aerosols exacerbate warming, with projections indicating continued temperature rise beyond mid-century, especially in high mountain regions.

1.1.2 Transitioning Precipitation Patterns in Mountain Areas

Persisting greenhouse gas emissions drive global temperature increases, intensifying evaporation and altering precipitation dynamics (IPCC, 2018a, b). Warmer temperatures shift precipitation from snow to rain, a pattern observed in regions like the Northern Hemisphere, Tibetan Plateau, and European Alps (Deng et al., 2017; Li et al., 2020; Quante et al., 2021). This shift disrupts snow-dependent hydrological processes, reducing winter and spring water storage, and impacting communities reliant on snowmelt for water supply (Mudryk et al., 2020; Safeeq et al., 2016). Global warming also accelerates snowmelt, further stressing water resources.

According to the IPCC (2022a, b), precipitation patterns in mountainous regions are influenced by seasonal and elevation-dependent variations. For example, the Himalayan–Tibetan Plateau experiences intensified and erratic monsoon precipitation (Palazzi et al., 2017; Sanjay et al., 2017), while studies in Europe and Central Asia reveal decreased snowfall at lower altitudes and increases at higher elevations (Li et al., 2020; Quante et al., 2021). Increased snowfall at high altitudes is attributed to insufficient warming to convert snow to rain (Kapnick & Delworth, 2013).

1.1.3 Climate Change Effects on Mountain Snow

Snow cover is a critical component of the Earth's surface, directly linked to the terrestrial climate system and hydrological processes (Li et al., 2018a, b). Mountainous regions are particularly sensitive to climate change, where snow cover significantly influences water availability, affecting agriculture, tourism, and hydropower (Bormann et al., 2018; Li et al., 2017, 2018a, b). Recent decades have witnessed a 78% decline in snow cover globally, with Asia, Europe, and South America experiencing faster-than-average losses (Notarnicola, 2020; Young, 2023). Despite global declines, some regions in the Northern Hemisphere show an 11% increase in winter snow cover.

Snow cover and phenological changes are elevation-dependent, with pronounced declines at lower and mid-elevations, while higher altitudes show greater stability due to cooler temperatures mitigating snowmelt (IPCC, 2022a, b). For instance, a projected 25% reduction in snow depth by mid-century under high-emission scenarios will affect regions such as the European Alps, Western North America, and the Himalayas (Fig. 1.2). Reduced snowpack leads to diminished streamflow and lower runoff efficiency, severely impacting water-dependent sectors including

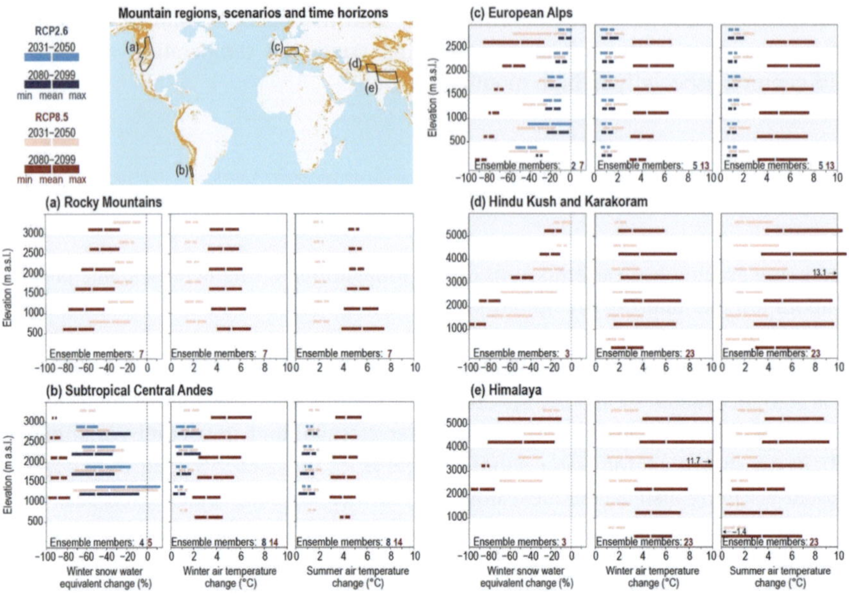

Fig. 1.2 Projected change (2031–2050 and 2080–2099 with respect to 1986–2005) of mean winter (December to May; June to August in Subtropical Central Andes) snow water equivalent, winter air temperature and summer air temperature (June to August; December to February in Subtropical Central Andes) in five high mountain regions for RCP8.5 (all regions) and RCP2.6 (European Alps and Subtropical Central Andes). Changes are averaged over 500 m (**a–c**) and 1,000 m (**d, e**) elevation bands. The numbers in the lower right of each panel reflect the number of simulations (note that not all models provide snow water equivalent) (IPCC, 2022a, b)

food production, hydropower, tourism, and drinking water supply (Beniston et al., 2018; Musselman et al., 2017; Notarnicola, 2020).

1.1.4 Glacier Dynamics and Implications

Mountain glaciers, excluding those in Antarctica, Greenland, and the Arctic regions, comprise approximately 170,000 glaciers covering 250,000 km^2 (RGI Consortium, 2017). These glaciers are retreating rapidly due to atmospheric warming, significantly affecting global sea level rise and regional hydrology (Fig. 1.3) (IPCC, 2022a, b). From 2000 to 2015, mass loss rates in regions such as the Southern Andes, European Alps, and High Mountain Asia ranged between -490 ± 100 and -150 ± 110 kg m^{-2} yr^{-1} (Brun et al., 2017). The primary driver is warming-induced glacier melt, although factors such as subsurface dynamics and meteorological variations also play a role (Thibert et al., 2018).

Climate-induced glacier loss is exacerbated by processes like increased precipitation variability, airflow changes, and the accumulation of light-absorbing particles

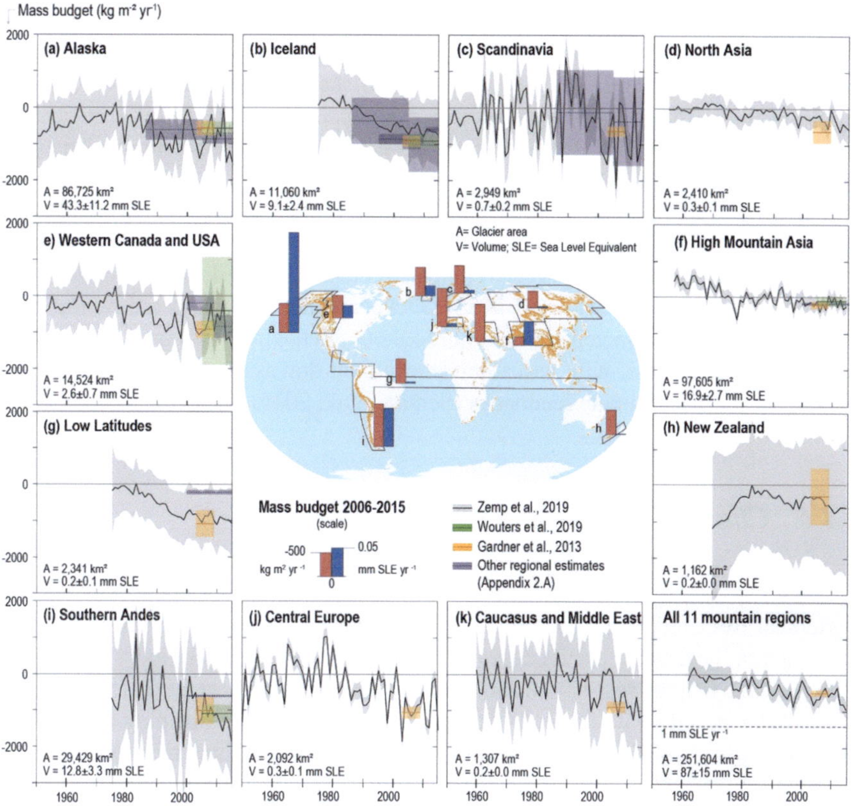

Fig. 1.3 Glacier mass budgets for the eleven mountain regions assessed. Red and blue bars on the map refer to regional budgets averaged over the period 2006–2015 in units of kg m^{-2} yr^{-1} and mm sea level equivalent (SLE) yr^{-1}, respectively, and are derived from each region's available mass-balance estimates (IPCC, 2022a, b)

that enhance surface melting (Duethmann et al., 2015; Painter et al., 2013; Williamson et al., 2019). Studies show that glacierized regions act as critical water towers, with their retreat affecting seasonal runoff, summer tourism, hydropower, and natural hazard frequency (Hanzer et al., 2018; Immerzeel et al., 2020). Without mitigation, glacier disintegration in this century could have lasting impacts on the sea level rise beyond 2100 (Zemp et al., 2019).

1.1.5 *Permafrost Degradation*

Permafrost, a subsurface phenomenon distinct from glaciers and snow, poses significant environmental and socio-economic concerns due to its sensitivity to warming.

In high mountain regions like the European Alps, Tibetan Plateau, and Tien Shan, rising temperatures since the 1980s have accelerated permafrost thaw, leading to ground ice depletion and increased instability (Biskaborn et al., 2019; IPCC, 2022a, b). While the warming rate varies by location, it consistently contributes to infrastructure risks, hydrological shifts, and increased greenhouse gas emissions (Cao et al., 2018; Zhao & Wu, 2019).

Permafrost degradation releases stored carbon, amplifying global warming and intensifying natural hazards. In regions with critical permafrost, the thickening of the active layer further disrupts ecosystems, accelerating land degradation and water resource vulnerability (Gao et al., 2023; Göckede et al., 2019). According to Chadburn et al. (2017), projected warming throughout the twenty-first century will exacerbate permafrost loss, with cascading impacts on downstream hydrology, carbon cycling, and global climate feedbacks (Schuur et al., 2015; Shen et al., 2023).

1.1.6 Changes in Lake and River Ice Cover

Globally, rivers cover more than one-third of the landmass, with many experiencing periodical freezing (Verpoorter et al., 2014). Lake and river ice plays a crucial role in regulating heat storage and ecosystem temperature, while also supporting human activities like fishing, recreation, and transportation (Brammer et al., 2015; Hampton et al., 2017; Knoll et al., 2019; O'Reilly et al., 2015). Recent warming trends have accelerated ice cover reduction, with future projections indicating worsening ice deterioration (Sharma et al., 2019, Sharma et al., 2021; Wang et al., 2021).

Although comprehensive data on lake ice is limited, both remote sensing and in situ studies have mapped ice thickness and duration. Wang et al. (2021) identified significant ice loss from 1985 to 2020 across Europe, northern United States, and central Asia, attributing it to rising air temperatures (Ades et al., 2020). Changes in ice cover alter the physical and biochemical properties of lakes, increasing solar absorption, warming water bodies, and boosting evaporation—potentially raising drought risks by 16% by 2100 (Stainsby et al., 2011; Williams et al., 2020).

The IPCC (2022a, b) highlights regional variations in ice patterns, with earlier ice melt trends on the Tibetan Plateau and Scandinavia (Du et al., 2017). Key factors influencing ice dynamics include altitude, lake depth, shoreline complexity, and solar radiation, while glacier-fed lakes exhibit unique responses driven by hydrodynamics and morphology (Sharma et al., 2019).

1.2 Climate-Driven Ecological and Environmental Challenges in High Mountain Regions

The resources found in mountain ecosystems, such as water and wood, are vital to the survival of nearly half of the global population. However, these ecosystems face escalating risks from climate change and natural disasters, which threaten their long-term sustainability. High mountain cold regions experience a range of environmental challenges, including changing glacier dynamics, fluctuating water availability, degraded water quality, landslides, unstable slopes, and flooding events such as snow avalanches and glacier lake outburst floods (GLOFs). Addressing these issues requires a comprehensive understanding of the interconnected processes shaping these environments.

1.2.1 Future Water Security in Glacier-Fed River Basins

Approximately 1.9 billion people in mountainous regions rely on glacier-fed river basins for freshwater, with this meltwater playing a crucial role in the global water cycle (Immerzeel et al., 2020b). While increased runoff from accelerated glacier melt has temporarily boosted water availability, this trend is unsustainable. The ongoing reduction in glaciated areas threatens long-term water supply stability. Recent studies emphasize that meltwater supports critical functions, such as food production in the Indus River basin, where nearly 40% of the population depends on glacier meltwater for irrigation (Lutz et al., 2022). Additionally, hydropower generation in regions like the European Alps and High Mountain Asia faces significant future declines due to glacial retreat, with mid-century projections indicating reduced energy output (Schaefli et al., 2019). The shift from snow- to rain-dominated precipitation reduces water storage capacity, exacerbating seasonal water scarcity (Laurent et al., 2020).

While human systems can implement adaptive measures, terrestrial and aquatic ecosystems are more vulnerable. Changes in discharge patterns disrupt habitat stability, reduce biodiversity, and alter ecosystem functions (Cauvy-Fraunié & Dangles, 2019). Future scenarios predict intensified droughts and economic challenges, highlighting the need for early intervention, glacier monitoring, and integrated resource management.

1.2.2 Water Quality Risks and Ecosystem Health in Glacier-Fed Basins

Glacial retreat exacerbates water quality issues as stored pollutants are released into downstream environments. These include black carbon, fallout radionuclides (FRNs), heavy metals, and persistent organic pollutants (POPs), all transported by atmospheric and cryospheric processes (Beard et al., 2022; Pawlak et al., 2021). Mercury contamination presents significant ecological and human health risks, particularly for communities reliant on glacier-fed fisheries (Hawkings et al., 2021). Furthermore, glacier meltwater increases acid rock drainage, raising acidity and toxicity in lakes and rivers (Colombo et al., 2018). Rising water temperature associated with glacier loss further degrade aquatic ecosystems, reducing species diversity and altering chemical properties (Michel et al., 2020). Addressing these impacts requires targeted contaminant monitoring, enhanced water quality regulations, and proactive conservation strategies.

1.2.3 *Permafrost Thaw and Mountain Slope Instability*

Permafrost degradation, along with increased water infiltration into glacial slopes, accelerates the movement of frozen debris and the loss of ground-level ice. The stability of frozen slopes is highly sensitive to surface temperature increases, ice content, and water availability (Kenner et al., 2017; Wirz et al., 2016). Evidence indicates a rise in rockfalls since the mid-1900s across regions like North America, New Zealand, and Europe due to permafrost thaw (Allen et al., 2011; Coe et al., 2018; Fischer et al., 2012; Ravanel & Deline, 2011). Rock avalanches, larger in scale than rockfalls, become more frequent with melting permafrost, as confirmed by field data and simulations (IPCC, 2022a, b). Glacier retreat further contributes to slope instability. For example, studies in the Himalayas have revealed a correlation between deglaciation and increased landslide activity (Liu et al., 2021). Between 1973 and 2003, one-third of significant landslides in Northern British Columbia were linked to glacier loss, which exposed cirque walls (Cloutier et al., 2016). Unique glacial instabilities, such as the complete collapse of low-angle slope glaciers, have led to catastrophic events. Notable examples include the Kolka Glacier collapse in 2002, which resulted in 130 deaths, and the twin collapses in Tibet in 2016, both linked to glacier bed melting and water infiltration (Gilbert et al., 2018; Kääb et al., 2018).

1.2.4 *Glacier-Related Flood Hazards*

Glacier-related floods pose significant threats to both ecosystems and human infrastructure, particularly in low-lying downstream areas. Glacier lake outburst floods (GLOFs), triggered by the sudden breach of moraine-dammed lakes, are among the most hazardous events (Carrivick & Tweed, 2016; Rawat et al., 2022). Glacier retreat creates conditions conducive to the formation of new lakes and the expansion of existing ones, increasing the likelihood of catastrophic GLOFs (Emmer, 2017; Vilca et al., 2021). Recent global studies highlight substantial growth in glacier lake volume and quantity since the 1990s, placing approximately 15 million people, primarily in countries like India, Pakistan, Peru, and China, at risk (Taylor et al., 2023).

Glacier surges and temporary ice dams further contribute to flood hazards. These surges block rivers and create large, unstable reservoirs that can lead to catastrophic floods upon collapse (Round et al., 2017; Stearns et al., 2015). Rain-on-snow events amplify these risks by rapidly melting snowpack, intensifying river discharge and increasing downstream impacts (Rawat et al., 2022).

Effective flood risk reduction strategies include comprehensive glacier lake inventories, predictive GLOF susceptibility modeling, and robust early warning systems. Integrated approaches are essential to protecting vulnerable populations, infrastructure, and ecosystems.

1.3 Sustainability and Vulnerability Challenges in the HKH Region

The Hindu Kush Himalayan (HKH) region is one of the world's most ecologically diverse and fragile landscapes. It provides vital ecosystem services, including freshwater resources, biodiversity, and climate regulation, for millions of people. However, this region faces significant sustainability challenges related to climate change, water resource management, agriculture, and biodiversity conservation.

1.3.1 Climate Vulnerability and Socioeconomic Challenges

The HKH's mountain cryosphere supports river flow regulation, groundwater recharge, and hydroelectric power generation (Scott et al., 2019). However, rising temperatures and altered precipitation patterns are disrupting the hydrological cycle, impacting agriculture and water availability (Huss & Hock, 2018; Kraaijenbrink et al., 2017; Krishnan et al., 2019). Increasing precipitation variability and climate hazards such as floods, landslides, and glacier lake outburst floods (GLOFs) have damaged irrigation infrastructure, forcing farmers to abandon lands (Dhakal et al., 2021). In Nepal, avalanches and rising temperatures have diminished food and water supplies for livestock (Tuladhar et al., 2021). The region's reliance on rainfed agriculture increases it's vulnerablility to erratic rainfall, with studies predicting declining crop yields due to rising temperatures (ICIMOD, 2019; World Bank, 2022).

Socioeconomic factors exacerbate vulnerability. Poverty, inadequate infrastructure, and reliance on natural resources hinder resilience (ICIMOD, 2019; Manandhar et al., 2018). Climate change intensifies resource competition and migration pressures, affecting social equity and livelihoods (Agnew, 2011; Hsiang et al., 2011; Bhusal & Subedi, 2014). Rapid glacial melt temporarily boosts river discharge but threatens long-term water and food security (Immerzeel et al., 2010; Nepal &

Shrestha, 2015). Proactive policies are required to mitigate cascading impacts on agriculture, economic stability, and ecosystem services.

1.3.2 Managing Water Scarcity and Resource Complexity

Known as Asia's Water Tower, the HKH sustains major rivers, including the Ganges and Yangtze. Rising water demand for agriculture and urbanization, combined with glacier retreat, has led to regional water scarcity (WBGU, 2022). Pollution from agricultural runoff and untreated wastewater further degrades water quality (ICIMOD, 2019). Unsustainable groundwater extraction in regions like Northwest India has depleted aquifers due to unregulated well drilling and free electricity policies (Mishra et al., 2018).

The HKH region has a hydropower potential exceeding 500 gigawatts yet retreating glaciers and changing river discharge are reducing generation capacity (Bolch, 2017). Seasonal variations increase energy costs and strain infrastructure (Huss & Hock, 2018; Kraaijenbrink et al., 2017). Meanwhile, degraded lakes suffer from pollution and eutrophication due to agricultural discharge and untreated sewage (Kamboj et al., 2020; Sharma et al., 2023; Singh et al., 2022). Addressing water scarcity requires integrated management of surface and groundwater, pollution control, and adaptive hydropower strategies.

1.3.3 Agriculture and Food Security Under Climate Stress

Agriculture is the backbone of livelihoods in the HKH, yet climate change and resource depletion pose significant damage to food security. Water stress, extreme weather, and inefficient practices reduce crop yields (Negi et al., 2017). Flash floods, landslides, and droughts degrades arable land, often leading to its abandonment (Dahal et al., 2022; Tiwari & Joshi, 2015). Unregulated urbanization further reduces agricultural space and strains resources (ICIMOD, 2019).

Studies associate declining snowfall and rising temperatures with reduced cereal crop yeilds in India's Himalayan regions (Bolch et al., 2019). Bhutan faces increased flood and landslide risks, endangering subsistence farmers reliant on rainfed agriculture (Chhogyel et al., 2020). Sustainable farming technologies and robust land-use planning are critical for resilience.

1.3.4 Land Degradation from Unsustainable Land-Use Practices

Unsustainable land-use practices, such as deforestation, overgrazing, monocropping, and excessive use of chemical fertilizers, are key drivers of land degradation in the HKH region (Halbac-Cotoara-Zamfir et al., 2020). Soil erosion and loss of fertile topsoil reduce agricultural productivity and exacerbate food insecurity (UNDP, 2020). The shift from traditional farming methods to intensive agriculture depletes soil nutrients, affecting crop yields and groundwater recharge (NITI Aayog, 2018).

Forest loss reduces manure availability, leading to a drop in soil fertility and productivity, as observed in Uttarakhand, where yields declined by 25% between 1980 and 2010 (ICIMOD, 2019). Furthermore, deforestation increases surface runoff, intensifying soil erosion and slope instability in mountainous regions (Nabiollahi et al., 2018; Zhang et al., 2017). Decreased plant cover worsens drought impacts, degrading grasslands and reducing livestock production (Kumari et al., 2020).

1.3.5 Biodiversity Loss and Ecosystem Decline Under Human and Climate Pressures

The HKH region, a global biodiversity hotspot, is under increasing strain from deforestation, overgrazing, habitat degradation, and illegal wildlife trade (ICIMOD, 2019). Unsustainable land-use practices and climate change threaten vital ecosystem services and traditional knowledge systems of indigenous communities (Chettri & Sharma, 2016). Species such as the Himalayan brown bear are experiencing habitat fragmentation, driven by both anthropogenic activities and climate-induced changes, leading to population declines (Dar et al., 2021; McLellan et al., 2017). Research by Su et al. (2018) identifies climate change as a key driver of these trends.

The HKH ecosystems also provide carbon storage, water regulation, and biodiversity maintenance. For instance, sustainable management of oak forests in Bhutan enhances socio-cultural ecosystem services (Dorji et al., 2019). However, biodiversity losses in Nepal's Shivapuri-Nagarjun National Park have reduced ecosystem services, including water quality and carbon sequestration, by up to 94% (Kattel, 2022; Peh et al., 2016). Overgrazing by livestock further depletes plant species critical for wildlife, such as bamboo forests essential to red pandas (Acharya et al., 2018; Bista et al., 2017; Lama et al., 2020).

Ecotourism, often promoted as sustainable, can exacerbate deforestation and ecosystem stress. In Nepal, poorly managed tourism results in habitat destruction and waste generation, undermining conservation efforts (Bhatta, 2019; Bhatta & Chan, 2023). Balanced policies are essential for aligning economic development with biodiversity protection.

1.4 Problems and Objectives Addressed in This Book

As discussed in previous sections, the Hindu Kush Himalayan (HKH) region stands as one of the world's most extensive and vital mountain systems, serving as a cornerstone for global ecological balance and human survival. Its vast glaciers supply freshwater to nearly two billion people through ten of Asia's major river systems, underscoring its critical role in sustaining regional livelihoods. Additionally, the HKH's rich biodiversity contributes to essential ecosystem services and climate regulation on a planetary scale. However, the region is increasingly vulnerable to sustainability threats driven by climate change and intensifying human activities. Glacial retreat, altered hydrological cycles, and the rising frequency of extreme events—floods, landslides, and droughts—are disrupting both ecosystems and human settlements. These changes endanger food security, water resources, and biodiversity, exacerbating socio-economic challenges such as poverty and environmental degradation.

To address these complex challenges, scientific research and policy action must be closely aligned. This book seeks to deepen understanding of the HKH's sustainability dynamics by integrating remote sensing technologies and advanced analytical frameworks. By tackling pressing scientific questions, it aims to equip policymakers and researchers with the tools needed to foster resilience and sustainable development.

1.4.1 Key Research Questions

The core scientific questions guiding this work are:

What are the advancements of the sustainable water management, agriculture, and ecological conservation in the HKH region?

How do human activities and climate change impact the hydrological systems that sustain the HKH region?

What roles do current agricultural practices play in advancing food security, and how can they be optimized?

How have human interventions and climate shifts affected ecosystem status in the HKH?

1.4.2 Objectives of the Study

This book has several interrelated objectives. First, it seeks to overcome critical data limitations that have historically impeded effective sustainability assessments in the HKH. Through satellite remote sensing, it aims to develop comprehensive frameworks for monitoring agricultural productivity, water resource dynamics, and ecological health. Second, it strives to enhance data integration and accessibility by leveraging geospatial technologies to provide timely, actionable insights. Finally, the book emphasizes evidence-based decision-making, offering strategies for adaptive management and promoting cooperation among HKH nations.

By fulfilling these objectives, the book aspires to contribute to the sustainable resource use, ecological conservation, and regional resilience, aligning local actions with broader global sustainability targets.

1.4.3 Methodological Approach

A multi-disciplinary, technology-driven framework underpins this study. Satellite imagery from missions such as Landsat, Sentinel, and Chinese Gaofen forms the foundation for analyzing glacier changes, land use shifts, and surface water variations. Machine learning models enhance classification accuracy and yield predictions for agricultural systems, while temporal and spatial analyses allow the integration of historical and real-time data to identify patterns and forecast risks. Cloud computing platforms like Google Earth Engine enable large-scale, real-time data processing and visualization.

This methodological blend ensures comprehensive monitoring and analysis, supporting robust sustainability assessments tailored to the HKH's unique environmental and socio-economic context.

1.4.4 Structure and Expected Outcomes

The structure of this book reflects a logical progression from problem identification to actionable solutions:

This Chapter introduces the HKH region and outlines its sustainability challenges.

Chapters 2–5 offer in-depth analyses of water resources, agriculture, and ecosystems.

Chapter 6 proposes a sustainable development indicator system, demonstrating remote sensing applications.

Chapter 7 synthesizes findings and provides policy recommendations.

The expected outcomes include:

Innovative Monitoring Tools: New frameworks for monitoring sustainable development indicators.

Enhanced Regional Datasets: Comprehensive, open-access geospatial data for broader use.

Policy-Relevant Insights: Strategies for mitigating climate impacts and enhancing resilience.

Strengthened SDG Monitoring Capacity: Practical tools to track progress toward sustainability goals.

Ultimately, this book aims to empower decision-makers with scientific insights and practical solutions, fostering sustainable development across the HKH region.

References

Acharya, K. P., Shrestha, S., Paudel, P. K., Sherpa, A. P., Jnawali, S. R., Acharya, S., & Bista, D. (2018). Pervasive human disturbance on habitats of endangered red panda Ailurus fulgens in the central Himalaya. *Global Ecology and Conservation, 15*, e00420. https://doi.org/10.1016/j.gecco.2018.e00420

Ades, M., Adler, R., Allan, R., Allan, R. P., Anderson, J., Argüez, A., Arosio, C., Augustine, J. A., Azorin-Molina, C., Barichivich, J., Barnes, J., Beck, H. E., Becker, A., Bellouin, N., Benedetti, A., Berry, D. I., Blenkinsop, S., Bock, Olivier., Bosilovich, M. G., … Ziese, M. (2020). Global Climate. *Bulletin of the American Meteorological Society, 101*(8), S9–S128. https://doi.org/10.1175/BAMS-D-20-0104.1

Agnew, J. (2011). Waterpower: Politics and the geography of water provision. *Annals of the Association of American Geographers, 101*(3), 463–476. https://doi.org/10.1080/00045608.2011.560053

Allen, S. K., Cox, S. C., & Owens, I. F. (2011). Rock avalanches and other landslides in the central Southern Alps of New Zealand: A regional study considering possible climate change impacts. *Landslides, 8*(1), 33–48. https://doi.org/10.1007/s10346-010-0222-z

Beard, D. B., Clason, C. C., Rangecroft, S., Poniecka, E., Ward, K. J., & Blake, W. H. (2022). Anthropogenic contaminants in glacial environments I: Inputs and accumulation. *Progress in Physical Geography: Earth and Environment, 46*(4), 630–648. https://doi.org/10.1177/03091333221107376

Beniston, M., Farinotti, D., Stoffel, M., Andreassen, L. M., Coppola, E., Eckert, N., Fantini, A., Giacona, F., Hauck, C., Huss, M., Huwald, H., Lehning, M., López-Moreno, J.-I., Magnusson, J., Marty, C., Morán-Tejéda, E., Morin, S., Naaim, M., Provenzale, A., … Vincent, C. (2018). The European mountain cryosphere: a review of its current state, trends, and future challenges. *The Cryosphere, 12*(2), 759–794. https://doi.org/10.5194/tc-12-759-2018

Bhatta, K. D. (2019). Exploring socio-cultural impacts of ecotourism in the Annapurna conservation area, Nepal. *Journal of Engineering Technology and Planning, 1*, 55–74. https://doi.org/10.3126/joetp.v1i0.38245

Bhatta, K. D., & Chan, R. C. K. (2023). Ecotourism, environmental impacts and sustainability in the Himalayan Settlements: Study of Sagarmatha (Mt. Everest) National Park, Nepal. *Himalayan Journal of Applied Science and Engineering, 4*(1), 15–31. https://doi.org/10.3126/hijase.v4i1.56129

Bhusal, J. K., & Subedi, B. P. (2014). Climate change induced water conflict in the Himalayas: A case study from Mustang, Nepal. *Ecopersia, 2*, 585–595. https://www.researchgate.net/publication/272790313

Biskaborn, B. K., Smith, S. L., Noetzli, J., Matthes, H., Vieira, G., Streletskiy, D. A., Schoeneich, P., Romanovsky, V. E., Lewkowicz, A. G., Abramov, A., Allard, M., Boike, J., Cable, W. L., Christiansen, H. H., Delaloye, R., Diekmann, B., Drozdov, D., Etzelmüller, B., Grosse, G., … Lantuit, H. (2019). Permafrost is warming at a global scale. *Nature Communications, 10*(1), 264. https://doi.org/10.1038/s41467-018-08240-4

Bista, D., Shrestha, S., Kunwar, A. J., Acharya, S., Jnawali, S. R., & Acharya, K. P. (2017). Status of gastrointestinal parasites in Red Panda of Nepal. *PeerJ, 5*, e3767. https://doi.org/10.7717/peerj.3767

Bolch, T. (2017). Asian glaciers are a reliable water source. *Nature, 545*(7653), 161–162. https://doi.org/10.1038/545161a

Bolch, T., Shea, J. M., Liu, S., Azam, F. M., Gao, Y., Gruber, S., Immerzeel, W. W., Kulkarni, A., Li, H., Tahir, A. A., Zhang, G., & Zhang, Y. (2019). Status and change of the cryosphere in the extended Hindu Kush Himalaya Region. In *The Hindu Kush Himalaya Assessment* (pp. 209–255). Springer International Publishing. https://doi.org/10.1007/978-3-319-92288-1_7

Bormann, K. J., Brown, R. D., Derksen, C., & Painter, T. H. (2018). Estimating snow-cover trends from space. *Nature Climate Change, 8*(11), 924–928. https://doi.org/10.1038/s41558-018-0318-3

Brammer, J. R., Samson, J., & Humphries, M. M. (2015). Declining availability of outdoor skating in Canada. *Nature Climate Change, 5*(1), 2–4. https://doi.org/10.1038/nclimate2465

Brun, F., Berthier, E., Wagnon, P., Kääb, A., & Treichler, D. (2017). A spatially resolved estimate of High Mountain Asia glacier mass balances from 2000 to 2016. *Nature Geoscience, 10*(9), 668–673. https://doi.org/10.1038/ngeo2999

C. Cloutier, J. Locat, M. Geertsema, M. Jakob, & M. Schnorbus. (2016). Potential impacts of climate change on landslides occurrence in Canada. In *Slope Safety Preparedness for Impact of Climate Change* (1st ed., p. 34). CRC Press.

Calvin, K., Dasgupta, D., Krinner, G., Mukherji, A., Thorne, P. W., Trisos, C., Romero, J., Aldunce, P., Barrett, K., Blanco, G., Cheung, W. W. L., Connors, S., Denton, F., Diongue-Niang, A., Dodman, D., Garschagen, M., Geden, O., Hayward, B., Jones, C., … Ha, M. (2023). *IPCC, 2023: Climate Change 2023: Synthesis Report. Contribution of Working Groups I, II and III to the Sixth Assessment Report of the Intergovernmental Panel on Climate Change* [Core Writing Team, H. Lee and J. Romero (eds.)]. IPCC. https://doi.org/10.59327/IPCC/AR6-9789291691647

Cao, B., Zhang, T., Peng, X., Mu, C., Wang, Q., Zheng, L., Wang, K., & Zhong, X. (2018). Thermal characteristics and recent changes of permafrost in the upper reaches of the Heihe River Basin, Western China. *Journal of Geophysical Research: Atmospheres, 123*(15), 7935–7949. https://doi.org/10.1029/2018JD028442

Carrivick, J. L., & Tweed, F. S. (2016). A global assessment of the societal impacts of glacier outburst floods. *Global and Planetary Change, 144*, 1–16. https://doi.org/10.1016/j.gloplacha.2016.07.001

Cauvy-Fraunié, S., & Dangles, O. (2019). A global synthesis of biodiversity responses to glacier retreat. *Nature Ecology & Evolution, 3*(12), 1675–1685. https://doi.org/10.1038/s41559-019-1042-8

Chadburn, S. E., Burke, E. J., Cox, P. M., Friedlingstein, P., Hugelius, G., & Westermann, S. (2017). An observation-based constraint on permafrost loss as a function of global warming. *Nature Climate Change, 7*(5), 340–344. https://doi.org/10.1038/nclimate3262

Chettri, N., & Sharma, E. (2016). Reconciling mountain biodiversity conservation and human wellbeing: Drivers of biodiversity loss and new approaches in the Hindu-Kush Himalayas. *Proceedings of the Indian National Science Academy, 82*(1). https://doi.org/10.16943/ptinsa/2016/v82i1/48378

Chhogyel, N., Kumar, L., & Bajgai, Y. (2020). Consequences of climate change impacts and incidences of extreme weather events in relation to crop production in Bhutan. *Sustainability, 12*(10), 4319. https://doi.org/10.3390/su12104319

Coe, J. A., Bessette-Kirton, E. K., & Geertsema, M. (2018). Increasing rock-avalanche size and mobility in Glacier Bay National Park and Preserve, Alaska detected from 1984 to 2016 Landsat imagery. *Landslides, 15*(3), 393–407. https://doi.org/10.1007/s10346-017-0879-7

Colombo, N., Salerno, F., Gruber, S., Freppaz, M., Williams, M., Fratianni, S., & Giardino, M. (2018). Review: Impacts of permafrost degradation on inorganic chemistry of surface fresh water. *Global and Planetary Change, 162*, 69–83. https://doi.org/10.1016/j.gloplacha.2017.11.017

Dahal, K. R., Dahal, P., Adhikari, R. K., Naukkarinen, V., Panday, D., Bista, N., Helenius, J., & Marambe, B. (2022). Climate change impacts and adaptation in a hill farming system of the Himalayan Region: Climatic trends, farmers' perceptions and practices. *Climate, 11*(1), 11. https://doi.org/10.3390/cli11010011

Dar, S. A., Singh, S. K., Wan, H. Y., Kumar, V., Cushman, S. A., & Sathyakumar, S. (2021). Projected climate change threatens Himalayan brown bear habitat more than human land use. *Animal Conservation, 24*(4), 659–676. https://doi.org/10.1111/acv.12671

Deng, H., Pepin, N. C., & Chen, Y. (2017). Changes of snowfall under warming in the Tibetan Plateau. *Journal of Geophysical Research: Atmospheres, 122*(14), 7323–7341. https://doi.org/10.1002/2017JD026524

Dhakal, D., O'Brien, D., & Mueser, P. (2021). Government policy and performance of agricultural cooperatives: A case study in Chitwan District, Nepal. *Sustainability, 13*(21), 12282. https://doi.org/10.3390/su132112282

Dorji, T., Brookes, J., Facelli, J., Sears, R., Norbu, T., Dorji, K., Chhetri, Y., & Baral, H. (2019). Socio-cultural values of ecosystem services from oak forests in the Eastern Himalaya. *Sustainability, 11*(8), 2250. https://doi.org/10.3390/su11082250

Du, J., Kimball, J. S., Duguay, C., Kim, Y., & Watts, J. D. (2017). Satellite microwave assessment of Northern Hemisphere lake ice phenology from 2002 to 2015. *The Cryosphere, 11*(1), 47–63. https://doi.org/10.5194/tc-11-47-2017

Duethmann, D., Bolch, T., Farinotti, D., Kriegel, D., Vorogushyn, S., Merz, B., Pieczonka, T., Jiang, T., Su, B., & Güntner, A. (2015). Attribution of streamflow trends in snow and glacier melt-dominated catchments of the Tarim River, Central Asia. *Water Resources Research, 51*(6), 4727–4750. https://doi.org/10.1002/2014WR016716

Emmer, A. (2017). Geomorphologically effective floods from moraine-dammed lakes in the Cordillera Blanca, Peru. *Quaternary Science Reviews, 177*, 220–234. https://doi.org/10.1016/j.quascirev.2017.10.028

Fischer, L., Purves, R. S., Huggel, C., Noetzli, J., & Haeberli, W. (2012). On the influence of topographic, geological and cryospheric factors on rock avalanches and rockfalls in high-mountain areas. *Natural Hazards and Earth System Sciences, 12*(1), 241–254. https://doi.org/10.5194/nhess-12-241-2012

Gao, J. (2020). *Global 1-km Downscaled Population Base Year and Projection Grids Based on the Shared Socioeconomic Pathways, Revision 01.* .

Gao, X., Lin, K., Liu, M., Dong, C., Yao, Z., Liu, Z., Xiao, M., Xie, X., & Huang, L. (2023). Dynamic changes in permafrost distribution over China and their potential influencing factors under climate warming. *Science of the Total Environment, 874*, 162624. https://doi.org/10.1016/j.scitotenv.2023.162624

Gao, Y., Chen, F., Lettenmaier, D. P., Xu, J., Xiao, L., & Li, X. (2018). Does elevation-dependent warming hold true above 5000 m elevation? Lessons from the Tibetan Plateau. *NPJ Climate and Atmospheric Science, 1*(1), 19. https://doi.org/10.1038/s41612-018-0030-z

Gilbert, A., Leinss, S., Kargel, J., Kääb, A., Gascoin, S., Leonard, G., Berthier, E., Karki, A., & Yao, T. (2018). Mechanisms leading to the 2016 giant twin glacier collapses, Aru Range, Tibet. *The Cryosphere, 12*(9), 2883–2900. https://doi.org/10.5194/tc-12-2883-2018

Göckede, M., Kwon, M. J., Kittler, F., Heimann, M., Zimov, N., & Zimov, S. (2019). Negative feedback processes following drainage slow down permafrost degradation. *Global Change Biology, 25*(10), 3254–3266. https://doi.org/10.1111/gcb.14744

Halbac-Cotoara-Zamfir, R., Smiraglia, D., Quaranta, G., Salvia, R., Salvati, L., & Giménez-Morera, A. (2020). Land degradation and mitigation policies in the Mediterranean Region: A brief commentary. *Sustainability, 12*(20), 8313. https://doi.org/10.3390/su12208313

Hampton, S. E., Galloway, A. W. E., Powers, S. M., Ozersky, T., Woo, K. H., Batt, R. D., Labou, S. G., O'Reilly, C. M., Sharma, S., Lottig, N. R., Stanley, E. H., North, R. L., Stockwell, J. D., Adrian, R., Weyhenmeyer, G. A., Arvola, L., Baulch, H. M., Bertani, I., Bowman, L. L., … Xenopoulos, M. A. (2017). Ecology under lake ice. *Ecology Letters, 20*(1), 98–111. https://doi.org/10.1111/ele.12699

Hanzer, F., Förster, K., Nemec, J., & Strasser, U. (2018). Projected cryospheric and hydrological impacts of 21st century climate change in the Ötztal Alps (Austria) simulated using a physically based approach. *Hydrology and Earth System Sciences, 22*(2), 1593–1614. https://doi.org/10.5194/hess-22-1593-2018

Hawkings, J. R., Linhoff, B. S., Wadham, J. L., Stibal, M., Lamborg, C. H., Carling, G. T., Lamarche-Gagnon, G., Kohler, T. J., Ward, R., Hendry, K. R., Falteisek, L., Kellerman, A. M., Cameron, K. A., Hatton, J. E., Tingey, S., Holt, A. D., Vinšová, P., Hofer, S., Bulínová, M., … Spencer, R. G. M. (2021). Large subglacial source of mercury from the southwestern margin of the Greenland Ice Sheet. *Nature Geoscience, 14*(7), 496–502. https://doi.org/10.1038/s41561-021-00753-w

Hsiang, S. M., Meng, K. C., & Cane, M. A. (2011). Civil conflicts are associated with the global climate. *Nature, 476*(7361), 438–441. https://doi.org/10.1038/nature10311

Huss, M., & Hock, R. (2018). Global-scale hydrological response to future glacier mass loss. *Nature Climate Change, 8*(2), 135–140. https://doi.org/10.1038/s41558-017-0049-x

ICIMOD. (2019). *The Hindu Kush Himalaya Assessment* (P. Wester, A. Mishra, A. Mukherji, & A. B. Shrestha, Eds.). Springer International Publishing. https://doi.org/10.1007/978-3-319-92288-1

Immerzeel, W. W., Lutz, A. F., Andrade, M., Bahl, A., Biemans, H., Bolch, T., Hyde, S., Brumby, S., Davies, B. J., Elmore, A. C., Emmer, A., Feng, M., Fernández, A., Haritashya, U., Kargel, J. S., Koppes, M., Kraaijenbrink, P. D. A., Kulkarni, A. V., Mayewski, P. A., … Baillie, J. E. M. (2020). Importance and vulnerability of the world's water towers. *Nature, 577*(7790), 364–369. https://doi.org/10.1038/s41586-019-1822-y

Immerzeel, W. W., van Beek, L. P. H., & Bierkens, M. F. P. (2010). Climate change will affect the Asian water towers. *Science, 328*(5984), 1382–1385. https://doi.org/10.1126/science.1183188

IPCC. (2018a). *Human, impacts of 1.5 °C of global warming on natural and systems.*

IPCC. (2018b). Summary for Policymakers. In V. Masson-Delmotte, P. Zhai, H.-O. Pörtner, D. Roberts, J. Skea, P. R. Shukla, A. Pirani, W. Moufouma-Okia, R. P. C. Péan, S. Connors, J. B. R. Matthews, Y. Chen, X. Zhou, M.I. Gomis, E. Lonnoy, T. Maycock, M. Tignor, & T. Waterfield (Eds.), *Global Warming of 1.5°C.* (pp. 3–24). Cambridge University Press. https://doi.org/10.1017/9781009157940.001

IPCC. (2022a). Climate Change 2022: Impacts, Adaptation and Vulnerability. In H.-O. Pörtner, D. C. Roberts, M. Tignor, E. S. Poloczanska, K. Mintenbeck, A. Alegría, M. Craig, S. Langsdorf, S. Löschke, V. Möller, A. Okem, & B. Rama (Eds.), *Contribution of Working Group II to the Sixth Assessment Report of the Intergovernmental Panel on Climate Change* (p. 3056). Cambridge University Press, Cambridge, UK. https://doi.org/10.1017/9781009325844

IPCC. (2022b). High mountain areas. In *The Ocean and Cryosphere in a Changing Climate* (pp. 131–202). Cambridge University Press. https://doi.org/10.1017/9781009157964.004

Jones, B., & O'Neill, B. C. (2016). Spatially explicit global population scenarios consistent with the shared socioeconomic pathways. *Environmental Research Letters, 11*(8), 084003. https://doi.org/10.1088/1748-9326/11/8/084003

Kääb, A., Leinss, S., Gilbert, A., Bühler, Y., Gascoin, S., Evans, S. G., Bartelt, P., Berthier, E., Brun, F., Chao, W.-A., Farinotti, D., Gimbert, F., Guo, W., Huggel, C., Kargel, J. S., Leonard, G. J., Tian, L., Treichler, D., & Yao, T. (2018). Massive collapse of two glaciers in western Tibet in 2016 after surge-like instability. *Nature Geoscience, 11*(2), 114–120. https://doi.org/10.1038/s41561-017-0039-7

Kamboj, V., Kamboj, N., & Sharma, N. (2020). A review on general characteristics, classification and degradation of river systems. In *Environmental Degradation: Causes and Remediation Strategies* (pp. 47–62). Agro Environ Media—Agriculture and Ennvironmental Science Academy, Haridwar, India. https://doi.org/10.26832/aesa-2020-edcrs-04

Kapnick, S. B., & Delworth, T. L. (2013). Controls of global snow under a changed climate. *Journal of Climate, 26*(15), 5537–5562. https://doi.org/10.1175/JCLI-D-12-00528.1

Kattel, G. R. (2022). Climate warming in the Himalayas threatens biodiversity, ecosystem functioning and ecosystem services in the 21st century: Is there a better solution? *Biodiversity and Conservation, 31*(8–9), 2017–2044. https://doi.org/10.1007/s10531-022-02417-6

Kenner, R., Phillips, M., Beutel, J., Hiller, M., Limpach, P., Pointner, E., & Volken, M. (2017). Factors controlling velocity variations at short-term, seasonal and multiyear time scales, ritigraben rock glacier, western Swiss alps. *Permafrost and Periglacial Processes, 28*(4), 675–684. https://doi.org/10.1002/ppp.1953

Knoll, L. B., Sharma, S., Denfeld, B. A., Flaim, G., Hori, Y., Magnuson, J. J., Straile, D., & Weyhenmeyer, G. A. (2019). Consequences of lake and river ice loss on cultural ecosystem services. *Limnology and Oceanography Letters, 4*(5), 119–131. https://doi.org/10.1002/lol2.10116

Kraaijenbrink, P. D. A., Bierkens, M. F. P., Lutz, A. F., & Immerzeel, W. W. (2017). Impact of a global temperature rise of 1.5 degrees Celsius on Asia's glaciers. *Nature, 549*(7671), 257–260. https://doi.org/10.1038/nature23878

Krishnan, R., Shrestha, A. B., Ren, G., Rajbhandari, R., Saeed, S., Sanjay, J., Syed, Md. A., Vellore, R., Xu, Y., You, Q., & Ren, Y. (2019). Unravelling climate change in the Hindu Kush Himalaya: Rapid warming in the mountains and increasing extremes. In *The Hindu Kush Himalaya Assessment* (pp. 57–97). Springer International Publishing. https://doi.org/10.1007/978-3-319-92288-1_3

Kumari, R., Banerjee, A., Kumar, R., Kumar, A., Saikia, P., & Latif Khan, M. (2020). Deforestation in India: Consequences and Sustainable Solutions. In *Forest Degradation around the World*. IntechOpen. https://doi.org/10.5772/intechopen.85804

Lama, S., Shrestha, S., Koju, N. P., Sherpa, A. P., & Tamang, M. (2020). Assessment of the impacts of livestock grazing on endangered red panda (*Ailurus fulgens*) habitat in Eastern Nepal. *Open Journal of Ecology, 10*(03), 97–110. https://doi.org/10.4236/oje.2020.103008

Laurent, L., Buoncristiani, J.-F., Pohl, B., Zekollari, H., Farinotti, D., Huss, M., Mugnier, J.-L., & Pergaud, J. (2020). The impact of climate change and glacier mass loss on the hydrology in the Mont-Blanc massif. *Scientific Reports, 10*(1), 10420. https://doi.org/10.1038/s41598-020-67379-7

Li, J., Yuan, G.-L., Wu, M.-Z., Sun, Y., Han, P., & Wang, G.-H. (2017). Evidence for persistent organic pollutants released from melting glacier in the central Tibetan Plateau, China. *Environmental Pollution, 220*, 178–185. https://doi.org/10.1016/j.envpol.2016.09.037

Li, T., Zhou, X. C., Ikhumhen, H. O., & Difei, A. (2018a). Research on the optimization of air quality monitoring station layout based on spatial grid statistical analysis method. *Environmental Technology (United Kingdom), 39*(10). https://doi.org/10.1080/09593330.2017.1327557

Li, X., Ding, Y., Xu, J., He, X., Han, T., Kang, S., Wu, Q., Mika, S., Yu, Z., & Li, Q. (2018b). Importance of mountain glaciers as a source of dissolved organic carbon. *Journal of Geophysical Research: Earth Surface, 123*(9), 2123–2134. https://doi.org/10.1029/2017JF004333

Li, Z., Chen, Y., Li, Y., & Wang, Y. (2020). Declining snowfall fraction in the alpine regions, Central Asia. *Scientific Reports, 10*(1), 3476. https://doi.org/10.1038/s41598-020-60303-z

Liu, J., Wu, Y., & Gao, X. (2021). Increase in occurrence of large glacier-related landslides in the high mountains of Asia. *Scientific Reports, 11*(1), 1635. https://doi.org/10.1038/s41598-021-81212-9

Lutz, A. F., Immerzeel, W. W., Siderius, C., Wijngaard, R. R., Nepal, S., Shrestha, A. B., Wester, P., & Biemans, H. (2022). South Asian agriculture increasingly dependent on meltwater and groundwater. *Nature Climate Change, 12*(6), 566–573. https://doi.org/10.1038/s41558-022-01355-z

Manandhar, S., Xenarios, S., Schmidt-Vogt, D., Hergarten, C., & Foggin, M. (2018). *Climate Vulnerability & Adaptive Capacity of Mountain Societies in Central Asia*. www.msri.ucentralasia.org

McLellan, B. N., Proctor, M. F., Huber, D., & Michel, S. (2017). *Ursus arctos. The IUCN red list of threatened species.*

Michel, A., Brauchli, T., Lehning, M., Schaefli, B., & Huwald, H. (2020). Stream temperature and discharge evolution in Switzerland over the last 50 years: Annual and seasonal behaviour. *Hydrology and Earth System Sciences, 24*(1), 115–142. https://doi.org/10.5194/hess-24-115-2020

Mishra, S. K., Hayse, J., Veselka, T., Yan, E., Kayastha, R. B., LaGory, K., McDonald, K., & Steiner, N. (2018). An integrated assessment approach for estimating the economic impacts of climate change on River systems: An application to hydropower and fisheries in a Himalayan River, Trishuli. *Environmental Science & Policy, 87*, 102–111. https://doi.org/10.1016/j.envsci.2018.05.006

Mudryk, L., Santolaria-Otín, M., Krinner, G., Ménégoz, M., Derksen, C., Brutel-Vuilmet, C., Brady, M., & Essery, R. (2020). Historical Northern Hemisphere snow cover trends and projected changes in the CMIP6 multi-model ensemble. *The Cryosphere, 14*(7), 2495–2514. https://doi.org/10.5194/tc-14-2495-2020

Musselman, K. N., Clark, M. P., Liu, C., Ikeda, K., & Rasmussen, R. (2017). Slower snowmelt in a warmer world. *Nature Climate Change, 7*(3), 214–219. https://doi.org/10.1038/nclimate3225

Nabiollahi, K., Golmohamadi, F., Taghizadeh-Mehrjardi, R., Kerry, R., & Davari, M. (2018). Assessing the effects of slope gradient and land use change on soil quality degradation through digital mapping of soil quality indices and soil loss rate. *Geoderma, 318*, 16–28. https://doi.org/10.1016/j.geoderma.2017.12.024

Negi, V. S., Maikhuri, R. K., Pharswan, D., Thakur, S., & Dhyani, P. P. (2017). Climate change impact in the Western Himalaya: People's perception and adaptive strategies. *Journal of Mountain Science, 14*(2), 403–416. https://doi.org/10.1007/s11629-015-3814-1

Nepal, S., & Shrestha, A. B. (2015). Impact of climate change on the hydrological regime of the Indus, Ganges and Brahmaputra river basins: A review of the literature. *International Journal of Water Resources Development, 31*(2), 201–218. https://doi.org/10.1080/07900627.2015.1030494

NITI Aayog. (2018). *Inventory and Revival of Springs in the Himalayas for Water Security. Report of Working Group I.*

Notarnicola, C. (2020). Hotspots of snow cover changes in global mountain regions over 2000–2018. *Remote Sensing of Environment, 243*, 111781. https://doi.org/10.1016/j.rse.2020.111781

O'Reilly, C. M., Sharma, S., Gray, D. K., Hampton, S. E., Read, J. S., Rowley, R. J., Schneider, P., Lenters, J. D., McIntyre, P. B., Kraemer, B. M., Weyhenmeyer, G. A., Straile, D., Dong, B., Adrian, R., Allan, M. G., Anneville, O., Arvola, L., Austin, J., Bailey, J. L., … Zhang, G. (2015). Rapid and highly variable warming of lake surface waters around the globe. *Geophysical Research Letters, 42*(24). https://doi.org/10.1002/2015GL066235

Painter, T. H., Flanner, M. G., Kaser, G., Marzeion, B., VanCuren, R. A., & Abdalati, W. (2013). End of the little ice age in the alps forced by industrial black carbon. *Proceedings of the National Academy of Sciences, 110*(38), 15216–15221. https://doi.org/10.1073/pnas.1302570110

Palazzi, E., Filippi, L., & von Hardenberg, J. (2017). Insights into elevation-dependent warming in the Tibetan Plateau-Himalayas from CMIP5 model simulations. *Climate Dynamics, 48*(11–12), 3991–4008. https://doi.org/10.1007/s00382-016-3316-z

Pawlak, F., Koziol, K., & Polkowska, Z. (2021). Chemical hazard in glacial melt? The glacial system as a secondary source of POPs (in the Northern Hemisphere). A systematic review. *Science of the Total Environment, 778*, 145244. https://doi.org/10.1016/j.scitotenv.2021.145244

Peh, K.S.-H., Thapa, I., Basnyat, M., Balmford, A., Bhattarai, G. P., Bradbury, R. B., Brown, C., Butchart, S. H. M., Dhakal, M., Gurung, H., Hughes, F. M. R., Mulligan, M., Pandeya, B., Stattersfield, A. J., Thomas, D. H. L., Walpole, M., & Merriman, J. C. (2016). Synergies between biodiversity conservation and ecosystem service provision: Lessons on integrated ecosystem service valuation from a Himalayan protected area, Nepal. *Ecosystem Services, 22*, 359–369. https://doi.org/10.1016/j.ecoser.2016.05.003

Quante, L., Willner, S. N., Middelanis, R., & Levermann, A. (2021). Regions of intensification of extreme snowfall under future warming. *Scientific Reports, 11*(1), 16621. https://doi.org/10.1038/s41598-021-95979-4

Ravanel, L., & Deline, P. (2011). Climate influence on rockfalls in high-Alpine steep rockwalls: The north side of the Aiguilles de Chamonix (Mont Blanc massif) since the end of the 'Little Ice Age.' *The Holocene, 21*(2), 357–365. https://doi.org/10.1177/0959683610374887

Rawat, M., Pandey, A., & Gupta, P. K. (2022). *Geospatial Analysis of Glacial Lake Outburst Flood (GLOF)* (pp. 141–160). https://doi.org/10.1007/978-3-030-98981-1_6

RGI Consortium. (2017). *Randolph Glacier Inventory (RGI)—A dataset of global glacier outlines version 6.0.* Technical report, Global Land Ice Measurements from Space, Boulder, Colorado, USA.

Round, V., Leinss, S., Huss, M., Haemmig, C., & Hajnsek, I. (2017). Surge dynamics and lake outbursts of Kyagar Glacier, Karakoram. *The Cryosphere, 11*(2), 723–739. https://doi.org/10.5194/tc-11-723-2017

Safeeq, M., Shukla, S., Arismendi, I., Grant, G. E., Lewis, S. L., & Nolin, A. (2016). Influence of winter season climate variability on snow–precipitation ratio in the western United States. *International Journal of Climatology, 36*(9), 3175–3190. https://doi.org/10.1002/joc.4545

Sanjay, J., Krishnan, R., Shrestha, A. B., Rajbhandari, R., & Ren, G.-Y. (2017). Downscaled climate change projections for the Hindu Kush Himalayan region using CORDEX South Asia regional climate models. *Advances in Climate Change Research, 8*(3), 185–198. https://doi.org/10.1016/j.accre.2017.08.003

Schaefli, B., Manso, P., Fischer, M., Huss, M., & Farinotti, D. (2019). The role of glacier retreat for Swiss hydropower production. *Renewable Energy, 132*, 615–627. https://doi.org/10.1016/j.renene.2018.07.104

Schuur, E. A. G., McGuire, A. D., Schädel, C., Grosse, G., Harden, J. W., Hayes, D. J., Hugelius, G., Koven, C. D., Kuhry, P., Lawrence, D. M., Natali, S. M., Olefeldt, D., Romanovsky, V. E., Schaefer, K., Turetsky, M. R., Treat, C. C., & Vonk, J. E. (2015). Climate change and the permafrost carbon feedback. *Nature, 520*(7546), 171–179. https://doi.org/10.1038/nature14338

Scott, C. A., Zhang, F., Mukherji, A., Immerzeel, W., Mustafa, D., & Bharati, L. (2019). Water in the Hindu Kush Himalaya. In *The Hindu Kush Himalaya Assessment* (pp. 257–299). Springer International Publishing. https://doi.org/10.1007/978-3-319-92288-1_8

Sharma, A. K., Sharma, M., Sharma, A. K., Sharma, M., & Sharma, M. (2023). Mapping the impact of environmental pollutants on human health and environment: A systematic review and meta-analysis. *Journal of Geochemical Exploration, 255*, 107325. https://doi.org/10.1016/j.gexplo.2023.107325

Sharma, S., Blagrave, K., Magnuson, J. J., O'Reilly, C. M., Oliver, S., Batt, R. D., Magee, M. R., Straile, D., Weyhenmeyer, G. A., Winslow, L., & Woolway, R. I. (2019). Widespread loss of lake ice around the Northern Hemisphere in a warming world. *Nature Climate Change, 9*(3), 227–231. https://doi.org/10.1038/s41558-018-0393-5

Sharma, S., Blagrave, K., Filazzola, A., Imrit, M. A., & Hendricks Franssen, H. (2021). Forecasting the permanent loss of lake ice in the northern hemisphere within the 21st century. *Geophysical Research Letters, 48*(1). https://doi.org/10.1029/2020GL091108

Shen, T., Jiang, P., Ju, Q., Yu, Z., Chen, X., Lin, H., & Zhang, Y. (2023). Changes in permafrost spatial distribution and active layer thickness from 1980 to 2020 on the Tibet Plateau. *Science of the Total Environment, 859*, 160381. https://doi.org/10.1016/j.scitotenv.2022.160381

Singh, A. K., Kumar, A., & Chandra, R. (2022). Environmental pollutants of paper industry wastewater and their toxic effects on human health and ecosystem. *Bioresource Technology Reports, 20*, 101250. https://doi.org/10.1016/j.biteb.2022.101250

Stainsby, E. A., Winter, J. G., Jarjanazi, H., Paterson, A. M., Evans, D. O., & Young, J. D. (2011). Changes in the thermal stability of Lake Simcoe from 1980 to 2008. *Journal of Great Lakes Research, 37*, 55–62. https://doi.org/10.1016/j.jglr.2011.04.001

Stearns, L. A., Hamilton, G. S., van der Veen, C. J., Finnegan, D. C., O'Neel, S., Scheick, J. B., & Lawson, D. E. (2015). Glaciological and marine geological controls on terminus dynamics of Hubbard Glacier, southeast Alaska. *Journal of Geophysical Research: Earth Surface, 120*(6), 1065–1081. https://doi.org/10.1002/2014JF003341

Su, J., Aryal, A., Hegab, I. M., Shrestha, U. B., Coogan, S. C. P., Sathyakumar, S., Dalannast, M., Dou, Z., Suo, Y., Dabu, X., Fu, H., Wu, L., & Ji, W. (2018). Decreasing brown bear (*Ursus arctos*) habitat due to climate change in Central Asia and the Asian Highlands. *Ecology and Evolution, 8*(23), 11887–11899. https://doi.org/10.1002/ece3.4645

Taylor, C., Robinson, T. R., Dunning, S., Rachel Carr, J., & Westoby, M. (2023). Glacial lake outburst floods threaten millions globally. *Nature Communications, 14*(1), 487. https://doi.org/10.1038/s41467-023-36033-x

Thibert, E., Dkengne Sielenou, P., Vionnet, V., Eckert, N., & Vincent, C. (2018). Causes of glacier melt extremes in the alps since 1949. *Geophysical Research Letters, 45*(2), 817–825. https://doi.org/10.1002/2017GL076333

Tiwari, P. C., & Joshi, B. (2015). Environmental changes, rural out-migration and social dynamics in Himalayas. *Studia Obszarów Wiejskich, 38*, 77–96. https://doi.org/10.7163/SOW.38.5

Tuladhar, S., Pasakhala, B., Maharjan, A., & Mishra, A. (2021). Unravelling the linkages of cryosphere and mountain livelihood systems: A case study of Langtang, Nepal. *Advances in Climate Change Research, 12*(1), 119–131. https://doi.org/10.1016/j.accre.2020.12.004

UNDP. (2020). Human development report 2020. In *UNDP (United Nations Development Programme).* http://report2020.archive.s3-website-us-east-1.amazonaws.com/

Verpoorter, C., Kutser, T., Seekell, D. A., & Tranvik, L. J. (2014). A global inventory of lakes based on high-resolution satellite imagery. *Geophysical Research Letters, 41*(18), 6396–6402. https://doi.org/10.1002/2014GL060641

Vilca, O., Mergili, M., Emmer, A., Frey, H., & Huggel, C. (2021). The 2020 glacial lake outburst flood process chain at Lake Salkantaycocha (Cordillera Vilcabamba, Peru). *Landslides, 18*(6), 2211–2223. https://doi.org/10.1007/s10346-021-01670-0

Wang, X., Feng, L., Gibson, L., Qi, W., Liu, J., Zheng, Y., Tang, J., Zeng, Z., & Zheng, C. (2021). High-resolution mapping of ice cover changes in over 33,000 lakes across the north temperate zone. *Geophysical Research Letters, 48*(18). https://doi.org/10.1029/2021GL095614

WBGU. (2022). *Mountains matter: Water tower of Asia—The state of the Hindukush Karakoram Himalaya Region's water resources.*

Williams, A. P., Cook, E. R., Smerdon, J. E., Cook, B. I., Abatzoglou, J. T., Bolles, K., Baek, S. H., Badger, A. M., & Livneh, B. (2020). Large contribution from anthropogenic warming to an emerging North American megadrought. *Science, 368*(6488), 314–318. https://doi.org/10.1126/science.aaz9600

Williamson, C. J., Cameron, K. A., Cook, J. M., Zarsky, J. D., Stibal, M., & Edwards, A. (2019). Glacier algae: A dark past and a darker future. *Frontiers in Microbiology, 10*, 524. https://doi.org/10.3389/fmicb.2019.00524

Wirz, V., Geertsema, M., Gruber, S., & Purves, R. S. (2016). Temporal variability of diverse mountain permafrost slope movements derived from multi-year daily GPS data, Mattertal, Switzerland. *Landslides, 13*(1), 67–83.

World Bank. (2022). *Population data in the world..* Https://Data.Worldbank.Org/Indicator/SP.POP.TOTL

Xie, A., Zhu, J., Qin, X., Wang, S., Xu, B., & Wang, Y. (2023). Surface warming from altitudinal and latitudinal amplification over Antarctica since the International Geophysical Year. *Scientific Reports, 13*(1), 9536. https://doi.org/10.1038/s41598-023-35521-w

Young, S. S. (2023). Global and regional snow cover decline: 2000–2022. *Climate, 11*(8), 162. https://doi.org/10.3390/cli11080162

Zemp, M., Huss, M., Thibert, E., Eckert, N., McNabb, R., Huber, J., Barandun, M., Machguth, H., Nussbaumer, S. U., Gärtner-Roer, I., Thomson, L., Paul, F., Maussion, F., Kutuzov, S., & Cogley, J. G. (2019). Global glacier mass changes and their contributions to sea-level rise from 1961 to 2016. *Nature, 568*(7752), 382–386. https://doi.org/10.1038/s41586-019-1071-0

Zhang, S., Fan, W., Li, Y., & Yi, Y. (2017). The influence of changes in land use and landscape patterns on soil erosion in a watershed. *Science of the Total Environment, 574*, 34–45. https://doi.org/10.1016/j.scitotenv.2016.09.024

Zhao, D., & Wu, S. (2019). Projected changes in permafrost active layer thickness over the Qinghai-Tibet Plateau under climate change. *Water Resources Research, 55*(9), 7860–7875. https://doi.org/10.1029/2019WR024969

Zheng, J., Zuo, Z., Lin, Z., Xiao, D., & Liang, Q. (2022). Linkage of the surface air temperature over Tibetan Plateau and Northeast hemisphere in winter at interannual timescale. *Atmospheric Research, 274*, 106229. https://doi.org/10.1016/j.atmosres.2022.106229

Chapter 2
Overview of the Hindu Kush-Himalayan (HKH) Region

2.1 Geographical Scope

Spanning over 4.5 million km^2, the Hindu Kush-Himalayan (HKH) region lies between latitudes 39.32°N and 15.96°S and longitudes 105.05°E and 60.85°W, covering eight countries: Afghanistan, Pakistan, India, China, Nepal, Bhutan, Bangladesh, and Myanmar (Fig. 2.1).

The HKH region includes almost all of Afghanistan's provinces, which are located between latitudes 38.49°N and 29.36°S and longitudes 74.89°E and 60.87°W. These Afghan provinces make up 60.4% of the country's total area, representing 8.8% of the entire HKH region (approximately 394,278.2 km^2).

Bangladesh's Chittagong Hill Tracts, located at the HKH boundary zone, account for 11.5% of the country's land area, or roughly 0.4% of the total HKH region, situated between 92.67°E to 88.39°W and 26.63°N to 20.74°S. Nearly the entire territory of Bhutan and Nepal—comprising about 0.9% and 3.3% of the HKH area, respectively—lies within the HKH region.

China, the largest country in the HKH region, is located between longitudes 105.04°E and 73.94°W and latitudes 39.08°N and 26.86°S. Its portion of the HKH region spans approximately 2,685,115 km^2 (about 60%), which accounts for roughly 27.3% of China's total land area. This includes provinces such as Gansu, Qinghai, Sichuan, Xinjiang Uygur Autonomous Region, Xizang Autonomous Region, and Yunnan.

Pakistan, the second-largest country in the HKH region after China, has 10.8% of its total area covered by the HKH. This includes the two centrally administered areas of Azad Jammu and Kashmir and Gilgit-Baltistan, as along with the provinces of Balochistan, Khyber-Pakhtunkhwa, and the Federally Administered Tribal Areas (FATA).

India, the third-largest country in the HKH region, covers around 12.8% of its total land area, or about 9.4% of the HKH region (roughly 345,898.82 km^2). The

S. Lu et al., *Remote Sensing Assessment of Sustainable Development in the Hindu Kush-Himalayan (HKH) Region*, https://doi.org/10.1007/978-981-96-8675-9_2

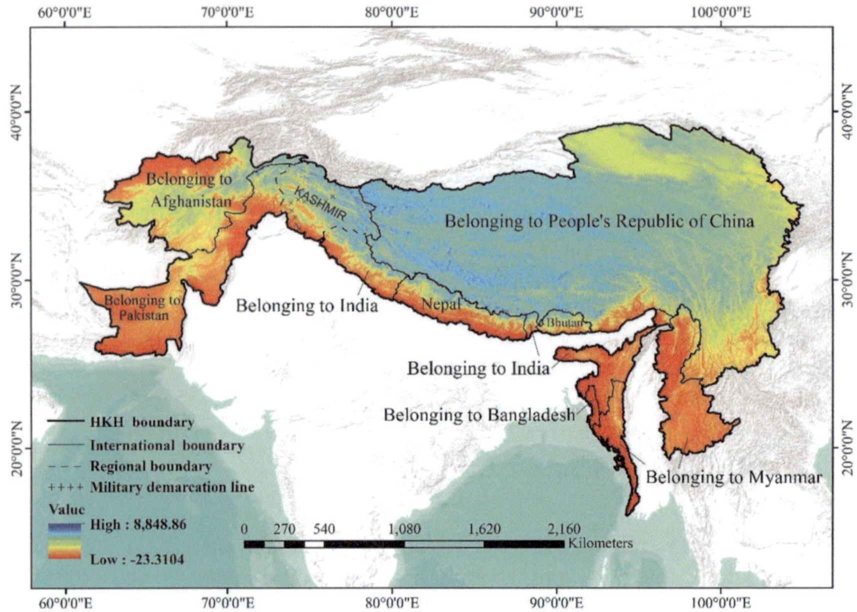

Fig. 2.1 The geographical scope of the HKH region

country is situated between longitudes 97.16°E and 73.78°W and latitudes 35.48°N and 21.97°S. This region encompasses about 11 mountain states.

Myanmar, which occupies 8.2% of the HKH region, is located between 101.2°E and 92.2°W and 28.5°N and 16°S. The HKH portion of Myanmar includes six mountain states, covering more than half of the country's total land area (365,778.9 km^2).

2.2 Topography and Landform

The HKH region is renowned for its towering mountain ranges, jagged peaks, vast valleys, alpine glaciers, boulders, and plateaus. The highest elevation in the region reaches approximately 8,848.86 m, with some slopes approaching an 88-degree incline. This area is home to the world's highest mountains, including Mount Everest, which stands at about 8,848.86 m. In Nepal, it is called "Sagarmatha," while the Chinese refer to it as "Qomolangma Feng." Located to the north of the Himalayas, the Tibetan Plateau is also home to numerous mountains, valleys, and plateaus. Notably, the area between the Nyaingentanggula, Kunlun, Tanggula, Kangdese, and other mountain ranges constitutes about two-thirds of the Northern Tibetan Plateau. The southern and eastern parts of the plateau consist of smaller river and lake valleys.

Among the HKH nations, China and Nepal boast the highest peak of Mount Everest, which has a slope of approximately 84 degrees. In contrast, Bangladesh has the lowest peak in the HKH region, standing at 1,034.4 m above sea level with a slope of 74.6 degrees (Fig. 2.2a). Bangladesh is the only HKH country with mountains below 2,000 m in altitude, placing it within the outer/sub-Himalayan category (sometimes referred to as the Siwalik Range). This region, with elevations ranging between 900 and 1,500 m, is highly vulnerable to heavy rainfall, deforestation, and erosion, leading to sediment runoff onto the plains.

The remaining HKH countries fall into one of three Himalayan classifications—Outer/Sub-Himalaya, Lower Himalayas, or the Great Himalayas—based on their elevation range, which spans from 5,000 to 8,849 m (Fig. 2.2b). The Lower Himalayas, also known as the Middle Himalayas, range from 3,600 to 4,600 m, while the Great Himalayas, which rise above 4,900 m, contain some of the world's tallest peaks, including Mount Everest (8,848.86 m), India's Nanda Devi (7,817 m), Trisul (7,120 m) in Uttarakhand, and Kamet (7,755 m).

The high altitude of the Himalayas ensures that their summits are perpetually covered with snow, which sustains most of the glaciers in the region. The northern parts of the HKH region feature thick soil cover, supporting alpine grasses and lowland forests. In the eastern Himalayas, the damp highland soils are rich in humus, making them ideal for growing tea. In contrast, the higher elevations of the Himalayas are characterized by lithosols—infertile, low-productivity soils composed of inadequately weathered rock fragments.

2.3 Climate Characteristics and Meteorological Changes

The HKH region is characterized by diverse climatic conditions, which vary depending on elevation and geographical location. As shown in Fig. 2.3, between 2000 and 2020, the region had an average annual precipitation of 1,249.4 mm, a mean temperature of 13.8 °C, and potential evaporation of 1,083.1 mm. During this period, the region experienced a decrease in precipitation of approximately 141.6 mm. Notably, years with precipitation over 1,300 mm occurred in 2000, 2011, 2015, 2017, and 2020, with the peak of 1,392.3 mm recorded in 2017. In contrast, precipitation fell below 1,200 mm in 2009, 2014, 2018, and 2019, with 2009 recording the lowest at 1,096.2 mm. Despite a modest increase in temperature (0.1 °C) and potential evaporation (6.5 mm) over the two decades, the lowest precipitation in 2009 was closely tied to the highest recorded average temperature of 14.26 °C. Additionally, temperature increases above 14 °C were observed in 2006, 2010, and 2016.

As previously noted, the climate in the HKH region varies by location. Between 2000 and 2020, countries like Bangladesh, Bhutan, India, Myanmar, and Nepal experienced significantly higher annual precipitation, ranging above 700 mm, with Bangladesh receiving the highest average of 3,104.9 mm and Nepal 718.4 mm (Fig. 2.4). On the other hand, higher-altitude regions such as Afghanistan, China, and Pakistan recorded lower precipitation, below 700 mm. Afghanistan had the highest

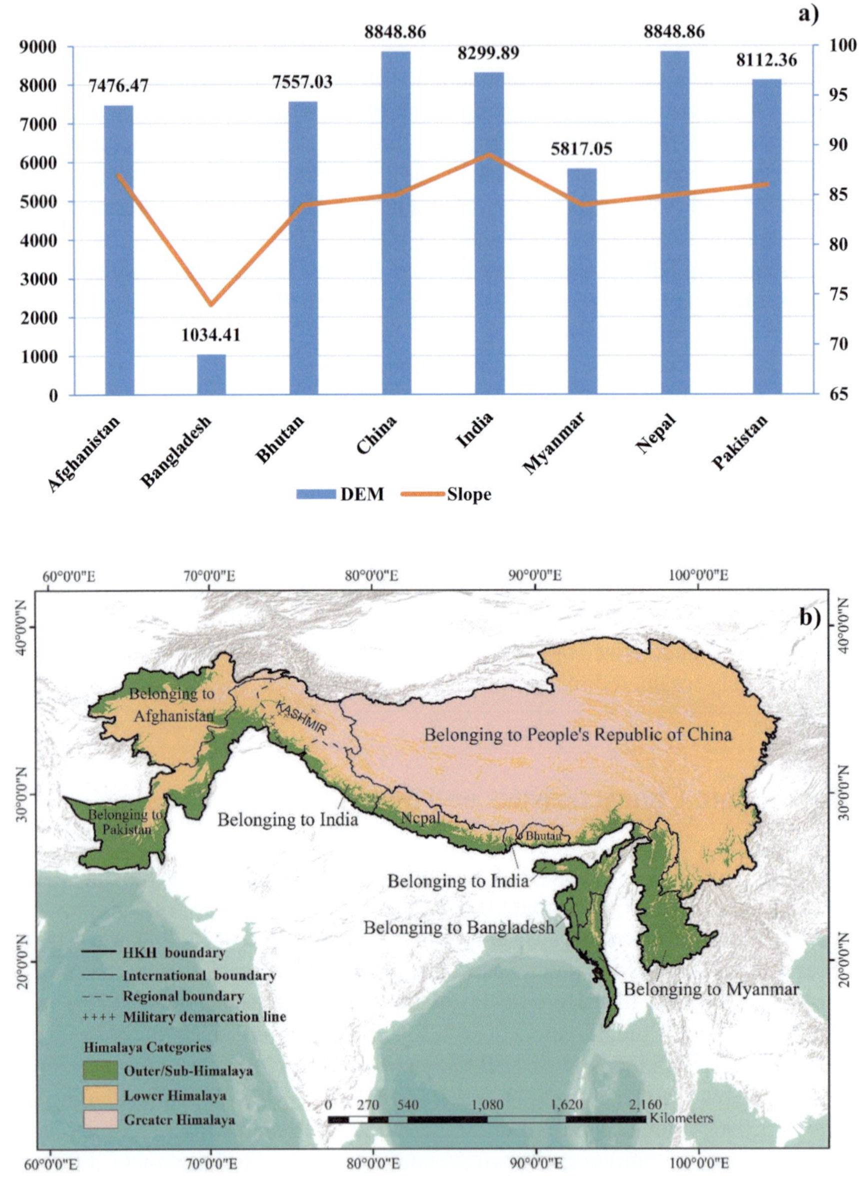

Fig. 2.2 Topographic features of the HKH countries. **a** The graphical presentation of the slope and elevation; **b** the spatial distribution map Himalaya categories

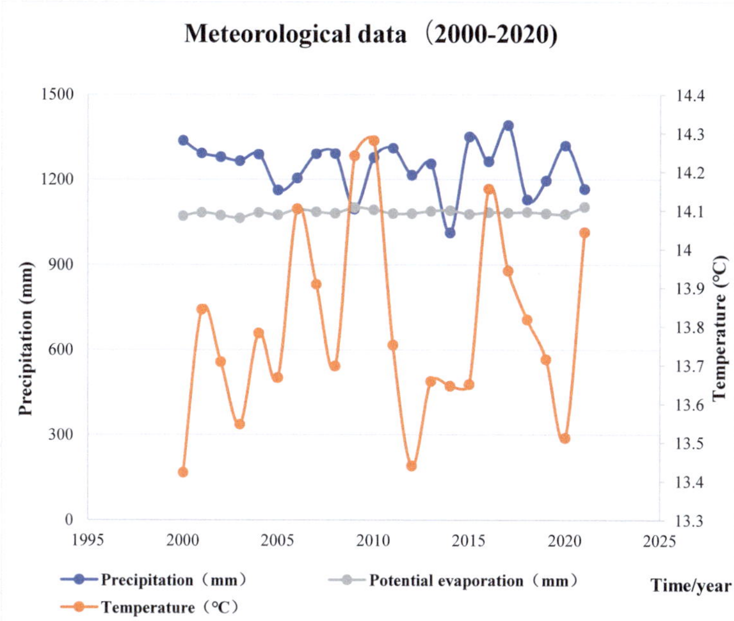

Fig. 2.3 Overall meteorological information of the HKH region from 2000 to 2020

precipitation in this category, with 557.8 mm recorded in 2015, while Pakistan recorded the lowest at 203.8 mm in 2018, making it the driest part of the HKH. While precipitation in Pakistan and Afghanistan increased by 183.6 and 195.2 mm, respectively, the other countries in the region saw a decline in total precipitation. Bangladesh, despite having the highest precipitation in the HKH, saw fluctuations over the period, with significant highs (3,104.9 mm in 2017) and lows (1,874.5 mm in 2014). As previously discussed, the low elevation of Bangladesh (sub-Himalaya) makes it more susceptible to heavy rainfall. Bhutan, with a peak precipitation of 2,145.6 mm in 2008 and a low of 1,463.3 mm in 2009, ranked second in precipitation after Bangladesh. Myanmar, with the lowest rainfall recorded in 2014 (1,455.6 mm) and the highest in 2016 (2,118.4 mm), ranked third.

Over the past two decades, the HKH region experienced an average temperature increase of 0.1 °C, with the greatest rises seen in Bangladesh (0.1 °C), China (0.6 °C), and Myanmar (0.6 °C) (Fig. 2.4). In contrast, India showed no temperature change, and countries like Afghanistan (−0.7 °C), Nepal (−0.1 °C), and Pakistan (−0.3 °C) saw slight decreases. Bangladesh and Myanmar recorded the highest temperatures, with 26.6 and 22.5 °C, respectively, in 2009 and 2010. Conversely, countries like China, Afghanistan, and Bhutan had average temperatures below 11 °C, with China recording the lowest temperature of −0.2 °C in 2000. The sub-Himalaya and lower Himalaya regions, with elevations under 6,000 m, experienced milder temperatures, as evidenced by Bangladesh and Myanmar. The higher-altitude Greater Himalaya,

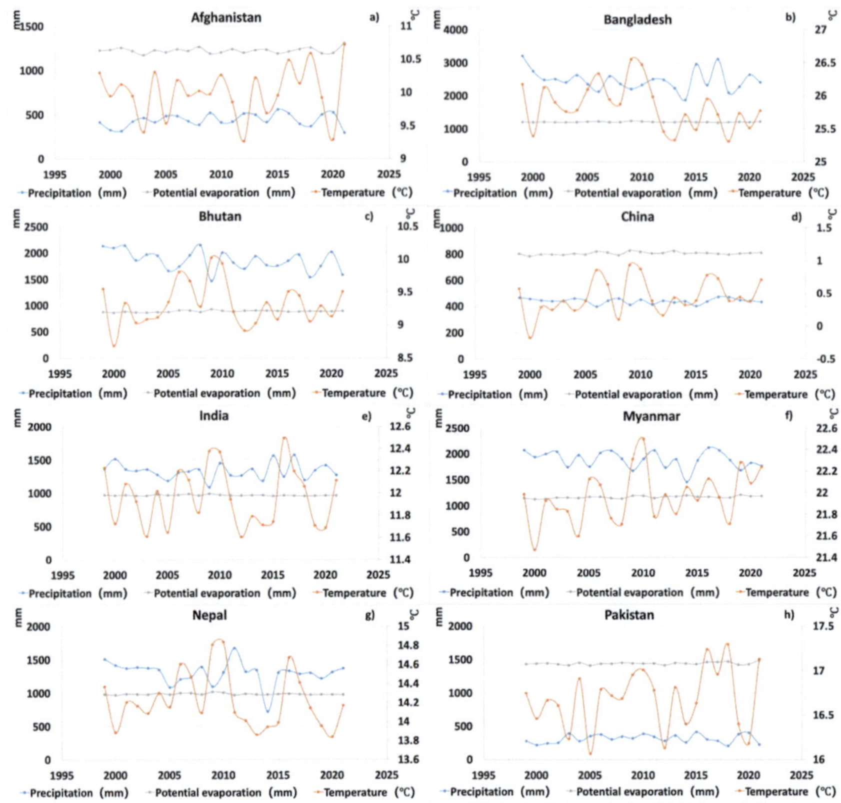

Fig. 2.4 Meteorological information of the different HKH countries from 2000 to 2020

on the other hand, had colder temperatures. These temperature variations, along with precipitation differences, contributed to the discrepancies in evaporation rates across the region. For example, China's low evaporation rate of 782 mm in 2000 can be attributed to its low temperature and reduced precipitation. Conversely, Pakistan, which has a warmer climate and lower precipitation, recorded the highest evaporation rates, reaching 1,461.7 mm in 2017 and 1,467.8 mm in 2018.

Apart from Pakistan, which had the highest evaporation rates (above 1,400 mm), other countries such as Afghanistan, Bangladesh, Myanmar, and Bhutan recorded average evaporation rates of approximately 1,100 mm. Between 2000 and 2020, countries like Bhutan (15.5 mm), China (25.5 mm), Myanmar (54 mm), and Nepal (8.7 mm) experienced a general rise in evaporation. In contrast, Pakistan, despite its high evaporation rates, saw a decrease of 10.3 mm over the two-decade period. India and Bangladesh also experienced slight declines in evaporation, with decreases of 1.2 and 3.1 mm, respectively.

2.4 Variations in Glaciers and Snow Cover

2.4.1 Variation in Glacier

As discussed in the previous section, the HKH region experienced a modest temperature increase of 0.1 °C over the past two decades. This warming, though slight, exhibited varying regional patterns, which significantly impacted the glaciers in the area. Between 2000 and 2020, although the regional annual glacier mass balance showed a slight improvement of 2.29 mm.w.e., the overall glacier mass balance remained negative, shifting from −407.19 mm.w.e in 2000 to −404.90 mm.w.e in 2020 (Fig. 2.5a). Additionally, the volume and area of glaciers in the HKH experienced a slight reduction, with losses of −0.006 Gt and −0.01 km^2, respectively. These reductions in glacier volume and size are major contributors to increased runoff, as evidenced by the rise in runoff from 1.13 to 1.16 Mt during the 2000–2020 period, driven by progressive glacier melt.

This ongoing glacier melt culminated in the HKH region reaching its lowest glacier area and volume in 2020—0.93 km^2 and 0.055 Gt, respectively—resulting in significantly higher runoff. In contrast, the greatest glacier extent (0.95 km^2) was recorded in 2011, while the largest glacier volume (0.061 Gt) and the lowest annual runoff (0.86 mm.w.e.) occurred in 2000 (Fig. 2.5b). Furthermore, glacier loss across the HKH was disproportionately greater than the accumulated mass, with the mass balance fluctuating between −84.33 mm.w.e in 2010 (the highest) and −507.39 mm.w.e in 2018 (the lowest).

Geographically, all countries in the HKH region possess glaciers, except Bangladesh, which has a subtropical climate with elevations below 1100 m. Over the past two decades, glacier size and volume have varied across the region (Fig. A1). While all countries saw a reduction in glacier volume of less than 0.012 Gt, glacier area declined significantly in all countries except Pakistan, where an increase of 0.023 km^2 was observed. Only a few countries experienced an increase in runoff due to glacier melt: Bhutan (0.218 Mt), China (0.236 Mt), Myanmar (0.005 Mt), and Nepal (0.244 Mt). Among these, Bhutan exhibited the most pronounced increase in runoff, peaking at 1.55 Mt in 2020, signaling the severe impact of glacier loss in the region.

While most HKH countries observed negative glacier mass balances from 2000 to 2020, Afghanistan, India, and Pakistan recorded minor increases in their mass balances, with gains of 410.5 mm.w.e, 69.86 mm.w.e, and 319.61 mm.w.e, respectively (Fig. A2).

Several regions exhibited positive mass balances over specific periods, indicating a temporary reduction in glacier melt and an increase in snow and ice accumulation. Notable positive mass balances occurred in Afghanistan (2002–2005, 2009–2010, 2012, and 2015) and Pakistan (2003–2005, 2009–2010, 2012, 2015, and 2019), with peak values recorded between 2009 and 2010 (Afghanistan: 560.72 mm.w.e and 676.58 mm.w.e; Pakistan: 294.15 mm.w.e and 352.84 mm.w.e). India also saw occasional positive balances, especially in 2010 (50.51 mm.w.e) and 2015 (3.44 mm.w.e).

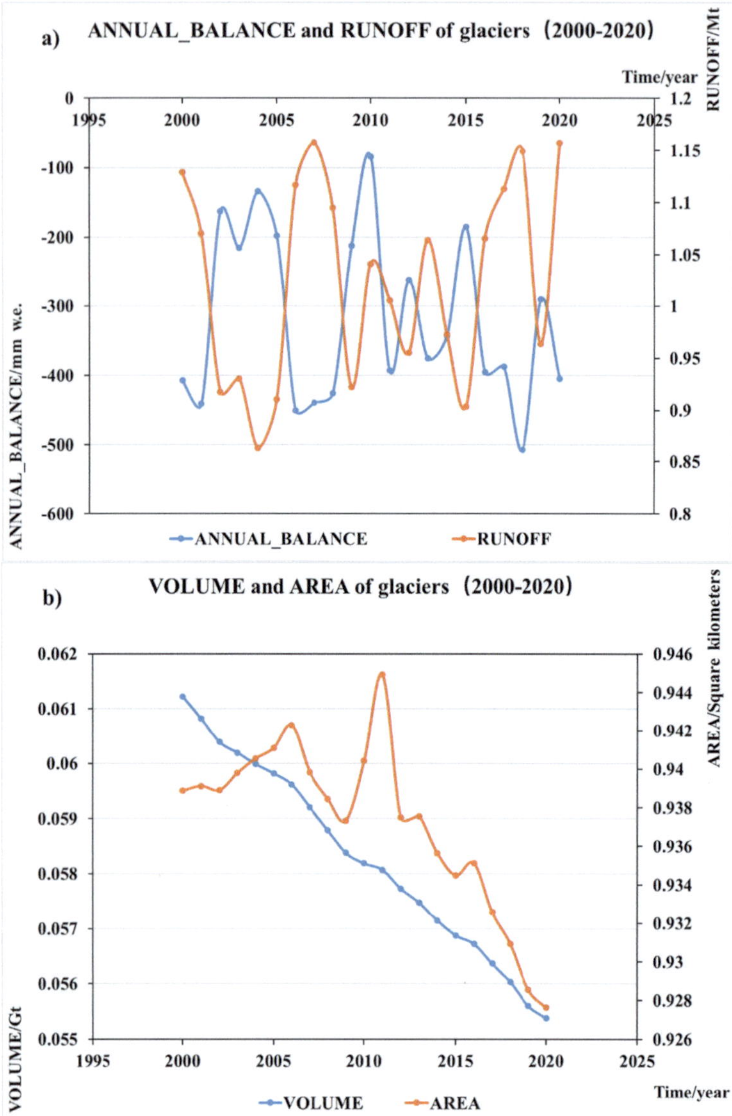

Fig. 2.5 A graphical representation of the HKH's. **a** Annual glacier balance and runoff (2000–2020); **b** Volume and area of glaciers (2000–2020)

Notably, India, Nepal, and Pakistan, with glacier areas exceeding 1.0 km^2 and volumes approaching 0.04 Gt, dominate the HKH in terms of glacier size. In contrast, Myanmar has the smallest glaciers, with areas of less than 0.4 km^2 and volumes below 0.01 Gt.

2.4.2 *Change in Snow Cover*

Over the same two-decade period (2000–2020), glaciers across the HKH region continued to melt, yet this did not result in a corresponding increase in snow cover. However, between 1985 and 2020, snow cover in the region did see a substantial increase, from 83,223.8 km^2 (1.86%) to 97,106.2km^2 (2.17%), a gain of 13,882.4 km^2 (Table 2.1). The greatest snow cover (2.21% of the total HKH area) was recorded in 1995. From a decadal perspective, the HKH region experienced a notable increase in snow cover, particularly from 2010 to 2020, which saw an additional 5,641.0 km^2 of snow cover. The slowest growth occurred between 2000 and 2010, with just 256.7 km^2 added during that period.

The overall increase in snow cover from 1985 to 2020 was felt across all countries in the region. China and Myanmar accounted for the greatest (over 45,000 km^2) and smallest (roughly 250 km^2) portions of total snow cover in the HKH, respectively. Snow cover in each country peaked at different times. For instance, Afghanistan and Pakistan recorded their highest snow cover in 2015 (4,537.5 km^2 and 23,575 km^2, respectively), while China's peak was in 1995 (43,151.1 km^2) and India's was in 1995 as well (25,468.2 km^2). Additionally, Nepal's highest snow cover (6,792.3 km^2) occurred in 2000, Bhutan's in 2005 (1,471.5 km^2), and Myanmar's in 2020 (250.0 km^2).

In general, all HKH countries, except Bhutan (which saw a slight decrease of 1.93% in snow cover, or a loss of 24.8 km^2 between 1990 and 1995), experienced a significant increase in snow cover in 1995 (Table 2.1). This rapid increase was responsible for the 11.9% overall growth in snow cover across the HKH, with particularly notable surges in Afghanistan (17.03%), India (19.79%), and Pakistan (12.97%).

2.5 Rivers and Lakes

As mentioned in subsequent sections, the HKH region has experienced increased runoff due to significant glacier melting. Glacier melt-induced runoff, combined with climate change, has significantly increased in the size of rivers and lakes in the region. The total surface area of rivers and lakes in the HKH expanded from 15,712.72 km^2 in 1987 to 81,322.97 km^2 in 2020, an increase of 65,610.24 km^2 (Fig. 2.6a).

When categorizing the water into permanent and seasonal types, a similar trend in water area changes was observed. From 1987, the water area steadily increased, with permanent water growing from 13,878.83 to 44,381.13 km^2 by 1992, and seasonal water expanding from 1,833.89 to 9,752.90 km^2 over the same period. However, between 1993 and 1996, a significant decline occurred, with the most dramatic reductions in 1995 ($-$12,883.17 km^2 for permanent water and $-$3,978.37 km^2 for seasonal water). By 1995, the total area of river and lake water was at its second-lowest level, only surpassing the 1987 figures.

Table 2.1 The snow coverage of the HKH and its changes from 1985 to 2020

Country	Area coverage (sq. km)									
	1985	1990	1995	2000	2005	2010	2015	2020	Growth % (1985–2020)	Differences (1985–2020)
Afghanistan	2404.3	2827.7	3309.5	2692.2	2994.2	3374.3	4537.5	3323.9	38.3	919.6
Bangladesh	0.0	0.0	0.0	0.0	0.0	0.0	0.0	0.0	0.0	0.0
Bhutan	1198.0	1282.3	1257.4	1309.0	1471.5	1237.6	1338.4	1213.2	1.3	15.2
China	41,603.3	42,236.7	45,750.9	45,062.3	45,052.6	42,170.3	41,892.2	43,619.0	4.9	2015.8
India	16,504.8	19,397.3	23,236.8	17,136.5	17,566.0	19,793.4	18,243.8	20,781.5	25.9	4276.7
Myanmar	219.3	231.4	236.0	228.9	214.9	217.5	228.1	250.0	14.0	30.7
Nepal	6015.8	5886.3	6446.5	6792.3	6291.9	6119.3	6002.9	6437.1	7.0	421.3
Pakistan	15,406.3	16,596.6	18,749.3	18,109.8	17,998.3	18,695.2	23,575.4	21,644.1	40.5	6237.8
Overall HKH coverage	83,202.4	88,260.8	98,722.9	91,186.6	91,424.6	91,436.3	95,668.2	97,080.1	16.7	13,877.7

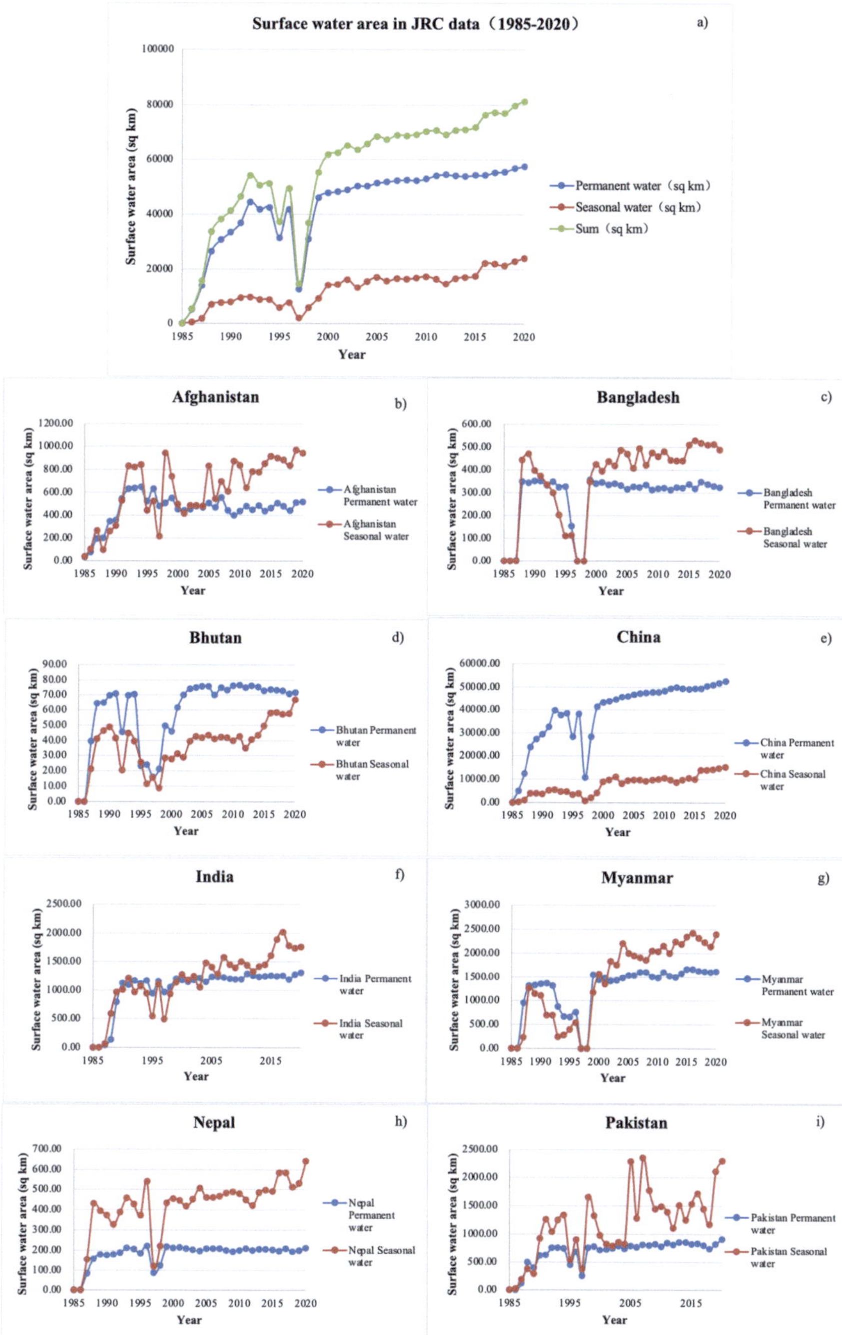

Fig. 2.6 A graphical representation of the surface water area. **a** The entire HKH region; **b–i** the 8 HKH countries

China, a major HKH country, accounts for a significant portion of the region's rivers and lakes, with a maximum water area capacity of 67,760.08 km^2 (Fig. 2.6e). In contrast, Bhutan has the smallest river and lake area, with a peak of 138.48 km^2.

From 1987 to 2020, rivers and lakes expanded across all HKH countries, although the trajectory of water area changes varied. In 1987, water areas were notably small across the region, with values ranging from 4.33 km^2 in Bangladesh to 13,337.31 km^2 in China. Over time, large fluctuations in water area occurred. For example, in 2020, the total water area was largest in Bhutan, China, Myanmar, Nepal, and Pakistan. However, Afghanistan, Bangladesh, and India did not see their highest water areas in 2020, with Afghanistan peaking at 1,488.15 km^2 in 1994, Bangladesh at 866.32 km^2 in 2017, and India at 3,282.25 km^2 in 2017.

Regarding permanent and seasonal water areas, each country showed varying patterns of peak values at different times. For instance, Afghanistan reached its largest permanent water area (649.24 km^2) in 1994, Nepal in 1999 (221.61 km^2), Bangladesh in 2011 (355.49 km^2), Bhutan in 1996 (76.63 km^2), and Myanmar in 2015 (1,668.66 km^2). In contrast, China, India, and Pakistan experienced their peak areas of permanent water in 2020, with China reaching 52,450.98 km^2, India 1,309.65 km^2, and Pakistan 910.08 km^2.

Notably, China holds the largest surface water area in the HKH, contributing over 80% of the entire region's water. Additionally, seasonal water size also peaked in 2020 for Bhutan (66.87 km^2), China (15,309.10 km^2), Myanmar (2,398.46 km^2), Nepal (641.40 km^2), and Pakistan (2,294.71 km^2) (Fig. 2.6).

2.6 Land Use and Land Cover

The HKH region's diverse environment is shaped by its complex climate, elevation, and terrain. Vegetation, comprising grasslands and forests, dominates over 60% of the region's landscape. Grasslands, the most widespread cover type, occupy between 2,232,433.42 km^2 (49.83%) and 2,769,855.43 km^2 (61.82%), while forests span between 581,988.61 km^2 (12.99%) and 1,126,629.60 km^2 (25.15%) (Fig. 2.7). From 1985 to 2020, total grassland and forest cover experienced slight declines of −0.34 and −0.06%, respectively, although fluctuations were observed within these categories. A significant finding was the sharp decrease in forest cover in 1995, down by −12.14% from 1990 levels, with much of this loss converting into grassland (up by 12.11%). This trend highlights deforestation driven by human activity and climate change.

Recovery of forest cover occurred by 2000, reaching 25.05%, but was followed by slower deforestation, resulting in a loss of −0.99% (−11,206.09 km^2). Vegetation patterns varied significantly across HKH countries. China, Pakistan, and Afghanistan were predominantly covered by grasslands (>45%), whereas Bhutan led the forest-rich countries with over 80% forest cover. Between 1985 and 2020, grassland area increased in Bhutan (7%) and Myanmar (1.29%), while other countries saw notable declines. Forest area trends also varied: Bangladesh and Nepal recorded increases of

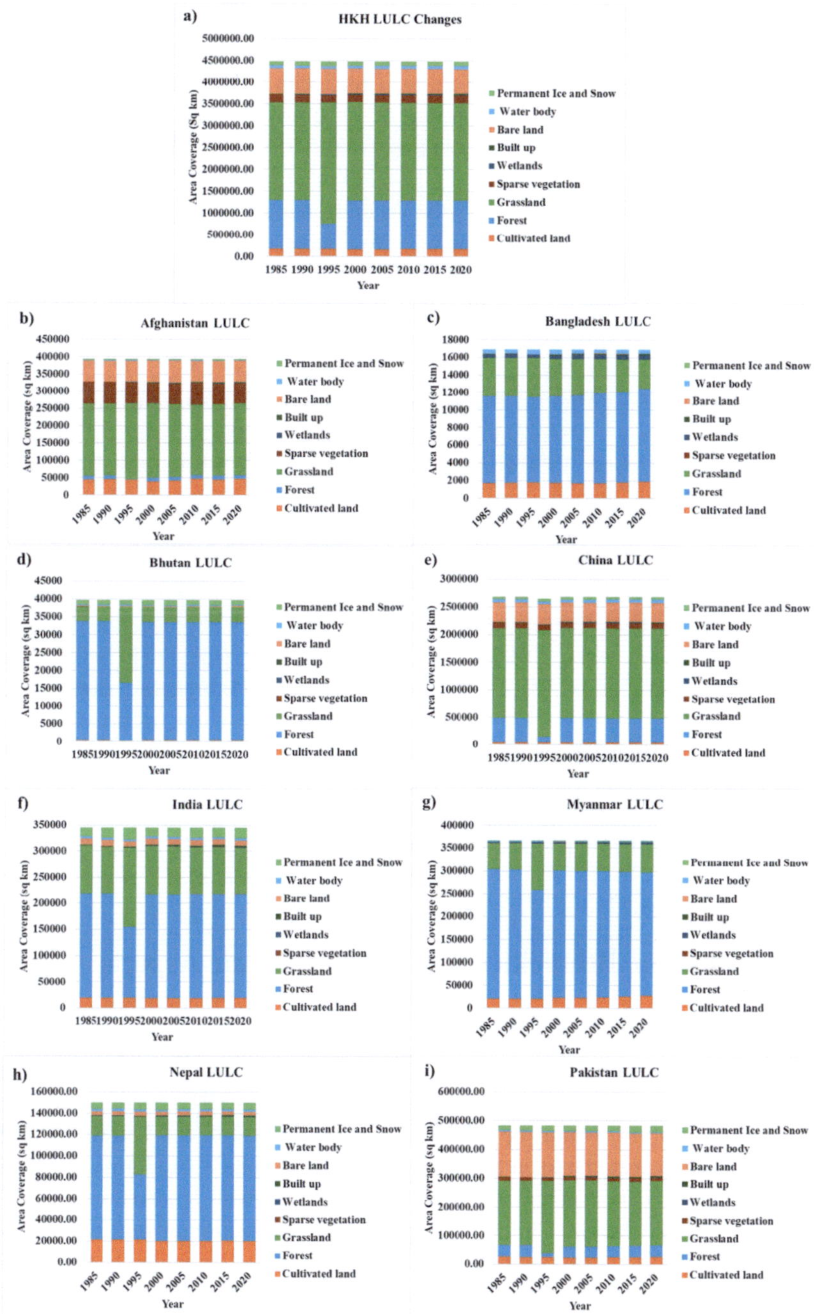

Fig. 2.7 A graphical representation of the HKH land use and land cover changes from 1985 to 2020

4.06% (686.06 km^2) and 0.58% (870.51 km^2), respectively, whereas other nations experienced losses. The forest decline in 1995, mirrored by grassland expansion, points to human influence on deforestation. This decline stabilized by 2000.

Despite agriculture being a key economic driver, cultivated land accounted for only a small fraction of the HKH's area, ranging from 3.75% to 4.13%. Between 1985 and 2020, cultivated land expanded by 7,104.24 km^2, from 177,761.92 km^2 (3.97%) to 184,866.16 km^2 (4.13%). This growth was uneven across countries: Afghanistan, Bangladesh, Bhutan, India, and Myanmar saw increases, while China (−2,252.90 km^2), Nepal (−1,073.33 km^2), and Pakistan (−38.08 km^2) experienced declines. Agricultural land exceeded 10% of total land in Nepal, Bangladesh, and Afghanistan. Peak cultivation years differed by country: India (19,430.46 km^2 in 1985), Nepal (21,659.02 km^2 in 1985), Pakistan (25,861.89 km^2 in 1985), China (42,757.12 km^2 in 1995), Bhutan (495.12 km^2 in 2015), and Afghanistan, Bangladesh, and Myanmar in 2020 (45,075.79 km^2, 1,903.27 km^2, and 28,583.52 km^2, respectively). Cultivation rates were lowest between 1990 and 2005, notably in 2005 for Bangladesh (1,728.44 km^2), China (39,593.28 km^2), India (18,858.29 km^2), and Nepal (20,323.22 km^2).

With rising global populations driving demand for food and infrastructure, the built-up area in the HKH increased significantly, from 4,781.98 km^2 (0.11%) in 1985 to 23,110.41 km^2 (0.52%) in 2020. This increase aligns with population growth and urban development. Infrastructure expansion (383.28%) outpaced the decline in bare land (−4.61%) between 1985 and 2020. China and Pakistan led in built-up area growth (8,611.15 km^2 and 4,321.25 km^2), while Bhutan and Bangladesh saw minimal increases (33.03 km^2 and 98.29 km^2). A sharp infrastructure surge occurred between 1995 and 2000, with a 185.69% increase in built-up area (10,880.80 km^2), tripling or quadrupling previous levels across countries. This development surge coincided with forest loss (−48.31%) during 1990–1995, indicating human-driven land use changes for infrastructure and economic purposes.

2.7 Population, Ethnicities, and Socioeconomic Aspects

Since the HKH region encompasses mountainous areas across eight countries, obtaining precise population statistics presents significant challenges. By analyzing the spatial distribution of population density within each HKH country, we estimated the population within the HKH boundaries using Eq. 2.1 below. In 2000, the HKH region was home to an estimated 164 million people, predominantly from China, India, and Pakistan. India had the largest share, with approximately 39.96 million people, followed by China and Pakistan, each exceeding 30 million. Bhutan, in contrast, had the smallest population, with around 577,000 inhabitants. Over time, the HKH population steadily increased, reaching 190 million in 2010 (a 16.21% rise from 2000) and 238 million in 2020 (a 24.81% increase). This represents a total population growth of 45.05% (73.95 million) between 2000 and 2020 (Table 2.2).

$$POP_n^x = PD_n^x \times A^x \tag{2.1}$$

where POP_n^x represents the population of country x in the HKH in the nth year, PD_n^x represents the nth year population density of country x within the HKH boundary, A^x represents the Area of country x within the HKH boundary.

In 2010, China, India, and Pakistan each had populations exceeding 38 million in the HKH, and by 2020, this figure rose to over 40 million in all three countries. India consistently led the region, with approximately 49 million people in 2010 and 58 million in 2020. Bhutan remained the least populous country in the region throughout this period despite its gradual population growth.

Table 2.2 Population and the ethnicity of the HKH countries

Countries	Population (million)			Population growth (%)	Ethnicity within the HKH
	2000	2010	2020		
Afghanistan	17	20.56	27.14	57.07	Pashtun (the largest ethnic group), Tajiks, Hazara, Uzbek, Aimaks, Baloch, Nuristani, Turkmen, and others
Bangladesh	3.62	4.39	5.71	57.73	Bengali group/ethnicity (Indo-Aryan ethnic group)
Bhutan	0.58	0.69	0.83	43.81	Ngalop, Sharchop, Lhotshampa, and the rural Bhutanese tribe
China	34.77	38.07	42.95	23.53	Tibetan, Qiang minority and the Mongolian ethnic minority
India	39.96	49.49	58.18	45.57	Assamese, Bengali (also practiced in Bangladesh), Gujarati, Hindi, Kashmiri, Konkani, Marathi and Punjabi
Myanmar	13.16	12.12	12.44	−5.48	Bamar, Karen, Shan, Kachin, Mon and Chine ethnic group
Nepal	23.26	25.71	29.35	39.74	Chhetri, Brahman-Hill, Magar, Tharu, Tamang, Kami, Musalman, Yadav and Rai ethic groups
Pakistan	31.52	39.74	50.25	59.40	Baloch and Pashtun ethnic groups, and others
TOTAL	164.15	190.77	238.10	45.05	

The substantial population of the HKH region is characterized by a diverse range of ethnic groups, which vary geographically. In Nepal, where the entire country lies within the HKH, prominent ethnic groups include Chhetri, Brahman-Hill, Magar, Tharu, Tamang, Kami, Musalman, Yadav, and Rai (Table 2.2). Afghanistan, a significant portion of which falls within the HKH, is home to various ethnic groups, including Pashtuns (the largest group), Tajiks, Hazara, Uzbek, Aimaks, Baloch, Nuristani, and Turkmen. Bangladesh, primarily associated with its Islamic majority, consists predominantly (98%) of Bengalis, a subgroup of the Indo-Aryan ethnic family extending into West Bengal, India.

In Bhutan, the ethnic composition is broadly categorized into four groups: Ngalop in the north and west, Sharchop in the east, Southern Lhotshampa, and indigenous rural tribes. China, which encompasses the largest geographical area of the HKH, is known for its ethnic diversity. The Tibetan, Qiang, and Mongolian minorities are the primary ethnic groups within its HKH portion. India, the second most populous country in the HKH, includes Indo-Aryan-speaking populations such as Assamese, Bengali, Gujarati, Hindi, Kashmiri, Konkani, Marathi, and Punjabi. Myanmar, part of which falls within the HKH, recognizes eight major ethnic groups; however, only six—Bamar, Karen, Shan, Kachin, Mon, and Chin—are significant in the HKH. Finally, Pakistan's HKH population comprises diverse ethnicities, including Baloch and Pashtuns, along with groups like Shina, Brahui, Kho, and Kohistani.

The population in the HKH region continues to grow, making it increasingly important to ensure the well-being of its inhabitants. Agriculture remains the primary source of income for people living in the mountainous areas, as they utilize the region's fertile soils to cultivate grains, fruits, and vegetables. Agricultural practices in the mountain regions of each HKH country make a significant contribution to national agricultural output, which in turn influences GDP growth. Due to limited data on the GDP of mountain-specific regions within HKH boundaries, an analytical approach was developed using Eq. 2.2 to estimate GDP outputs for each country's mountainous zones.

Between 2000 and 2020, China experienced a remarkable GDP increase of 1,112.5%, reaching a total of 14.69 trillion USD (Table 2.3). Agricultural production played a crucial role in this growth. According to our HKH GDP estimation, the agricultural output in China's HKH region grew by an astonishing 1,240%—from 33.36 billion USD in 2000 to 447 billion USD in 2020—significantly outpacing other HKH countries (Fig. 2.8). India ranked second among HKH nations, with GDP rising from 17.7 billion USD in 2000 to 111.3 billion USD in 2020, representing a 530% increase (Fig. 2.8).

$$GDP_x^n = GP_n \times PD_x^n \times A_x \tag{2.2}$$

where GDP_x^n represents the nth year GPD of country x within the HKH boundary, GP_n represents the nth year country per capita GDP, PD_x^n represents the nth year Population density of country x within the HKH boundary zone and A_x is the surface area of each HKH country.

Table 2.3 Annual GDP data of the HKH countries

Countries	GDP (Billion USD)					2000–2020 difference	2000–2020 change (%)
	2000	2005	2010	2015	2020		
AFG	0	6.20	15.86	19.13	19.96	19.96	0
BGD	53.37	69.44	28	195.08	373.90	320.53	600.59
BTN	0.42	0.80	1.55	2.00	2.46	2.03	479.01
CHN	1,211.33	2,285.96	6,087.19	11,061.57	14,687.74	13,476.41	1112.53
IND	468.39	820.38	1,675.62	2,103.59	2,671.60	2,203.20	470.37
MMR	8.91	11.99	49.54	59.69	79.05	70.14	787.65
NPL	5.49	8.13	16.00	24.36	33.43	27.94	508.52
PAK	99.48	145.21	196.71	299.96	300.43	200.94	201.98

Source World bank

Fig. 2.8 The GDP output of each HKH countries

Fruit production in Pakistan's mountainous regions—where crops such as apples, plums, cherries, figs, grapes, peaches, and pears are grown—has been a significant contributor to the country's overall fruit production, including exports (Hussain et al., 2016; ICIMOD, 2019). This sector has supported the country's GDP growth, from 99.5 billion USD in 2000 to 300.4 billion USD in 2020. Within the Hindu Kush-Himalayan (HKH) region, Pakistan's fruit production played a key role in the 227% GDP growth, increasing from 20.31 billion USD in 2000 to 66.44 billion USD in 2020. While all other countries in the HKH region also experienced substantial GDP growth, Bhutan showed the least GDP output, despite a notable 498% growth—from 0.42 billion USD in 2000 to 2.5 billion USD by 2020 (Fig. 2.8).

Livestock farming and the cultivation of neglected and underused species (NUS) also represent important livelihoods in the HKH region. These practices offer significant potential to diversify the food supply and increase farmer incomes. Cereal and cash crops considered as NUS, including barley, sorghum, millets, wheat, beans, and pulses, are gaining demand due to their nutritional benefits and resilience to climate change. For instance, the rising popularity of cakes made from wheat, millets, and oatmeal flour in Nepal's supermarkets has led to greater market demand and increased awareness of NUS, prompting greater interest among mountain dwellers in cultivating these crops (ICIMOD, 2019). Additionally, non-farming activities, such as tourism, handicrafts, and souvenir production, make significant contributions to the income of the HKH region.

Although the full potential of this concept remains underexplored due to a lack of comprehensive knowledge, the stunning landscapes of the HKH mountain areas make them highly suitable for non-agricultural industries, particularly tourism. The region stands to benefit significantly from the integration of various businesses, such as handicraft and souvenir production (including wood carving, blankets, baskets, jewelry, shawls, and embroidery). This integration could create numerous opportunities for local residents, enhancing their livelihoods and diversifying the regional economy (Rasul & Hussain, 2015).

Appendix

See Figs. A1 and A2.

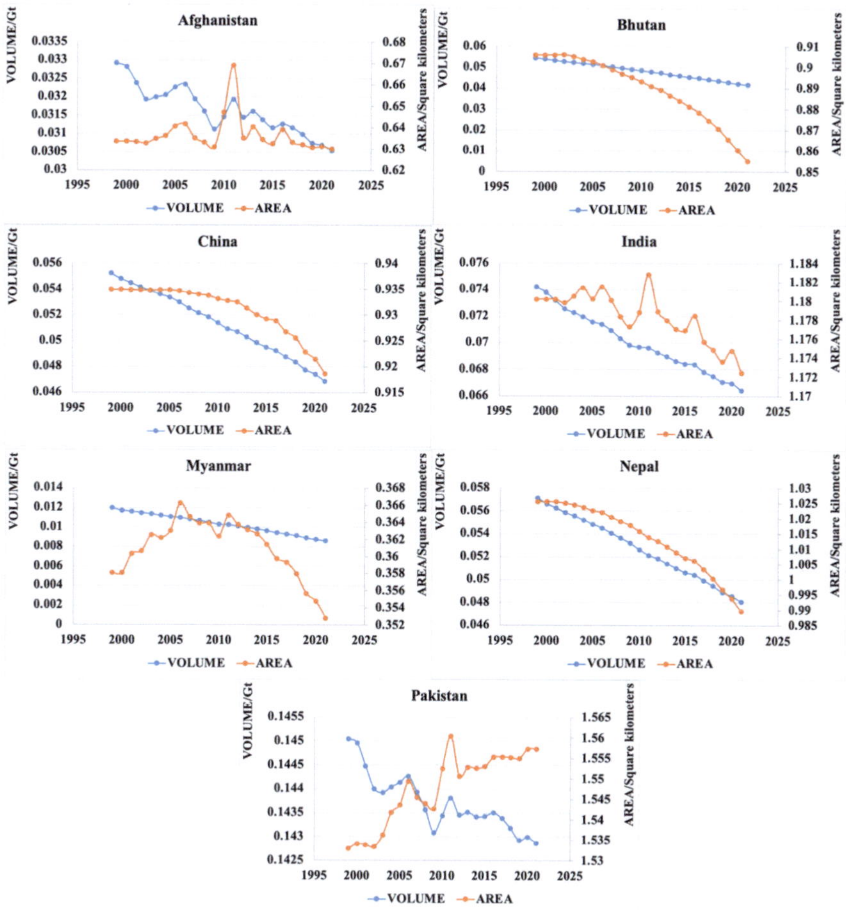

Fig. A1 The HKH countries' regional glacier area and volume

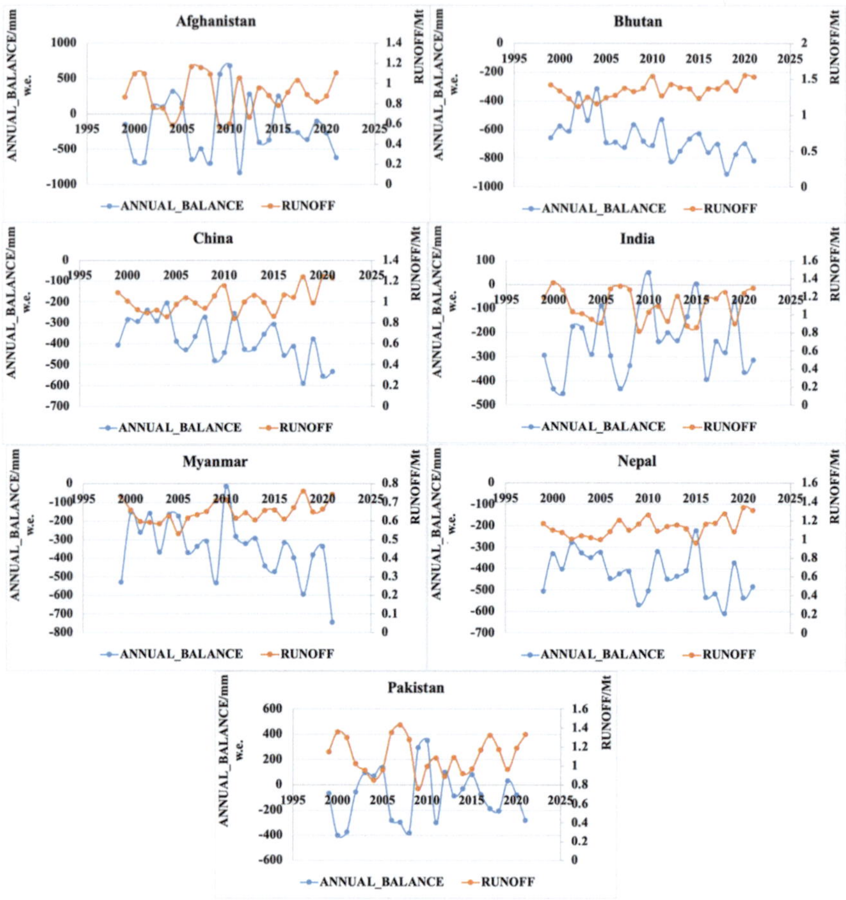

Fig. A2 Annual glacier mass balance of HKH countries by region

References

ICIMOD. (2019). The Hindu Kush Himalaya Assessment (P. Wester, A. Mishra, A. Mukherji, & A. B. Shrestha, Eds.). Springer International Publishing. https://doi.org/10.1007/978-3-319-92288-1

Hussain, A., Rasul, G., Mahapatra, B., & Tuladhar, S. (2016). Household food security in the face of climate change in the Hindu-Kush Himalayan region. *Food Security, 8*(5), 921–937. https://doi.org/10.1007/s12571-016-0607-5

Rasul, G., & Hussain, A. (2015). Sustainable food security in the mountains of Pakistan: Towards a policy framework. *Ecology of Food and Nutrition, 54*(6), 625–643. https://doi.org/10.1080/03670244.2015.1052426

Chapter 3
Remote Sensing Monitoring and Assessment of Sustainable Agricultural Development in the HKH Region

Food production and consumption systems have shaped human culture and the environment for millennia (Desor, 2017). Due to their central role, food and nutrition security have become top priorities in national and international development and environmental agendas (ICIMOD, 2019). In the Hindu Kush Himalayan (HKH) region, hunger and malnutrition are widespread. Of the 795 million undernourished people globally, 52% (415 million) are from HKH countries (FAO-IFAD-WFP, 2015). Achieving food and nutrition security is crucial for the HKH region, where agriculture serves as a primary means of subsistence. However, to meet food production demands, farmers in the region must contend with significant challenges, including climate change, water scarcity, and the depletion of agricultural land.

Climate change poses an increasing threat to agriculture, exacerbating food insecurity as it makes it harder to feed a rapidly growing population (Aryal et al., 2014; Rasul, 2010; Shrestha et al., 2017). In Nepal's dry Trans-Himalayan areas, for example, 50% of households face food insecurity due to poor agricultural practices and the high market costs of food transported from lower altitudes (Pandey, 2016). Himalayan warming has triggered several changes, including a shift from snow-dominated to rain-dominated precipitation systems, early flowering and fruiting, upslope treeline shifts, glacial lake expansion and outbursts, rapid snowmelt, and more frequent extreme events such as precipitation, drought, and desertification (Aryal et al., 2014; Kc et al., 2022; Lamsal et al., 2017). These climatic changes have resulted in water shortages, altered harvest timings, pest and disease outbreaks, changed phenology, and a shortage of feed and fodder, all of which impact food production (Rasul, 2010; Pandey, 2019).

At higher elevations, a significant concern is the reduced water supply caused by declining snowfall (ICIMOD, 2019). In Upper Mustang, Nepal, farmers must travel long distances to collect water due to the declining availability of water, as Mustang is an alpine desert region with an average annual rainfall of just 307 mm (Pandey et al., 2018). In India, water scarcity has led rice farmers to adjust their transplanting and harvesting schedules (Rymbai & Sheikh, 2018). Since 90% of agriculture in

S. Lu et al., *Remote Sensing Assessment of Sustainable Development in the Hindu Kush-Himalayan (HKH) Region*, https://doi.org/10.1007/978-981-96-8675-9_3

the Himalayas is rain-fed, water stress has a detrimental impact on both agricultural productivity and farmers' incomes (Kc et al., 2022). Reduced rainfall and more frequent, intense droughts are pushing farmers to cultivate crops in areas that were previously unsuitable for agriculture.

The forests in the agro-ecological system are critical for sustaining agricultural productivity. Historically, the residents of the HKH have relied on forests for their livelihoods, but several factors are increasing the vulnerability of these areas to risks and uncertainties. Population growth and subsequent changes in land use are leading to a reduction in forested areas, making agriculture in the HKH increasingly unsustainable (ICIMOD, 2019). Additionally, shifting land use patterns and declining annual rainfall have disrupted the hydrological regime, leading to reduced groundwater recharge and rapidly depleting water supplies (Tiwari & Joshi, 2012). Gaining insights into the resilience and sustainability of these crucial agricultural systems is essential, especially in light of the challenges faced by the HKH agricultural sector.

Based on distribution mapping data from the 2020 land use change product (Zhang et al., 2024), our study has identified six major food-producing zones within the Hindu Kush-Himalayan region: northern Afghanistan, southern Pakistan-northern India, Nepal, China's Qinghai and Yunnan provinces, and Myanmar. Figure 3.1 illustrates the geographical scope of the study, which encompasses various agro-ecological zones and provides a deeper understanding of agricultural productivity in the region. This broad coverage will allow us to examine land use changes in detail and their effects on food production, providing valuable new insights into the resilience and sustainability of these vital agricultural systems.

3.1 Monitoring and Analysis of Cultivated Land Distribution and Its Changing Trends

Monitoring the distribution and changes in cultivated land is crucial for ensuring food security, managing resources, protecting the environment, fostering economic development, and guiding policy formulation (Ntihinyurwa & de Vries, 2021; Wang, 2022). By tracking changes in farmland distribution, we can assess both the potential and risks associated with food production. This assessment aids governments and agricultural organizations in developing more effective food production strategies and response measures, which are essential for maintaining national and regional food security (Salois & Moss, 2010). Precise mapping of farmland allows for better planning and utilization of water and land resources, optimizing irrigation systems, enhancing resource use efficiency, and minimizing waste. Environmental protection is directly impacted by agricultural activities, which can lead to soil degradation, water pollution, and biodiversity loss. By monitoring changes in farmland, we can identify and address these environmental issues promptly, enabling the formulation of effective environmental protection measures. Accurate farmland monitoring also supports sustainable agricultural practices by providing the necessary data to mitigate

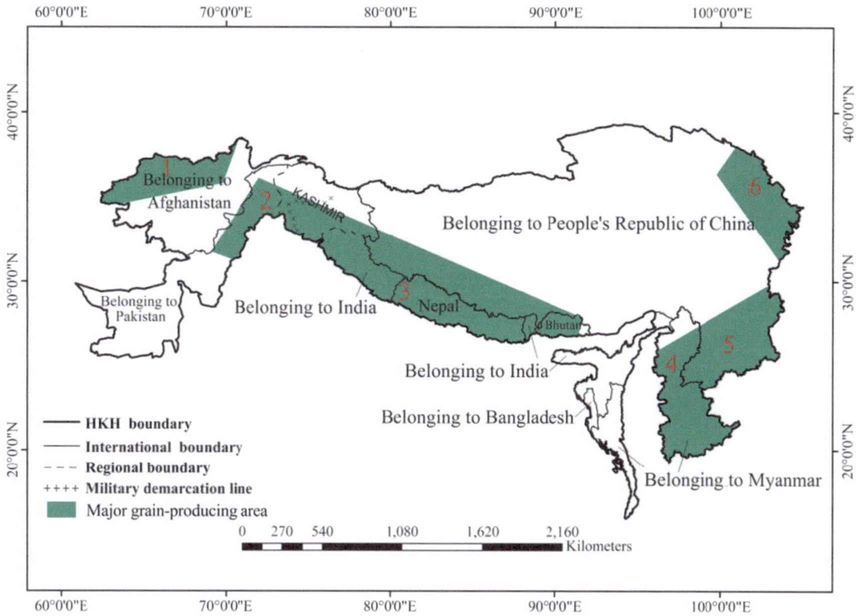

Fig. 3.1 Main grain-producing areas in the Hindu Kush-Himalayan region

negative environmental impacts and promote biodiversity conservation. Advancements in remote sensing technology, combined with machine learning, multi-source data fusion, and the analysis of specific variables, have revolutionized the monitoring of farmland distribution and trends. These technologies enable the monitoring of farmland from local to global scales, providing critical tools for understanding the environmental and socioeconomic impacts of agricultural activities. The integration of these advanced methods enhances our ability to manage agricultural landscapes effectively and sustainably, ensuring that agricultural production aligns with broader environmental and economic goals.

3.1.1 Methodology

(1) Data collection and processing

In the six major grain-producing areas of the Hindu Kush-Himalayan region, satellite image data spanning from 2000 to 2020 were collected. Given that no single satellite could cover this entire period, multiple remote sensing satellite images were utilized as data sources. The study incorporated Landsat-5, Landsat-7, Landsat-8 surface reflectance products, and Sentinel-2 Level 1C image data products from various regions. These datasets were essential for analyzing and identifying the distribution of cultivated land across different areas.

For tasks such as extracting farmland distribution and subsequently classifying major food crops, satellite images from different time periods were employed. Landsat-5 imagery was used to collect data for the years 2000 and 2010, providing a comprehensive perspective of the earlier years. Subsequently, Landsat-8 imagery and Sentinel-2 satellite imagery were used for the analysis in 2020.

By leveraging these different satellite images across various time periods, this research conducted a comprehensive analysis of the distribution and classification of major food crops in the grain-producing regions of the Hindu Kush-Himalayan area. This approach allowed for a detailed temporal analysis of farmland distribution and crop classification over the past two decades. Tables 3.1, 3.2, 3.3, and 3.4 show the satellite parameters for Landsat 5, 7, 8, and Sentinel 2, respectively.

Table 3.1 Landsat 5 TM remote sensing image parameters

Name	Description	Resolution (m)
B1	Blue	30
B2	Green	30
B3	Red	30
B4	Near infrared	30
B5	Shortwave infrared 1	30
B6	Thermal infrared 1. Resampled from 60 to 30 m	30
B7	Shortwave infrared 2	30

Table 3.2 Landsat 7 ETM + remote sensing image parameters

Name	Description	Resolution (m)
B1	Blue	30
B2	Green	30
B3	Red	30
B4	Near infrared	30
B5	Shortwave infrared 1	30
B6_ VCID_1	Low-gain thermal infrared 1. This band has expanded dynamic range and lower radiometric resolution (sensitivity), with less saturation at high Digital Number (DN) values. Resampled from 60 to 30 m	60
B6_ VCID_2	High-gain Thermal Infrared 1. This band has higher radiometric resolution (sensitivity), although it has a more restricted dynamic range. Resampled from 60 to 30 m	60
B7	Shortwave infrared 2	30
B8	Panchromatic	15

Table 3.3 Landsat 8 OLI Remote sensing image parameters

Name	Description	Resolution
B1	Coastal aerosol	30 m
B2	Blue	30 m
B3	Green	30 m
B4	Red	30 m
B5	Near infrared	30 m
B6	Shortwave infrared 1	30 m
B7	Shortwave infrared 2	30 m
B8	Band 8 Panchromatic	15 m
B9	Cirrus	30 m
B10	Thermal infrared 1, resampled from 100 to 30 m	30 m
B11	Thermal infrared 2, resampled from 100 to 30 m	30 m

Table 3.4 Sentinel-2 MSI remote sensing image parameters

Name	Description	Resolution (m)
B1	Aerosols	60
B2	Blue	10
B3	Green	10
B4	Red	10
B5	Red edge 1	20
B6	Red edge 2	20
B7	Red edge 3	20
B8	NIR	10
B8A	Red edge 4	20
B9	Water vapor	60
B10	Cirrus	60
B11	SWIR 1	20
B12	SWIR 2	20

(2) Farmland identification method

Sample Production: Manually producing training samples by overlaying remote sensing images with regional vector files is a crucial step in preparing for supervised classification algorithms. This process relies on ground survey data, historical data, or high-resolution images from the study area. By manually interpreting and labeling these images based on sample points from existing field surveys, we ensure the accuracy and representativeness of the training data. This meticulous approach guarantees that the training samples accurately reflect the conditions on the ground, thus enhancing the performance of the classification algorithms.

Feature Selection: The feature selection process considers a variety of attributes, including spectral features, vegetation index features, and terrain features at different phases. Spectral features encompass the blue, green, red, and near-infrared bands from Sentinel-2 and Landsat series satellites. For cultivated land classification data, Landsat-5 imagery was used to collect data for the years 2000 and 2010, providing a comprehensive perspective of the earlier years. Subsequently, Landsat-8 imagery and Sentinel-2 satellite imagery were used for the analysis in 2020. Additionally, remote sensing indexes and texture features are incorporated into the feature set. This comprehensive integration of features provides robust support for identifying and extracting farmland. Table 3.5 shows the features used.

Optimization of Feature Set: To determine the most effective feature set for farmland identification, we employ Recursive Feature Elimination (RFE). This method systematically evaluates and ranks the importance of various features, ensuring that the final selection optimally enhances the classification accuracy. By refining the feature set through RFE, we can identify the most relevant attributes for accurate farmland extraction, thereby improving the overall reliability and effectiveness of the remote sensing analysis.

In this study, we utilized four vegetation indices of Difference Vegetation Index (DVI), Normalized Difference Vegetation Index (NDVI), Enhanced Vegetation Index (EVI), and Ratio Vegetation Index (RVI) as classification features (Miura et al., 2000; Pearson, 1974). These indices were selected due to their effectiveness in distinguishing different types of vegetation. Additionally, we generated the GLCM (Gray Level Co-occurrence Matrix) feature image using the red band of the image. GLCM is a statistical method that captures the texture characteristics of an image by calculating the frequency of different gray levels occurring together in a specified direction. This technique was widely used in image processing and analysis to extract texture information (Haralick et al., 1973).

Table 3.5 The feature set to be optimized

Feature type	Feature	
Optical features	Blue Green Red Near infrared Red (NIR)	
Index features	Difference vegetation index (DVI) Normalized Difference Vegetation Index (NDVI) Enhanced Vegetation Index (EVI) Ratio Vegetation index (RVI)	
Textural features	Grey Level Co-occurrence Matrix (GLCM)	Contrast Dissimilarity Homogeneity Energy Correlation Mean Entropy

The GLCM matrix elements indicate how often pairs of pixels with specific values and in a specified spatial relationship occur in an image, thus providing detailed texture representation. This study calculated these texture features to enhance the accuracy of vegetation classification. The specific formulas used to generate each of the selected features are detailed below.

$$DVI = NIR - R \tag{3.1}$$

$$NDVI = \frac{NIR - R}{NIR + R} \tag{3.2}$$

$$EVI = \frac{NIR - R}{NIR + 6\text{R - }7.5\text{B} + 1} \times 2.5 \tag{3.3}$$

$$RVI = \frac{NIR}{R} \tag{3.4}$$

$$\begin{cases} Mean = \dfrac{\sum_i \sum_j P(i,j) * i}{n^2} \\ Contrast = \sum_i \sum_j (i-j)^2 P(i,j) \\ Dissmilarity = \sum_i \sum_j |i-j| P(i,j) \\ Homogeneity = \sum_i \sum_j \dfrac{1}{(a+b)^2+1} P(i,j) \\ Energy = \sum_i \sum_j P(i,j)^2 \\ Correlation = \dfrac{\sum_i \sum_j (i-\mu_i)(j-\mu_j) P(i,j)}{\sigma_i \sigma_j} \\ Entropy = \sum_i \sum_j P(i,j) \log_2 P(i,j) \end{cases} \tag{3.5}$$

After initially obtaining the feature set, this study employs recursive feature elimination (RFE) to optimize the features. RFE is a feature selection method used to identify the most relevant features improving model performance (Chen & Jeong, 2007). The specific process is detailed as follows:

Stepwise Feature Selection: RFE begins with stepwise feature selection. Initially, the model is trained using the entire set of features. Each feature is then ranked based on its importance to the model's performance.

Feature Combination Iteration: In each iteration, RFE eliminates the least important features identified in the current model. The model is then retrained with the remaining features. This iterative process continues, progressively narrowing down the feature set.

Accuracy Evaluation: After each iteration, the model's performance is evaluated using cross-validation or other performance assessment methods. This step determines which features make significant contributions to the model's accuracy and overall performance.

Determination of the Best Feature Set: The iterative process continues until either a specified number of features is reached, or the performance metric stabilizes. The features retained at the end of this process are selected as the final optimal subset.

By systematically removing less important features, RFE reduces the feature space, enhances the model's generalization ability, mitigates overfitting, and simplifies the interpretation of the model results.

Ultimately, we selected the blue, green, red, NDVI, RVI, and EVI features from each satellite image dataset as the input features for the random forest classifier. This selection was based on the optimization process facilitated by RFE, ensuring that the most relevant features were used to improve the model's predictive performance. Figure 3.2 shows the process of feature selection.

Model training and classification were conducted using the Random Forest algorithm, as outlined by Breiman (2001). This approach was implemented on the Google Earth Engine (GEE) platform, leveraging its powerful computational capabilities and extensive data repository. Key parameters were meticulously configured to optimize the performance of the Random Forest model. These parameters included 'n_estimators' (the number of trees in the forest), 'max_depth' (the maximum depth of each tree), 'min_samples_split' (the minimum number of samples required to split an internal node), 'min_samples_leaf' (the minimum number of samples required to be at a leaf node), 'max_features' (the number of features to consider when looking

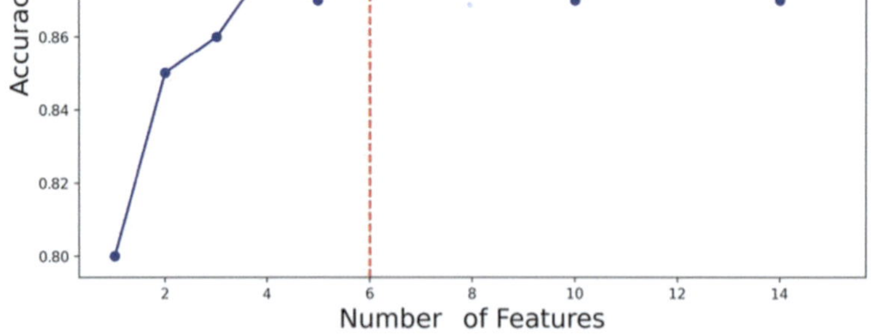

Fig. 3.2 Feature recursive elimination iteration curve

Table 3.6 Random forest parameter table

Random forest model parameters	Parameter values
n_estimators	200
max_depth	20
min_samples_split	2
min_samples_leaf	1
max_features	Sqrt (square root of the number of features)
bootstrap	TRUE

for the best split), and 'bootstrap' (whether bootstrap samples are used when building trees). The specific parameter values selected for this study are detailed in Table 3.6.

To enhance the model's accuracy and generalization ability, a comprehensive optimization process was undertaken. This involved fine-tuning the parameters and incorporating cross-validation techniques. Cross-validation helps in assessing the model's performance on different subsets of the training data, ensuring that the model is not overfitting and can generalize well to unseen data. Once optimized, the trained Random Forest model was deployed across the entire study area to identify and classify cultivated land with high precision.

The classification results obtained from the model were then rigorously validated using field sampling data. This ground-truth data provided a benchmark to evaluate the accuracy of the remote sensing image classification. By comparing the model's predictions with the actual field observations, we were able to assess the reliability and robustness of the classification results. This iterative process of model training, optimization, and validation ensured that the final output was both accurate and reliable, effectively supporting the study's objectives.

Figure 3.3 shows the specific distribution of arable land, while Figs. 3.4 and 3.5 illustrate the changes in arable land area statistics.

3.1.2 *Cultivated Land Identification Results and Analysis*

The classification results obtained through this method were validated for accuracy in 2020 using sample points acquired through visual interpretation, achieving an overall accuracy of 85.7%. Table 3.7 presents the confusion matrix.

Over the span of each decade, meticulous documentation has been undertaken to track the distribution and alterations in cultivated land across the study area (See Table 3.7). This comprehensive analysis culminates in the presentation of an area statistics table and a cultivated land distribution map, both of which serve as invaluable tools for understanding the evolving landscape. An overarching observation gleaned from regional statistical findings reveals a notable increase in the overall cultivated land area within the study area. Particularly striking is the significant 11% expansion of cultivated land in the Hindu Kush-Himalayan region from 2000 to 2020.

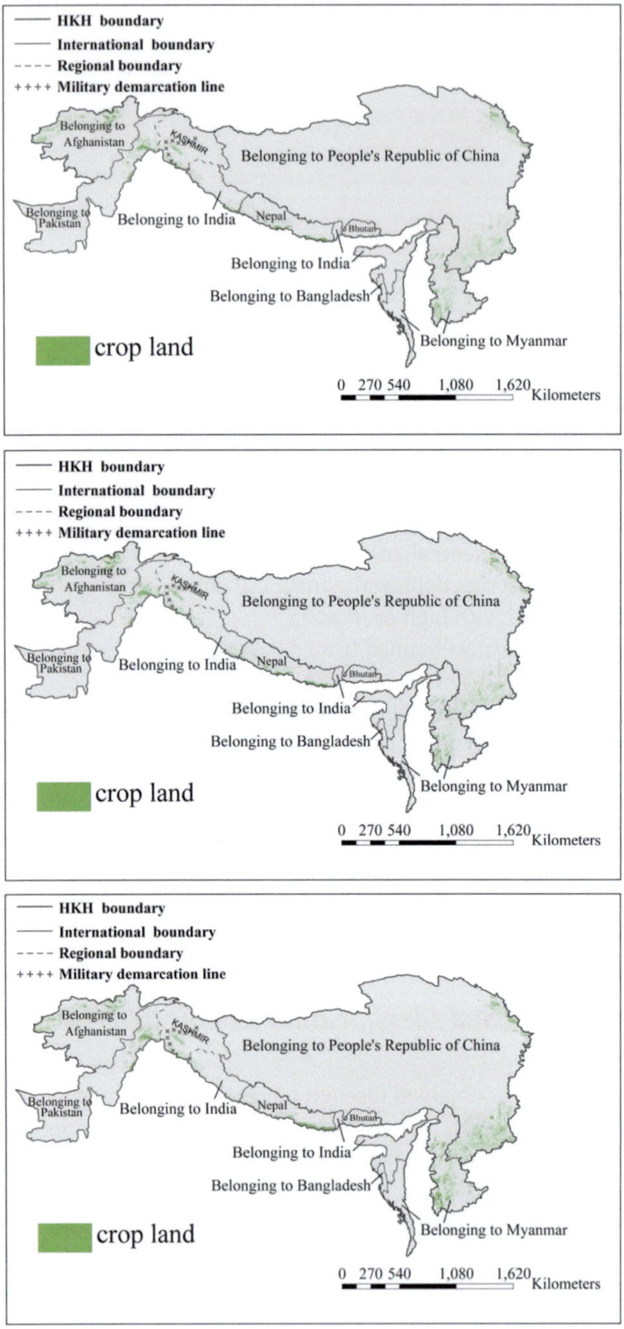

Fig. 3.3 Spatial distribution map of 2000–2020 grain producing areas in the HKH

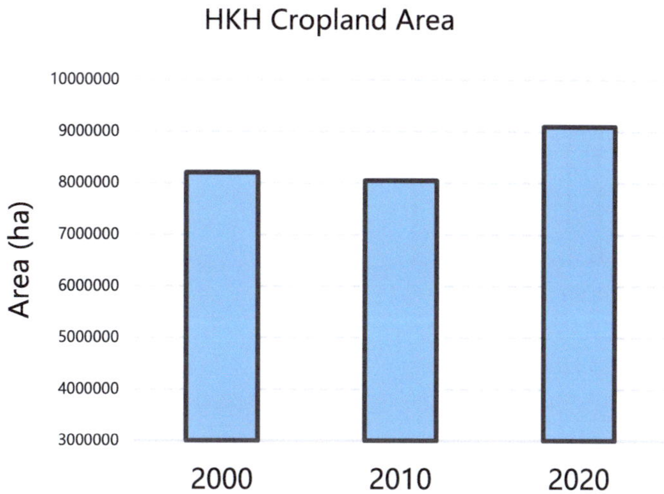

Fig. 3.4 Changes in overall cultivated land area (ha)

Noteworthy is the steady year-on-year growth in the main grain-producing regions of Myanmar and southern China, where cultivated land has surged by 25 and 20%, respectively. This dynamic landscape underscores the complex and evolving nature of agricultural development and cultivated land patterns in Myanmar.

The agricultural trajectory of Myanmar exhibits a multifaceted evolution marked by significant governmental emphasis on agriculture as a cornerstone of national economic advancement over the past two decades. Supported by scholarly insights (Zhi-jian, 2011), it's evident that Myanmar has witnessed a commendable annual agricultural added value growth rate averaging 11% from 2000 to 2009. However, a discernible deceleration is observed post-2010, with the growth rate tapering to a mere 2% annually (Tun et al., 2023; Wang et al., 2023), mirroring the trend observed in cultivated land area changes. Notably, following a substantial decline in cultivated land area in Pakistan and Afghanistan in 2010, the regions have shown signs of stabilization, with the cultivated land area by 2020 nearly restoring to the levels observed in 2000. Specifically, while Afghanistan's cultivated land area remained relatively stable from 2000 to 2020, Pakistan experienced a marginal 9% decrease in cultivated land area during the same period. Meanwhile, other regions have witnessed minimal fluctuations in cultivated land area, generally within a modest 2% range over the past two decades (Table 3.8).

The meticulous examination of cultivated land dynamics illuminates a nuanced narrative of agricultural evolution, intertwined with broader socioeconomic and political trajectories. Myanmar's agricultural landscape, shaped by governmental priorities and regional dynamics, stands as a testament to the intricate interplay between policy imperatives, economic forces, and environmental considerations. As cultivated land patterns continue to evolve, informed insights gleaned from rigorous

Fig. 3.5 Changes in cultivated land area in major grain-producing areas

Table 3.7 Confusion matrix of cultivated land classification

	Predict cultivated	Predict non-cultivated	Total
Actual cultivated	120	22	142
Actual non-cultivated	10	73	83
Total	130	95	225
Accuracy			0.8578

Table 3.8 Changes in cultivated land area in major grain-producing areas in each decade

Regions (ha)	Year		
	2000	2010	2020
Myanmar	1,252,204.09	1,515,017.99	1,573,632.11
Afghanistan	1,574,786.07	1,462,702.77	1,598,124.78
Nepal	273,518.46	278,018.46	281,168.46
Pakistan	2,972,132.28	2,576,323.71	2,727,533.97
South China	1,878,306.14	1,912,526.98	2,270,448.16
Qinghai, China	725,948.73	712,173.69	727,119.81

statistical analyses serve as invaluable tools for policymakers, researchers, and stakeholders alike in navigating the complex terrain of agricultural development and land management.

In the Hindu Kush-Himalaya (HKH) region, a meticulous approach was undertaken to assess land use dynamics. A systematic 100 km × 100 km grid was overlaid across the entire HKH area, facilitating a gridded examination of land use changes. This grid partitioned the region, allowing for precise analysis of variations in cultivated land area from 2000 to 2020. The observed changes were classified into five distinct levels based on percentage alterations: 'Significantly Reduced, 'Slightly Reduced', 'Remained Stable', 'Slightly Increased', and 'Significantly Increased' (Fig. 3.6). These categorizations corresponded to changes exceeding 5%, within the range of 2–5%, and within 2%, respectively.

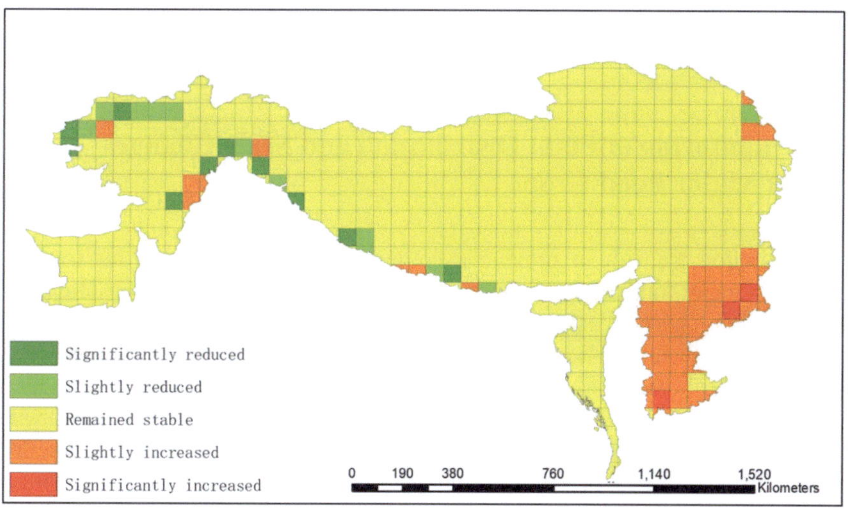

Fig. 3.6 2000–2020 change monitoring analysis of cultivated land area in main grain-producing areas

Upon scrutinizing the analytical outcomes, discernible trends emerged. Notably, certain southeastern regions, particularly the primary cultivation zones of Myanmar and southern China, exhibited a marked surge in cultivated land. Conversely, other areas demonstrated a mix of increments and decrements; however, the overall landscape portrayed a state of relative stability. This comprehensive analysis sheds light on the intricate dynamics of land use evolution within the HKH region, emphasizing the nuanced interplay between socio-economic factors, environmental influences, and agricultural practices.

3.1.3 Summary

The study undertook a meticulous examination of cultivated land dynamics over the span of each decade, providing a comprehensive understanding of the evolving landscape. This analysis revealed a notable 11% expansion of cultivated land in the Hindu Kush-Himalayan region from 2000 to 2020, alongside steady growth in main grain-producing regions like Myanmar and southern China. The agricultural trajectory of Myanmar exhibited a multifaceted evolution, shaped by government priorities and regional dynamics. While Pakistan and Afghanistan initially experienced declines in cultivated land area, both regions showed signs of stabilization by 2020. The study employed a systematic grid approach to assess land use dynamics in the Hindu Kush-Himalaya region, revealing discernible trends of surge in certain southeastern regions, while overall portraying a state of relative stability. These findings underscore the complex interplay between socio-economic factors, environmental influences, and agricultural practices in shaping cultivated land patterns.

3.2 Monitoring of Plant Distribution and Scale Changes of Major Food Crops

In this section we aim to provide a comprehensive methodology for tracking changes in major food crop distribution and scale within the Hindu Kush-Himalaya (HKH) region. This methodology is crucial for understanding agricultural dynamics, ensuring food security, and informing policy decisions in a region vulnerable to climate change and socio-economic pressures. Through the utilization of Landsat and Sentinel satellite data processed via Google Earth Engine (GEE), coupled with ground reference data collected through field surveys and high-resolution satellite images, the study employs advanced remote sensing techniques for crop identification and classification. By constructing training and feature sets and employing a random forest classifier, the methodology enables accurate identification of major food crops such as rice and wheat, considering regional phenological variations and terrain characteristics. This meticulous approach not only facilitates the monitoring of agricultural changes but also provides valuable insights into the complex interplay between land use, environmental factors, and agricultural practices in the HKH region.

3.2.1 Methodology

(1) Data collection and processing

To comprehensively gather data for each region in the HKH, we employed a meticulous process utilizing various satellite platforms and data sources. The collection effort encompassed Landsat-5, 7, and 8, as well as Sentinel-2 data with cloud cover below 20%, leveraging the capabilities of Google Earth Engine (GEE). Landsat 5 and 7 satellite top-of-atmosphere (TOA) data were utilized, ensuring a thorough temporal coverage of the regions under study. Landsat-5 imagery was used to collect data for the years 2000 and 2010, providing a comprehensive perspective of the earlier years. Subsequently, Landsat-8 imagery and Sentinel-2 satellite imagery were used for the analysis in 2020. From 2014 to 2017, Landsat 8 TOA data was incorporated, while Sentinel-2 data Level-1C was integrated from 2018 to 2020, specifically for cultivated land crop identification.

A critical aspect of our data collection methodology involved the preprocessing of Sentinel-2 Level-1C data. This process entailed several steps to ensure data accuracy and compatibility. Initially, radiometric and geometric corrections were applied to enhance the quality of the imagery. Subsequently, orthorectification techniques were implemented to rectify image distortion caused by terrain variations. Furthermore, radiometric offsets were added to refine the data, enhancing its precision and reliability. Finally, spatial registration with sub-pixel accuracy was conducted, aligning the imagery with a global reference system to facilitate consistent analysis across regions. This comprehensive preprocessing approach ensured that the collected data met the highest standards of quality and integrity, laying a solid foundation for subsequent analysis and interpretation.

(2) Main food crop identification methods

Training Set Construction: To build an effective training set, we embark on a meticulous process of data collection within the study area. This involves gathering ground reference data through various means, including meticulous field surveys and extraction of feature information from high-resolution satellite images. Through these efforts, we obtain sample point feature labels that serve as the foundation for our training set. It is imperative that the collected ground reference data accurately represent the diverse features present in the study area. Consequently, we meticulously select representative sample locations on the image for each feature class. These selections prioritize samples with conspicuous spectral characteristics, distinct ground object types, and a consistent absence of changes in ground object types over a span of 20 years. Each sample location is meticulously labeled to denote its corresponding category, ensuring that an ample number of sample points are selected for each ground object category to comprehensively capture their distinct spectral characteristics.

Feature Set Construction: Utilizing the robust capabilities of the Google Earth Engine (GEE) platform, we embark on the preprocessing of the collected Landsat and Sentinel-2 data to prepare our feature set. This preprocessing journey encompasses essential steps such as radiometric calibration and atmospheric correction, crucial for eliminating various errors and mitigating interference. Once the images are suitably processed, they undergo training and prediction phases utilizing a sophisticated random forest classifier. Our approach doesn't end here; we meticulously account for the diverse phenological stages exhibited by different crops across various regions. This entails an in-depth investigation into the crop growth processes specific to each region. Leveraging remote sensing time series images corresponding to these distinct growth stages, we construct a comprehensive classification feature set (Table 3.9). This feature set encompasses an array of parameters, including various vegetation indices such as the normalized vegetation index (NDVI), enhanced vegetation index (EVI), and Soil Adjusted Vegetation Index (SAVI), alongside spectral characteristics like blue, green, red, and near-infrared bands. Furthermore, considering the unique characteristics of rice crops, we incorporate the Normalized Water Index (NDWI) into our feature set. Through meticulous feature optimization methods, we ascertain the most effective feature sets tailored for identifying wheat and rice crops. By carefully crafting our training and feature sets, we lay the groundwork for robust and accurate classification models, which is pivotal for extracting valuable insights and supporting informed decision-making in agricultural and environmental contexts.

Analyzing the characteristic values of key phenological periods during crop growth within time series data provides a robust method for distinguishing crops from other vegetation types. By leveraging historical crop phenology data specific to each region, this approach allows for the targeted selection of remote sensing images at appropriate time phases, enhancing the accuracy of crop type identification. This strategic selection mitigates the risks associated with blindly choosing remote sensing data, as highlighted in prior studies (Bargiel, 2017; Hao et al., 2018). The methodology adopted in this study capitalizes on multi-temporal data to construct time series features, facilitating the identification and classification of crops with precision.

In the realm of model training and classification, recognizing the diverse landscape of crop categories and their growth patterns across different regions is paramount. To address this variability, separate models were developed for the primary food crops cultivated in six key production areas, including Pakistan, Myanmar, southern

Table 3.9 Collection of classification characteristics of major food crops

Crop type	Features
Wheat	Blue, Green, Red, NDVI, RVI, SAVI, DEM-SLOPE
Rice	Blue, Green, Red, NDVI, RVI, NDWI, EVI, DEM-SLOPE

Table 3.10 Confusion matrix of wheat and rice classification

	Predicted rice	Predicted wheat	Predicted non-cultivated land
Actual rice	123	3	10
Actual wheat	7	120	10
Actual non-cultivated land	26	15	151

China, Nepal, Afghanistan, and Qinghai, China. Tailoring the models to account for the unique planting structures f each region ensures optimal performance in crop classification. For instance, rice emerges as the predominant food crop in the primary production zones of southern China and Nepal, while wheat dominates in regions such as Afghanistan and Qinghai, China.

This tailored approach to model development not only acknowledges the regional nuances in crop cultivation but also underscores the significance of understanding local agricultural practices. By aligning the model with the specific crop portfolios and growth timelines prevalent in each region, the accuracy and reliability of crop classification are greatly enhanced. This nuanced strategy recognizes the complex interplay between environmental factors and agricultural practices, providing a foundation for more effective crop monitoring and management initiatives tailored to regional contexts.

3.2.2 *Study Outcome*

Data from 2020 was selected for accuracy verification, with an overall classification accuracy of 85% for wheat and rice. Table 3.10 shows the confusion matrix.

In the latest statistical analysis, a comprehensive overview of the distribution and dynamics of major crops in key grain-producing regions is presented. The data, spanning a decade, offers valuable insights into agricultural trends and land use patterns. Detailed area statistical tables and cultivated land distribution maps (Figs. 3.7, and 3.8) are appended, enriching the analysis and facilitating a deeper understanding of the findings.

Figures 3.9 and 3.10 show the changes in crop area. Regional statistical findings reveal notable shifts in cultivated land areas, particularly in major rice-producing regions. Overall, there has been a commendable 10% increase in the rice planting area, with significant expansions observed in eastern territories. Southern China, for instance, has seen a remarkable surge of 24% in rice cultivation, while neighboring Myanmar experienced a substantial 28% increase. Conversely, the rice planting area in Nepal has remained relatively stable since 2000, contrasting with a decline of approximately 8% observed in Pakistan. These variations underscore the diverse agricultural dynamics within the region.

Furthermore, the analysis highlights the stability of Wheat cultivation, with overall planting areas remaining consistent with historical levels. The fluctuations in Wheat

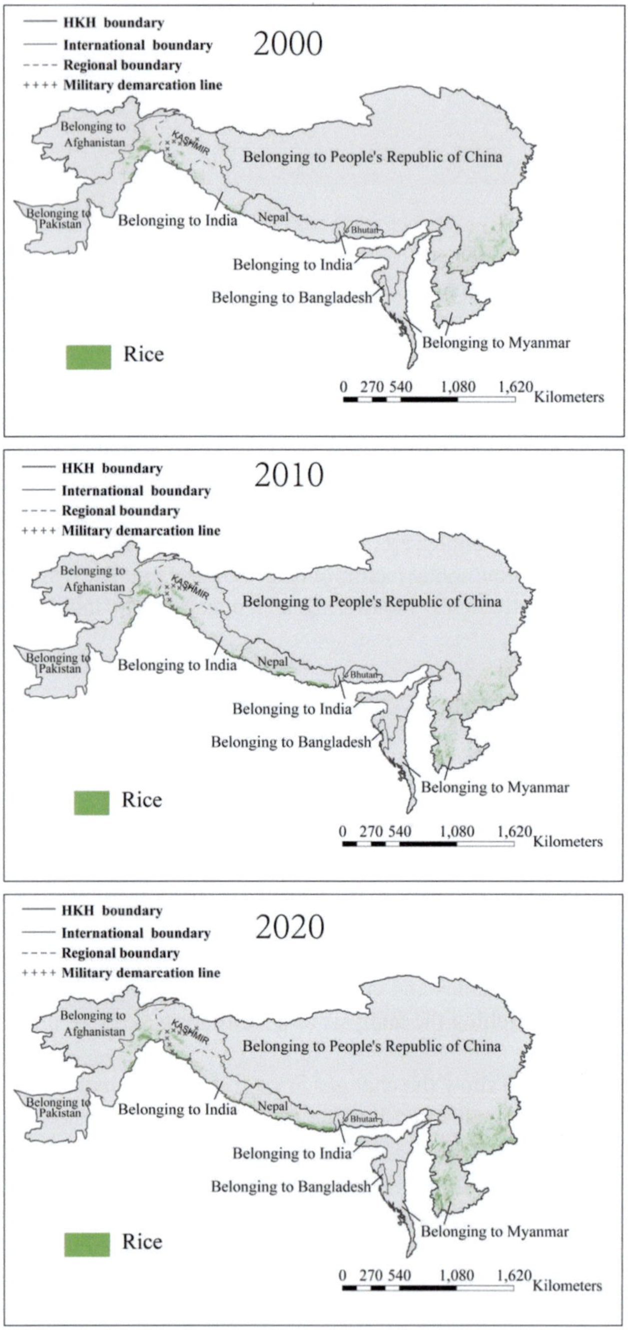

Fig. 3.7 2000–2020 distribution map of rice in major grain producing areas

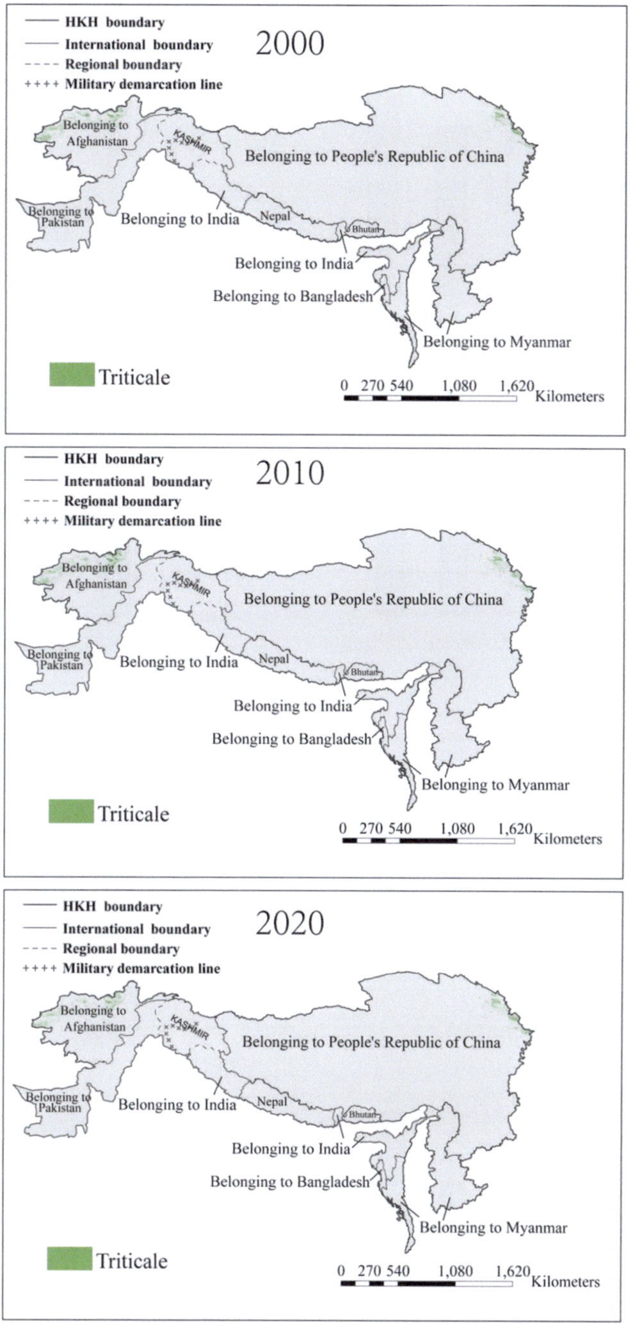

Fig. 3.8 2000–2020 distribution map of wheat in major grain producing areas

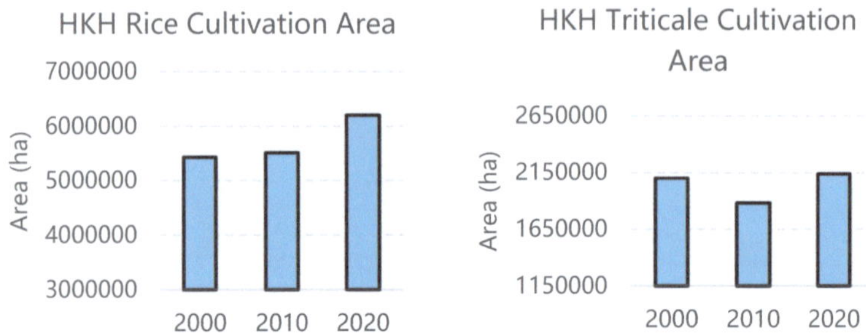

Fig. 3.9 Overall changes in crop area of rice and wheat in main grain-producing areas from 2000 to 2020 (ha)

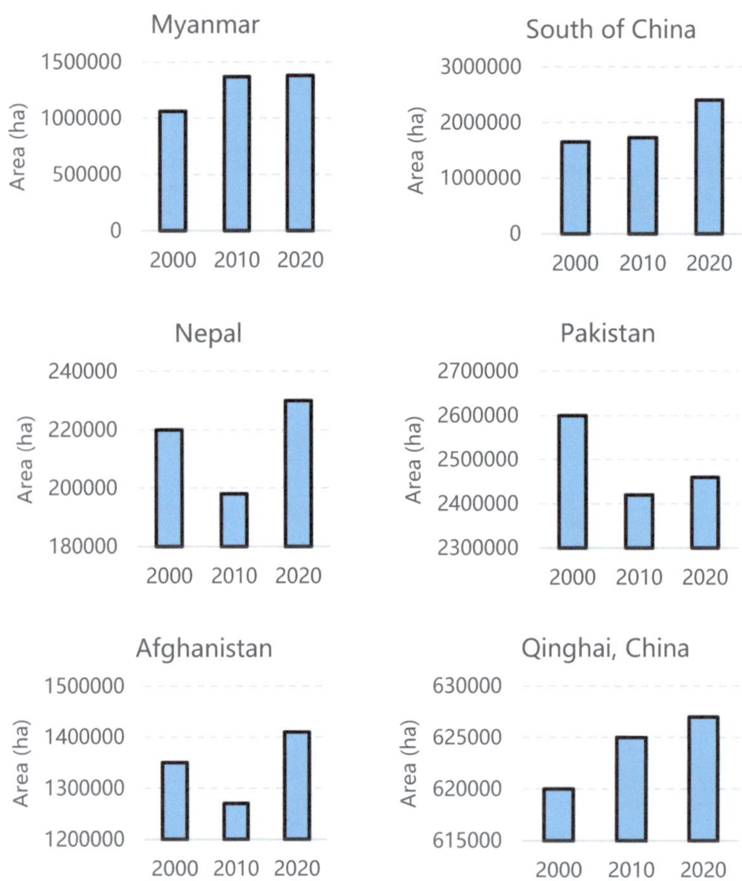

Fig. 3.10 Changes in crop area in major grain-producing areas from 2000 to 2020 (ha)

planting areas have been minimal, with changes staying within a narrow range of 5%. The resilience of wheat cultivation highlights its enduring significance in the agricultural landscape, contributing to food security and sustaining livelihoods across the studied regions.

Figures 3.11 and 3.12 show the changes in crop distribution from 2000 to 2020. Through detailed observation spanning the years 2000–2020, a discernible trend emerges regarding the rice cultivation landscape in southern China. Notably, there has been a marked increase in the rice planting area, particularly concentrated in the southeastern enclave of the Hindu Kush-Himalaya region. This expansion exhibits a notable outward trajectory, indicative of the dynamic shifts within the agricultural terrain. Conversely, while the primary rice-producing regions of Nepal and Pakistan display fluctuations in specific grid areas—some witnessing increases while others experiencing declines—the overall rice planting area has exhibited stability, reflecting a delicate balance amidst regional agricultural dynamics.

Turning attention to wheat cultivation, a nuanced pattern emerges across different regions. In Afghanistan, the wheat planting area exhibits a discernible downward trajectory in most regions, suggesting systemic challenges or shifts within the agricultural landscape. Conversely, pockets of Qinghai, China, reveal a contrasting trend with a slight uptick in wheat planting areas. This juxtaposition underscores the complex interplay of factors influencing agricultural practices, ranging from climatic conditions to socio-economic dynamics, and highlights the need for localized strategies in agricultural management and policy formulation.

The comprehensive analysis of these trends underscores the intricate dynamics shaping agricultural practices across the studied regions. It highlights the need for nuanced approaches in agricultural planning and policy formulation, considering

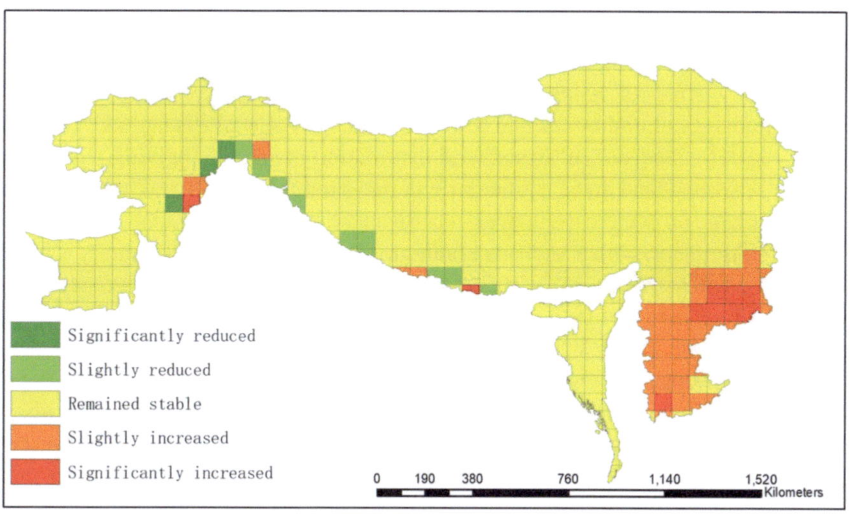

Fig. 3.11 Rice area change monitoring analysis in major grain-producing areas from 2000 to 2020

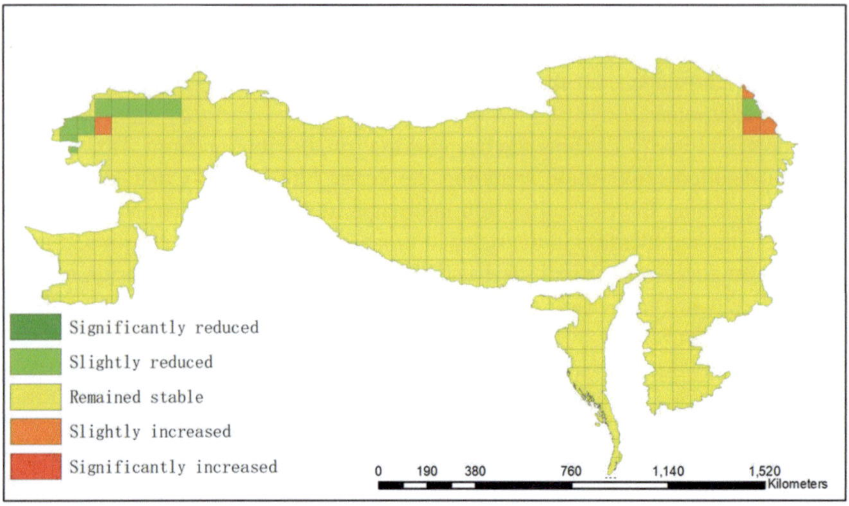

Fig. 3.12 Wheat area change monitoring analysis in major grain-producing areas from 2000 to 2020

the localized variations and systemic influences. Such insights are invaluable for fostering sustainable agricultural practices, enhancing food security, and addressing the evolving challenges posed by environmental and socio-economic factors.

3.2.3 Summary

In the latest statistical analysis, a decade-long comprehensive overview of the distribution and dynamics of major crops in key grain-producing regions reveals valuable insights into agricultural trends and land use patterns. The analysis includes detailed area statistical tables and cultivated land distribution maps, highlighting significant shifts in cultivated land areas, particularly in rice-producing regions. There has been a commendable 10% increase in overall cultivated land area, with substantial expansions in eastern territories such as southern China (24%) and Myanmar (28%), contrasting with stability in Nepal and an 8% decline in Pakistan. Wheat cultivation has remained stable, with minimal fluctuations within a narrow 5% range, underscoring its enduring significance in the agricultural landscape. The study identifies dynamic shifts in southern China's rice cultivation, while regions like Nepal and Pakistan maintain stability despite specific grid fluctuations. Conversely, wheat planting areas in Afghanistan show a downward trend, whereas Qinghai, China, experiences a slight increase, reflecting complex agricultural influences. These findings highlight the need for localized agricultural strategies to address environmental and socio-economic factors, fostering sustainable practices and enhancing food security across the studied regions.

3.3 Analysis of Major Grain Crop Yield Trends

The analysis of major grain crop yield trends in the Hindu Kush-Himalaya (HKH) region is a crucial effort in understanding the intricate dynamics of agricultural systems in this ecologically diverse and economically significant area. This comprehensive study delves into the shifting patterns of grain yield, examining the spatial and temporal variations in rice and wheat cultivation spanning the years 2000–2020. By comprehensively observing changes in planting areas and yield levels, the research aims to elucidate the underlying drivers of agricultural transformation, including climatic shifts, socio-economic factors, and land management practices. Through a multidisciplinary approach integrating remote sensing, GIS analysis, and socio-economic data, this study seeks to provide valuable insights into the sustainability of agricultural practices, the optimization of land use, and the enhancement of production efficiency in the HKH region, thus contributing to informed decision-making and policy formulation for agricultural development and environmental conservation.

3.3.1 Methodology

(1) Data collection and processing

To facilitate our research, we utilized a combination of satellite imagery and existing datasets to gather comprehensive and accurate information. Landsat-5, 7, and 8, along with Sentinel-2 data, were meticulously selected, ensuring a cloud cover of less than 20% in each region of interest. Leveraging the powerful capabilities of Google Earth Engine (GEE), we accessed and processed these datasets efficiently. This comprehensive approach to data collection ensured that our research was founded on robust and reliable information, providing a solid basis for our analyses and Summaries.

(2) Yield estimation method

In this study, we employed a method to derive the leaf area index (LAI) time series characteristics of crop growth stages by integrating the band characteristics of remote sensing images with empirical formulas. This approach allowed us to capture the dynamics of LAI over time, crucial for understanding crop growth. Subsequently, these time-series LAI crop parameters were integrated into the established GRU yield estimation model to predict wheat and rice yields. Given the diverse agricultural contexts across the HKH areas, applying a single GRU model proved challenging. Thus, we devised a strategy to map the model output results to FAO statistics of yield data per mu, accounting for regional disparities in agricultural productivity. This involved comparing the model outputs with FAO yield data for the same area and period, providing a benchmark for refining our yield estimations.

To train the GRU model, we utilized a dataset generated through simulations by the WOFOST(van Diepen et al., 1989) crop growth model, renowned for its accuracy in simulating crop growth dynamics and its relationship with environmental conditions (de Wit et al., 2019; Jin et al., 2015; S. Wu et al., 2021; Zhuo et al., 2022). Leveraging over 350,000 simulated data points across multiple scenarios, we calibrated the GRU model and established the LAI-spectrum relationship through empirical formulas. By integrating advanced modeling techniques with empirical data, we aimed to provide a comprehensive framework for estimating regional winter wheat yields. This approach not only leveraged the strengths of established crop growth models like WOFSOT but also utilized cutting-edge remote sensing technology to capture spatial and temporal variations in LAI. The detailed calibration and validation processes undertaken underscored our commitment to delivering robust and reliable yield estimations, essential for informing agricultural policies and practices in the HKH areas.

$$SeLI = \frac{R_{865} - R_{705}}{R_{865} + R_{705}} \tag{3.6}$$

$$LAI = 5.405 * SeLI - 0.114 \tag{3.7}$$

In a study conducted by Pasqualotto et al. (2019) the authors employed the Sentinel-2 LAIgreen Index (SeLI) to compare different fitting methods for the inversion of Leaf Area Index (LAI) using ground data. Specifically, they evaluated the linear, exponential, and second-order polynomial fitting results and ultimately opted for the linear fitting formula due to its superior performance. The accuracy assessment of the inversion process revealed a coefficient of determination (R2) of 0.732 and a Root Mean Square Error (RMSE) of 0.69, indicating a satisfactory level of accuracy in predicting LAI values. Subsequently, the calculated LAI distribution maps for each region were input into the yield model, allowing for the derivation of corresponding crop yields, thus demonstrating the practical utility of the LAI inversion approach within agricultural applications.

For Landsat satellite sensors, an alternative method proposed by Curran et al. (1992) was utilized for LAI inversion. This method involves the calculation of Normalized Difference Vegetation Index (NDVI) values from satellite imagery, followed by linear regression analysis to estimate LAI. By correlating NDVI with LAI through linear regression, this approach provides a means to indirectly infer LAI values from remote sensing data. This method offers a convenient and efficient means of LAI estimation at a broader spatial scale, leveraging the capabilities of satellite sensors to monitor vegetation dynamics over large areas. The utilization of such remote sensing techniques underscores the importance of integrating advanced technological approaches into ecological and agricultural research to enhance our understanding of ecosystem dynamics and improve management practices.

$$LAI = -5.4 + 15.95NDVI \tag{3.8}$$

The structure of the Gated Recurrent Unit (GRU) is fundamental yet sophisticated, comprising three main layers: input, hidden, and output. The input layer serves as the recipient of the input sequence, while the hidden layer houses the GRU units, the core elements of the model. These GRU units are distinguished by their update and output gates, which play pivotal roles in information retention. Unlike traditional recurrent neural networks, GRUs excel in preserving crucial information over extended periods, mitigating the common issue of vanishing gradients and allowing relevant data to persist throughout the learning process. This unique ability ensures that historical context remains accessible, enhancing the model's predictive accuracy. In our application, we harness the power of the GRU model to discern intricate relationships between Leaf Area Index (LAI) sequence characteristics and crop yields, focusing on staple crops like wheat and rice. By integrating GRU-based analyses with comprehensive datasets from the Food and Agriculture Organization (FAO), we can finely tune and validate our predictive models through linear corrections, thereby providing robust estimations of crop yields with practical applications in agricultural planning and resource management.

Figure 3.13 shows the framework of the estimation model. To validate the accuracy of our GRU yield estimation model, we conducted an extensive field survey in the Qinghai region in 2023 to measure actual wheat yields. The results from our model indicated a commendable level of accuracy, with an R^2 value of 0.45 and an RMSE of 20.42 kg/mu (Fig. 3.14). Figure 3.15 shows the estimated yield situation of the region. These metrics demonstrate a strong alignment between the model's estimations and the observed yields, reflecting its reliability in predicting wheat yields in the Qinghai region. Our verification process underscores the model's robustness and its ability to perform at a level consistent with established standards in yield estimation research.

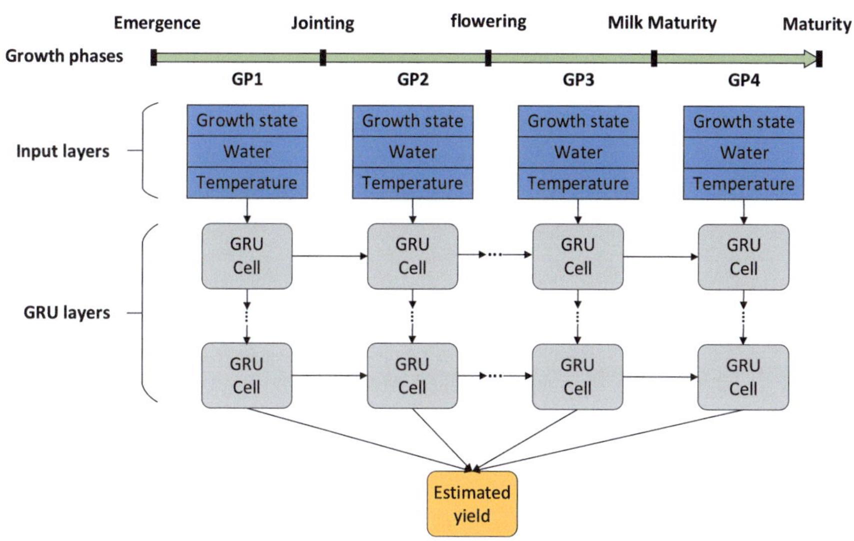

Fig. 3.13 GRU production estimation model structure

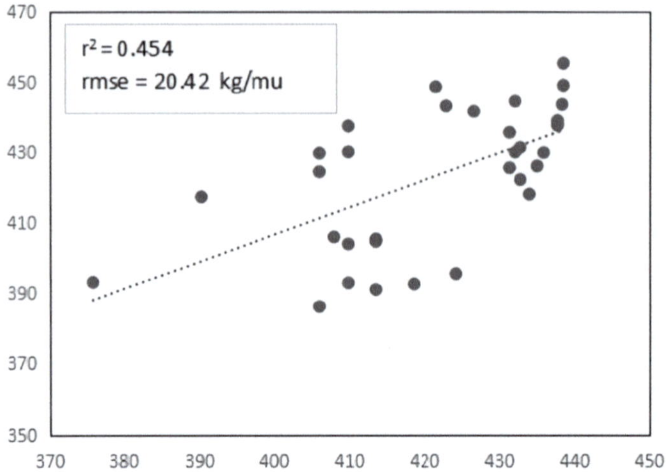

Fig. 3.14 Scatter plot of the predicted yield versus the actual yield

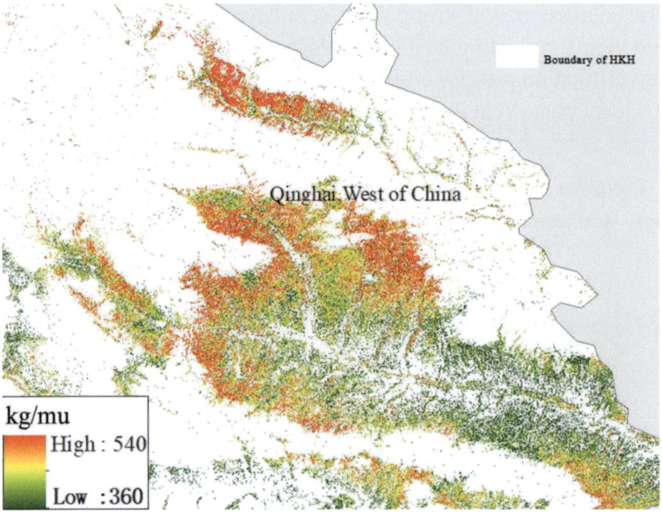

Fig. 3.15 Local regional yield estimation results (Qinghai region)

3.3.2 Results and Analysis of Crop Yield Trends and Variability in the Hindu Kush-Himalayan Region (2000–2020)

The objective of this research is to forecast the fluctuations in yield for key crops across the Himalayan region from 2000 to 2020. Our methodology involved acquiring spatial distribution data for diverse crops through the overlay of classification masks representing different crop types. Following this, we employed the GRU yield estimation model, utilizing time series data of Leaf Area Index (LAI), to project crop yields within each distinct region. Figures 3.16 and 3.17 show the yield distribution of rice and wheat in the HKH region. Our analysis reveals a discernible trend over the two-decade period, indicating a general uptrend in agricultural output across the region. The findings underscore a notable increase in yields for most crops, underscoring the overall positive trajectory in agricultural productivity within the Himalayan landscape.

Based on the estimations derived from our model, we conducted an in-depth statistical analysis of the yield changes per decade spanning from 2000 to 2020. Over this period, the Hindu Kush-Himalayan region, which serves as a pivotal agricultural production zone, has demonstrated a consistent and notable upward trajectory in grain crop output per unit area. This compelling finding underscores the ongoing enhancement of agricultural production capabilities within the region, with a concurrent rise in food crop yields per unit area. Nevertheless, juxtaposed with conventional agricultural hubs situated in plains, the crop yield per unit area in the Hindu Kush-Himalaya region remains notably lower. This disparity may be attributed to a combination of natural factors such as topography, climate, and soil conditions unique to the Hindu Kush-Himalayan region, alongside disparities in agricultural production techniques and management practices. These findings illuminate the multifaceted dynamics influencing agricultural productivity and underscore the imperative of tailored strategies to address region-specific challenges and optimize agricultural output sustainably.

Figure 3.18 specifically shows the changes in output across different regions. Between 2000 and 2020, the average per-acre rice yield in Myanmar rose from 220 to 254 kg/mu, marking a 15% increase. In southern China, the per-acre rice yield increased from 411 to 461 kg/mu, marking a 12% rise. Nepal saw a significant increase in rice yield, from 179 to 256 kg/mu, representing a 43% surge. In Pakistan, the rice yield went up from 201 to 251 kg/mu, reflecting a 24% increase. The wheat yield in Afghanistan saw a remarkable surge, rising from 48 to 135 kg/mu, a 180% increase. Similarly, in Qinghai, China, Wheat yields increased from 249 to 381 kg/mu, marking a 53% growth. These figures highlight the diverse rates of yield improvement across different regions within the HKH area, reflecting varying local conditions and agricultural practices.

3.3.3 Summary

This research provides a comprehensive analysis of the fluctuations in crop yields across the Himalayan region from 2000 to 2020, utilizing advanced GRU yield estimation models and spatial distribution data of diverse crops. Our findings reveal a significant upward trend in agricultural productivity throughout the region, marked

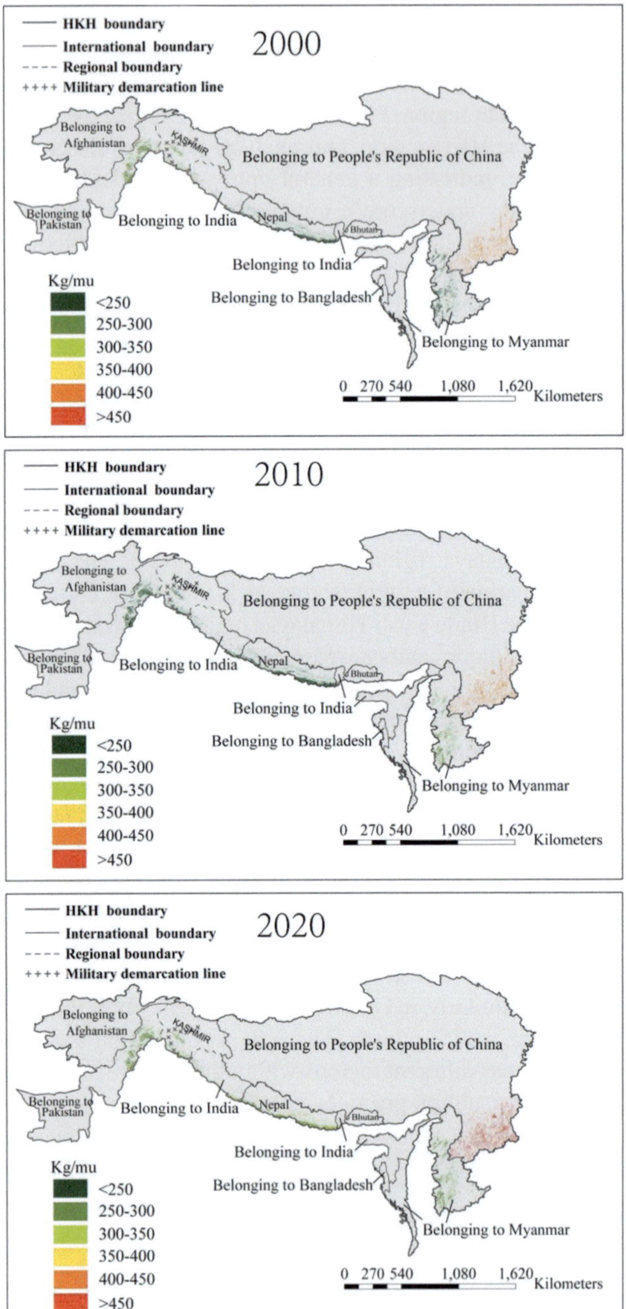

Fig. 3.16 2000–2020 rice yield distribution in the HKH region

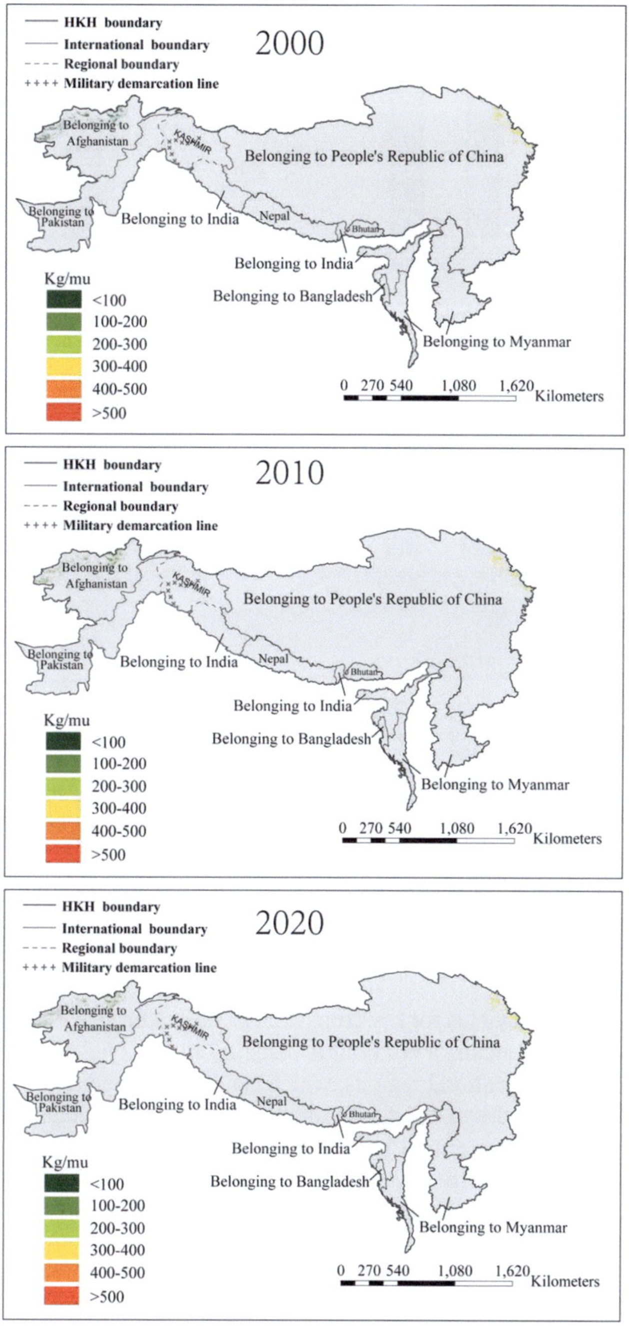

Fig. 3.17 2000–2020 wheat yield distribution in the HKH region

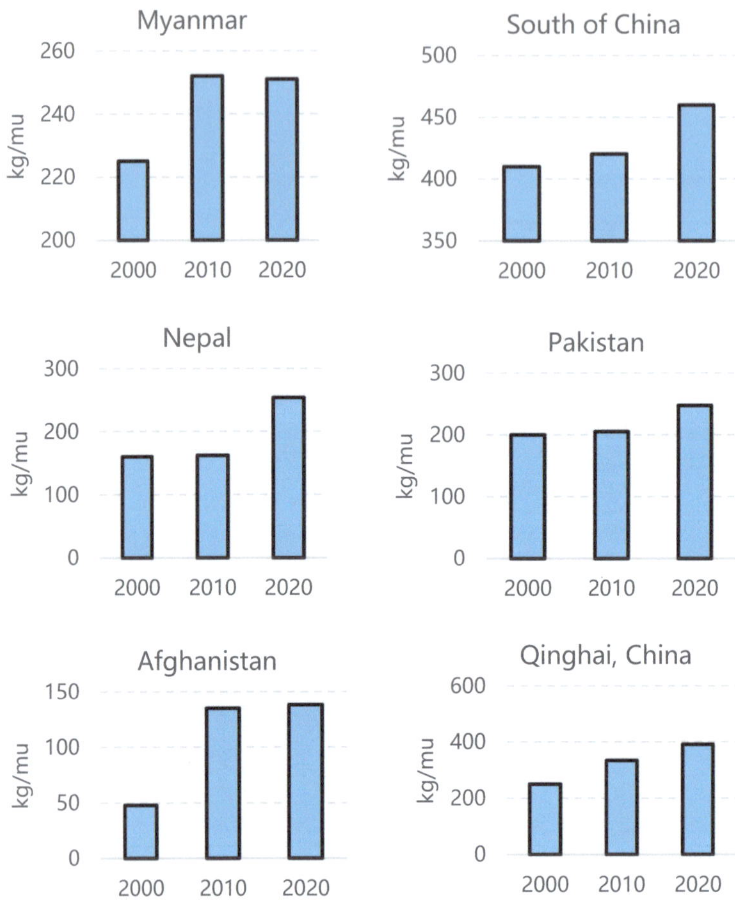

Fig. 3.18 Changes in yield per region from 2000 to 2020

by notable increases in yields for key crops such as rice and wheat. Despite this positive trajectory, there remains a discernible disparity in crop yields when compared to traditional plain agricultural regions, which can be attributed to the unique topographical, climatic, and soil conditions of the Hindu Kush-Himalayan region, as well as variations in agricultural practices and management levels. This research underscores the importance of developing tailored strategies to address these region-specific challenges and optimize agricultural output sustainably. By highlighting the diverse rates of yield improvement and the multifaceted dynamics influencing agricultural productivity, this study provides critical insights that can inform future efforts to enhance food security and agricultural resilience in the Himalayan landscape. The data and trends presented here provide a valuable foundation for policymakers, researchers, and practitioners working to foster sustainable agricultural development in this pivotal region.

3.4 Summary

The study meticulously analyzed the evolution of cultivated land in the Hindu Kush-Himalayan (HKH) region from 2000 to 2020, revealing an 11% expansion in cultivated areas over the two-decade period. This growth was particularly notable in key grain-producing regions such as Myanmar and southern China, which experienced significant increases of 28 and 24%, respectively. Despite initial declines in cultivated land in Pakistan and Afghanistan, both regions demonstrated signs of stabilization by 2020. The systematic grid approach employed in this study highlighted emerging trends in southeastern areas of the HKH region, while also portraying a general state of stability across the region. These findings emphasize the complex interplay of agricultural practices in shaping land use dynamics nexus with socioeconomic factors and environmental conditions.

The detailed statistical analysis further reveals key shifts in crop cultivation patterns across major grain-producing regions. The study identified a commendable 10% increase in the overall cultivated land area, with substantial expansions in eastern regions such as southern China and Myanmar. Conversely, while Nepal's rice planting area remained stable, Pakistan experienced an 8% decline. Wheat cultivation, however, exhibited stability with minimal fluctuations, highlighting its enduring role in the agricultural landscape. The analysis points to dynamic shifts in rice cultivation in southern China and stable patterns in Nepal and Pakistan, while also noting the downward trend in wheat planting in Afghanistan and a slight increase in Qinghai, China. These variations underscore the necessity for localized agricultural strategies to address diverse environmental and socio-economic influences.

In synthesizing the crop yield data from 2000 to 2020 using advanced GRU yield estimation models, the study reveals a significant upward trend in agricultural productivity throughout the HKH region. Notable increases in yields for key crops such as rice and Wheat reflect positive developments in agricultural output. However, disparities persist when compared to traditional plain agricultural regions, owing to the unique topographical, climatic, and soil conditions of the HKH region. This research underscores the need for region-specific strategies to optimize agricultural productivity and address challenges. By providing critical insights into the diverse rates of yield improvement and the complex factors influencing agricultural productivity, the study offers a valuable foundation for policymakers, researchers, and practitioners working to enhance food security and sustainability in the Himalayan region.

References

Aryal, A., Brunton, D., & Raubenheimer, D. (2014). Impact of climate change on human-wildlife-ecosystem interactions in the Trans-Himalaya region of Nepal. *Theoretical and Applied Climatology, 115*(3–4), 517–529. https://doi.org/10.1007/s00704-013-0902-4

Bargiel, D. (2017). A new method for crop classification combining time series of radar images and crop phenology information. *Remote Sensing of Environment, 198*, 369–383. https://doi.org/10.1016/j.rse.2017.06.022

Breiman, L. (2001). Random forests. *Machine Learning, 45*(1), 5–32. https://doi.org/10.1023/A:1010933404324

Chen, X., & Jeong, J. C. (2007). Enhanced recursive feature elimination. In *Sixth International Conference on Machine Learning and Applications (ICMLA 2007)* (pp. 429–435). https://doi.org/10.1109/ICMLA.2007.35

Curran, P. J., Dungan, J. L., & Gholz, H. L. (1992). Seasonal LAI in slash pine estimated with Landsat TM. *Remote Sensing of Environment, 39*(1), 3–13. https://doi.org/10.1016/0034-4257(92)90136-8

Desor, S. (2017). Ideas and initiatives towards an alternative food system in India. In *Kalpavriksh: Deccan Gymkhana.*

de Wit, A., Boogaard, H., Fumagalli, D., Janssen, S., Knapen, R., van Kraalingen, D., Supit, I., van der Wijngaart, R., & van Diepen, K. (2019). 25 years of the WOFOST cropping systems model. *Agricultural Systems, 168*, 154–167. https://doi.org/10.1016/j.agsy.2018.06.018

FAO-IFAD-WFP. (2015). The state of food insecurity in the world. Meeting the 2015 international hunger targets: Taking stock of uneven progress. A Joint publication of Food and Agriculture Organization (FAO) of the United Nations, International Fund for Agriculture Development (IFAD) and World Food Programme (WFP).

Hao, P., Wu, M., Niu, Z., Wang, L., & Zhan, Y. (2018). Estimation of different data compositions for early-season crop type classification. *PeerJ, 6*, e4834. https://doi.org/10.7717/peerj.4834

Haralick, R. M., Shanmugam, K., & Dinstein, I. (1973). Textural features for image classification. *IEEE Transactions on Systems, Man, and Cybernetics, SMC-3*(6), 610–621. https://doi.org/10.1109/TSMC.1973.4309314

ICIMOD. (2019). *The Hindu Kush Himalaya Assessment* (P. Wester, A. Mishra, A. Mukherji, & A. B. Shrestha, Eds.). Springer International Publishing. https://doi.org/10.1007/978-3-319-92288-1

Jin, M., Liu, X., Wu, L., & Liu, M. (2015). An improved assimilation method with stress factors incorporated in the WOFOST model for the efficient assessment of heavy metal stress levels in rice. *International Journal of Applied Earth Observation and Geoinformation, 41*, 118–129. https://doi.org/10.1016/j.jag.2015.04.023

Kc, K. B., Tzadok, E., & Pant, L. (2022). Himalayan ecosystem services and climate change driven agricultural frontiers: a scoping review. *Discover Sustainability, 3*(1), 35. https://doi.org/10.1007/s43621-022-00103-9

Lamsal, P., Kumar, L., Atreya, K., & Pant, K. P. (2017). Vulnerability and impacts of climate change on forest and freshwater wetland ecosystems in Nepal: A review. *Ambio, 46*(8), 915–930. https://doi.org/10.1007/s13280-017-0923-9

Miura, T., Huete, A. R., & Yoshioka, H. (2000). Evaluation of sensor calibration uncertainties on vegetation indices for MODIS. *IEEE Transactions on Geoscience and Remote Sensing, 38*(3), 1399–1409. https://doi.org/10.1109/36.843034

Ntihinyurwa, P. D., & de Vries, W. T. (2021). Farmland fragmentation, farmland consolidation and food security: Relationships, research lapses and future perspectives. *Land, 10*(2), 129. https://doi.org/10.3390/land10020129

Pandey, R. (2016). The state of food (in)security in the Trans-Himalaya, Upper-Mustang, Nepal. *Dhaulagiri Journal of Sociology and Anthropology, 10*, 92–122. https://doi.org/10.3126/dsaj.v10i0.15882

Pandey, R. (2019). Climate change impact and adaptation response in the Trans-Himalaya, Upper-Mustang, Nepal. In *The Annual Kathmandu Conference on Nepal and the Himalaya (First).* Himal Books for Social Science Baha: Kathmandu.

Pandey, R., Kumar, P., Archie, K. M., Gupta, A. K., Joshi, P. K., Valente, D., & Petrosillo, I. (2018). Climate change adaptation in the western-Himalayas: Household level perspectives on impacts and barriers. *Ecological Indicators, 84*, 27–37. https://doi.org/10.1016/j.ecolind.2017.08.021

Pasqualotto, N., Delegido, J., Van Wittenberghe, S., Rinaldi, M., & Moreno, J. (2019). Multi-crop green LAI estimation with a new simple Sentinel-2 LAI Index (SeLI). *Sensors, 19*(4), 904. https://doi.org/10.3390/s19040904

Pearson, R. L. (1974). Remote Mapping of Standing Crop Biomass for Estimation of the Productivity of the Shortgrass Prairie, Pawnee National Grasslands, Colorado. Department of Watershed Sciences, College of Forestry and Natural Resources, Colorado State University,

Rasul, G. (2010). The role of the Himalayan Mountain systems in food security and agricultural sustainability in South Asia. *International Journal of Rural Management, 6*(1), 95–116. https://doi.org/10.1177/097300521100600105

Rymbai, D., & Sheikh, F. M. (2018). The insight of agricultural adaptation to climate change: A case of rice growers in Eastern Himalaya, India. *International Journal of Biometeorology, 62*(10), 1833–1845. https://doi.org/10.1007/s00484-018-1586-3

Salois, M. J., & Moss, C. B. (2010). An information approach to the dynamics in farm income: Implications for farmland markets. *Entropy, 13*(1), 38–52. https://doi.org/10.3390/e13010038

Shrestha, H., Bhandari, T., Karky, B., & Kotru, R. (2017). Linking soil properties to climate change mitigation and food security in Nepal. *Environments, 4*(2), 29. https://doi.org/10.3390/environments4020029

Tiwari, P. C., & Joshi, B. (2012). Natural and socio-economic factors affecting food security in the Himalayas. *Food Security, 4*(2), 195–207. https://doi.org/10.1007/s12571-012-0178-z

Tun, P. T., Nguyen, T. T., & Buerkert, A. (2023). Transformation of agricultural landscapes and its consequences for natural forests in Southern Myanmar within the last 40 years. *Remote Sensing, 15*(6), 1537. https://doi.org/10.3390/rs15061537

van Diepen, C. A., Wolf, J., van Keulen, H., & Rappoldt, C. (1989). WOFOST: A simulation model of crop production. *Soil Use and Management, 5*(1), 16–24. https://doi.org/10.1111/j.1475-2743.1989.tb00755.x

Wang, X. (2022). Managing land carrying capacity: Key to achieving sustainable production systems for food security. *Land, 11*(4), 484. https://doi.org/10.3390/land11040484

Wang, Y., Hu, Y., Niu, X., Yan, H., & Zhen, L. (2023). Myanmar's land cover change and its driving factors during 2000–2020. *International Journal of Environmental Research and Public Health, 20*(3), 2409. https://doi.org/10.3390/ijerph20032409

Wu, S., Yang, P., Ren, J., Chen, Z., & Li, H. (2021). Regional winter wheat yield estimation based on the WOFOST model and a novel VW-4DEnSRF assimilation algorithm. *Remote Sensing of Environment, 255*, 112276. https://doi.org/10.1016/j.rse.2020.112276

Zhang, X., Zhao, T., Xu, H., Liu, W., Wang, J., Chen, X., & Liu, L. (2024). GLC_FCS30D: The first global 30 m land-cover dynamics monitoring product with a fine classification system for the period from 1985 to 2022 generated using dense-time-series Landsat imagery and the continuous change-detection method. *Earth System Science Data, 16*(3), 1353–1381. https://doi.org/10.5194/essd-16-1353-2024

Zhi-jian, K. (2011). A primary study on the current situation and prospects for Myanmar's agricultural development. *Journal of Nanning Polytechnic.*

Zhuo, W., Fang, S., Gao, X., Wang, L., Wu, D., Fu, S., Wu, Q., & Huang, J. (2022). Crop yield prediction using MODIS LAI, TIGGE weather forecasts and WOFOST model: A case study for winter Wheat in Hebei, China during 2009–2013. *International Journal of Applied Earth Observation and Geoinformation, 106*, 102668. https://doi.org/10.1016/j.jag.2021.102668

Chapter 4
Remote Sensing Monitoring and Evaluation of Hydrology and Water Resources

This chapter delves into the significance of using remote sensing technologies to monitor and assess the changes in hydrology and water resources in the Hindu Kush-Himalayan (HKH) region. As one of the most ecologically sensitive areas on the planet, the HKH is home to vast glaciers, snowfields, and permafrost, which are crucial for the region's water resources and ecosystems. The rapid environmental changes driven by climate change and human activities have profound impacts on these natural resources, necessitating scientific monitoring and evaluation to better understand their dynamics.

We begin by focusing on the changes in glaciers, snow, and permafrost. Over the past few decades, rising temperatures have accelerated glacier melt, reduced snow cover, and led to the degradation of permafrost, all of which have significantly disrupted regional hydrological cycles and water resource sustainability. These changes not only threaten the stability of natural ecosystems but also affect the communities that rely on these water resources. Therefore, monitoring these changes is vital—not only to understand future trends in water resources but also to provide scientific support for effective regional water management.

The chapter then addresses the monitoring of surface water body distribution and water quality changes. The complex terrain and climate conditions of the HKH region pose significant challenges to water monitoring. Particularly in mountainous areas, monitoring the dynamic distribution and quality of water bodies requires high-resolution remote sensing data. By utilizing the latest satellite data, we have successfully developed a 10-m resolution surface water dynamic distribution product (DWH10), providing a reliable dataset for long-term water monitoring in the region. Additionally, water transparency, as a key water quality parameter, will be a focus of the study to fill the existing gaps in water quality monitoring in the HKH.

As global climate change intensifies, we also assess its impact on regional water resources. Climate change has not only accelerated the melting of glaciers and snow but also altered precipitation patterns and increased the frequency of extreme climate events. These changes directly affect the availability and quality of water resources.

S. Lu et al., *Remote Sensing Assessment of Sustainable Development in the Hindu Kush-Himalayan (HKH) Region*, https://doi.org/10.1007/978-981-96-8675-9_4

By comprehensively assessing the impacts of climate factors, we aim to identify key drivers of water resource changes and provide scientific guidance for water management strategies.

Finally, the chapter explores the impacts of human activities on water quality changes. With the expansion of agriculture, urbanization, and industrialization, water quality in the HKH region is facing growing threats. Agricultural runoff, deforestation, and over-exploitation of groundwater are key contributors to the deterioration of water quality. Additionally, climate change exacerbates water quality issues, including increased pollution from accelerated glacier melt and altered precipitation patterns. Addressing these complex challenges requires a multifaceted approach that takes into account both natural and human-driven factors, with collaborative efforts to safeguard water quality.

4.1 Monitoring of Glacial, Snow and Permafrost Changes

The term 'Himalaya', derived from ancient Indian pilgrims and meaning 'abode of snow', highlights the significance of snow in the region. In recent times, the importance of snow has shifted from being merely aesthetic to scientific, primarily due to its impact on climate change and its crucial role as a source of fresh water, which directly influences economic growth and social well-being (Gurung et al., 2017). For instance, two billion people, accounting for a quarter of the world's population, reside in the HKH region. The livelihoods of these communities depend on changing runoff patterns that supply essential ecosystem services (ICIMOD, 2019). Climate change, which exacerbates glacier melting, reduces snow cover reduction, and degrades permafrost in the HKH, is widely recognized for its profound impact on regional hydrological systems including runoff, groundwater, lakes, and rivers. As these impacts intensify, they increasingly disrupt hydrological cycles, affecting water availability and ecosystem stability. Monitoring changes in ice, snow, and permafrost dynamics is therefore crucial for effective water resources management and for ensuring the resilience of communities that depend on them.

4.1.1 Methodology

To perform this analysis, the global-scale land cover types were sourced from the Moderate Resolution Imaging Spectroradiometer (MODIS) annual land cover product MCD12Q1 (Friedl & Sulla-Menashe, 2019). This dataset is generated using a supervised classification algorithm applied to MODIS Terra and Aqua reflectance data, offering a spatial resolution of 500 m and covering the period from 2000 to 2020. The MCD12Q1 product includes five classification systems: The International Geosphere-Biosphere Program (IGBP), University of Maryland (UMD), Leaf Area Index (LAI), BIOME-Biogeochemical Cycles (BGC), and Plant Functional Types

(PFT), as well as the Food and Agriculture Organization (FAO) classification system (Friedl & Sulla-Menashe, 2019). For this study, the FAO classification system, which provides categories for tundra and permanent snow and ice, was utilized to monitor temporal changes in permafrost, ice, and snow in the HKH region from 2000 onwards using MCD12Q1 data.

4.1.2 Monitoring Outcomes of Glacial, Snow and Permafrost Changes

The distribution map below illustrates that the ice and snow cover in the HKH region is primarily concentrated in the Himalayas, at the border of China and Nepal, India and Pakistan, and in the central areas of China's Tibetan Plateau (Fig. 4.1). Notably, there was little to no ice and snow recorded in Bangladesh. Analyzing the changes over time since 2000 (Fig. 4.2), the ice and snow coverage in the HKH area has displayed inconsistent changes, but overall, there has been a significant growth of 10.9% by 2020, covering 3645.2 km^2 area. This increase in snow coverage, as depicted in Fig. 4.2, was particularly notable in Nepal, Pakistan, and Afghanistan, with growth rates of 10.2, 53.6, and 247.9% respectively. Conversely, China, Bhutan, and Myanmar experienced declines in snow coverage during the same period, at rates of −11.4, −21.9, and −77.0% respectively. Although China's overall trend showed a decline, it fluctuated significantly, displaying a notable decrease between 2000 and 2014 followed by an increase until 2020. While Pakistan and Afghanistan showed a consistent upward trend in ice and snow coverage, Nepal's coverage fluctuated, initially growing before 2008 (reaching a peak coverage of 2567.9 km^2), then declining until 2018 before rising again by 2020 (Fig. 4.2).

As for permafrost, it was mainly distributed in the central area of the Qinghai-Tibet Plateau in China and the northern area of Afghanistan (Fig. 4.3). Similarly, there is basically no permafrost distribution in Bangladesh due to its tropical climate. However, while other HKH countries exhibited a significant amount of permafrost coverage, Myanmar was observed to have exhibited the least coverage of permafrost with areas ranging within 1200 km^2. Judging from the time series changes since 2000 (Fig. 4.4), it was observed that the entire HKH region showed a significant downward trend with a −17.9% loss (−49,431.45 km^2) experienced by the year 2020. The downward trend in permafrost coverage experienced in the entire HKH was observed to be mainly prompted by changes in all the countries except for Myanmar which experienced a considerable growth of 8.8%. Although most HKH countries exhibited a decline in permafrost coverage, the rate and pattern of this trend varied. Since 2000, while permafrost trends have been inconsistent across the region, China, Pakistan and Afghanistan experienced significant losses of approximately 15.7, 86.9 and 35.2%, respectively. This was not the case for Bhutan and Nepal, which, despite experiencing slight permafrost losses of 1.4 and 2.1%, respectively, displayed notable fluctuations. Both countries initially saw a significant decline in

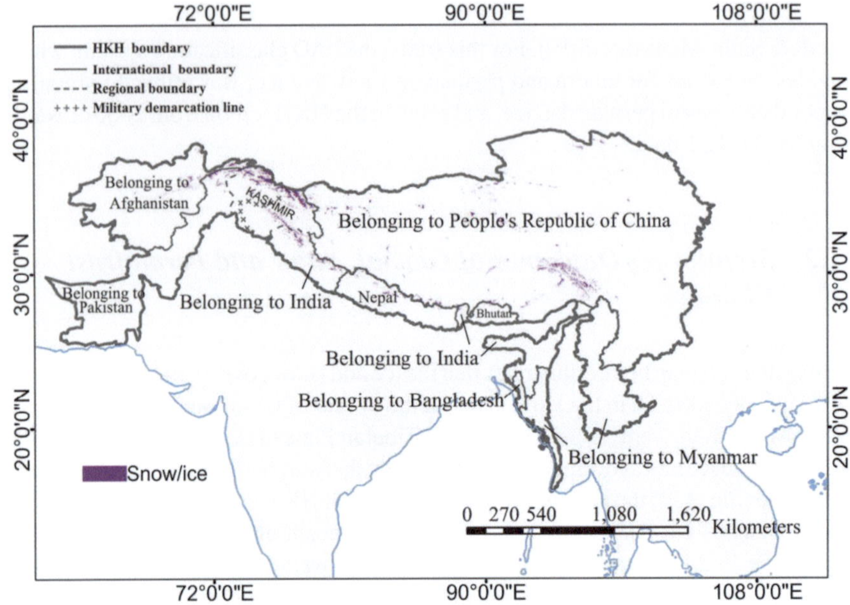

Fig. 4.1 Ice and snow coverage area of the HKH region in 2020

permafrost coverage, followed by periods of recovery. From 2000 onward, permafrost in these regions dropped sharply in 2008, then increased between 2008 and 2011, declined again by 2017, and ultimately showed an upward trend after 2017.

4.1.3 Summary

The study on the distribution of ice, snow, and permafrost in the Hindu-Kush Himalayan (HKH) region highlights distinct spatial and temporal patterns. Spatially, ice and snow were primarily found in the Himalayas, parts of the central Tibetan Plateau in China, and some northern regions of Myanmar. Permafrost, on the other hand, was predominantly located in the central Qinghai-Tibet Plateau in China, and the northern areas of Afghanistan and Myanmar. Over the past 20 years, temporal changes in ice and snow coverage have varied across different countries within the HKH region. In China, Bhutan, and Myanmar, ice and snow coverage generally decreased. Contrastingly, in Pakistan, Afghanistan, and India, ice and snow coverage have increased significantly. Nepal exhibited a pattern of initial increase followed by a decrease in ice and snow coverage. Regarding permafrost, the study found a significant decline in permafrost areas in China, Pakistan, and Afghanistan over the past two decades. In contrast, Nepal, Bhutan, Myanmar, and India showed no substantial overall change in permafrost coverage. However, within these countries, there

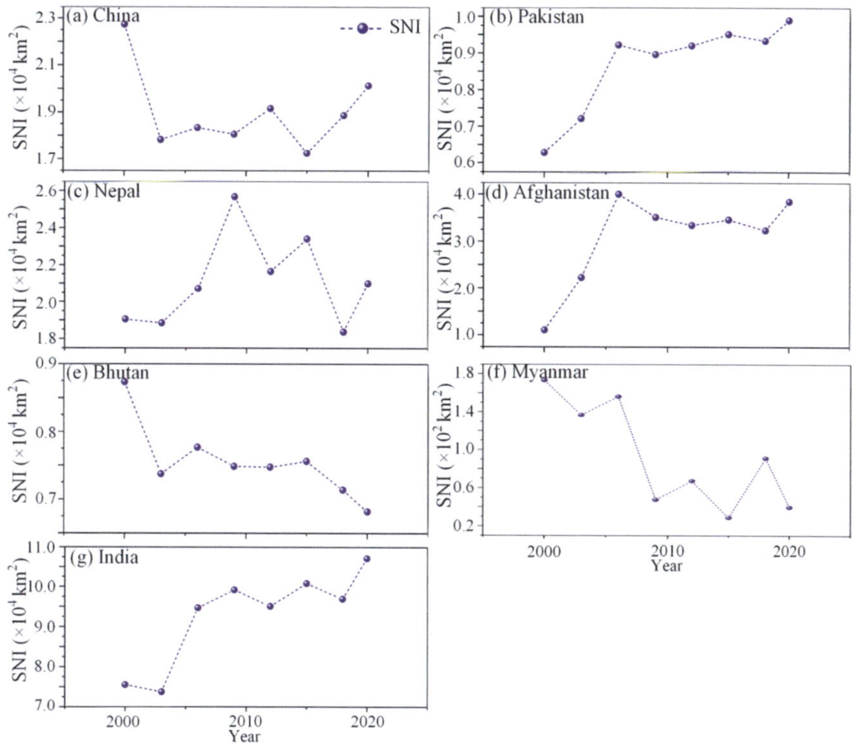

Fig. 4.2 Changes in ice and snow area within the HKH region (excluding Bangladesh due to minimal ice and snow coverage)

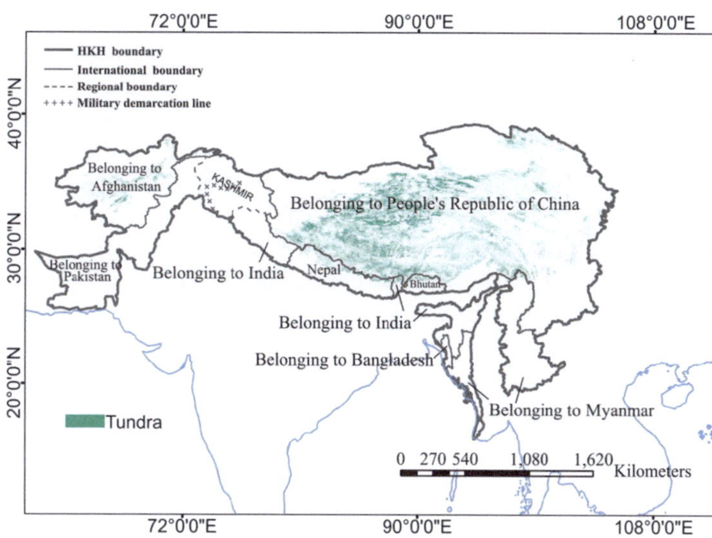

Fig. 4.3 HKH permafrost coverage area in 2020

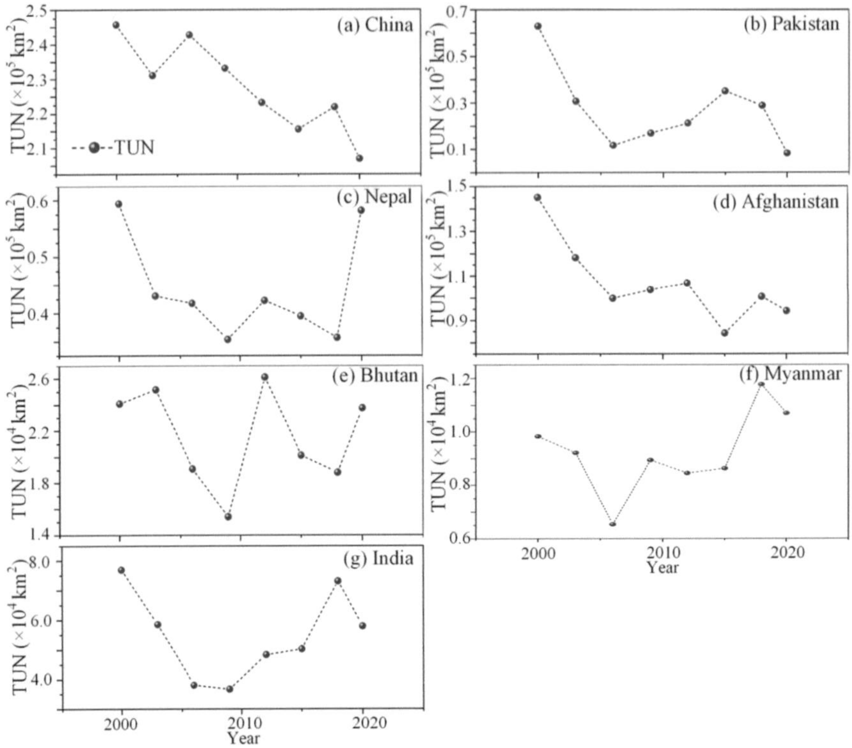

Fig. 4.4 Changes in permafrost area within the HKH region (excluding Bangladesh due to minimal permafrost coverage)

were notable fluctuations: Nepal, Myanmar, and India experienced initial decrease followed by increase in permafrost areas, whereas Bhutan showed significant fluctuations in permafrost coverage, with a marked increase around the year 2000, followed by ongoing variability through 2020. To conclude, the study reveals that while there are clear spatial concentrations of ice, snow, and permafrost within specific regions of the HKH, temporal changes over the past two decades have been highly variable, reflecting complex regional responses to climatic and environmental factors. These findings underscore the need for localized climate adaptation and mitigation strategies to address the differing trends in ice, snow, and permafrost across the HKH region.

4.2 Surface Water Body Distribution and Water Quality Change Monitoring

Surface water encompasses both inland freshwater and some saltwater, forming a crucial component of terrestrial ecosystems (Klein et al., 2017; Palmer et al., 2015). However, the Hindu Kush-Himalayan (HKH) region, in recent years have seen an increase in disasters such as floods and droughts due to climate change and human activities, along with significant water quality deterioration. These issues pose serious threats to regional water resource security and sustainable development (Yang et al., 2019; Zhang et al., 2022). Consequently, accurately mapping and monitoring the distribution and quality of surface water bodies in the HKH has become a top priority for water resource management and is crucial for achieving Sustainable Development Goal (SDG 6.6.1), which focuses on "ecosystem-wide monitoring of water-related areas" (Lu et al., 2021; McNicoll, 2019). The HKH's complex terrain, characterized by many terrain shadows and small water bodies nestled between mountains, poses significant challenges to accurate surface water monitoring (McNicoll, 2019). Current mainstream surface water products in the HKH are plagued by significant uncertainties in monitoring accuracy due to terrain shadows and the scale of water body observation (Pekel et al., 2016; Pickens et al., 2020), leaving the true distribution of water bodies unclear. Additionally, there is a notable scarcity of high-precision monitoring products that effectively capture the dynamic distribution of water bodies in the HKH. Fine-resolution water quality parameter products are also lacking, with most studies focusing on large lakes and reservoirs, thus failing to provide a comprehensive assessment of overall lake water quality. Furthermore, the development of long-term fine-resolution water quality products for the HKH region remains an unmet need.

To address these challenges and achieve long-term monitoring of surface water distribution and water quality in the HKH region, this study will leverage SDGSAT-1 nighttime thermal infrared and Sentinel-2 multispectral satellite data to develop a high-precision water monitoring method for mountainous areas. The goal is to create a 10-m resolution water dynamic distribution product for the HKH region by 2022 and build a long-term water distribution product based on Landsat data from 2000 to 2020. Additionally, the study will focus on water transparency as a key water quality parameter, developing a long-term water transparency product for the HKH region. This will involve using the constructed water distribution data and Landsat images to monitor and evaluate changes in water quality over time in the HKH region.

4.2.1 Methodology

(1) ***Dynamic change monitoring of surface water bodies distribution in HKH using SDGSAT-1 and Sentinel-2 satellite data***

The topography of mountainous areas causes interference from mountain shadow during the water extraction process. This leads to several challenging problems that need to be resolved, such as the creation of complex water body types and a lack of data that accurately reflects the distribution characteristics of water bodies in the area. To address these issues, a study was conducted in the Hindu Kush-Himalayan region specifically using thermal infrared (TIR) and Sentinel-2 optical satellite data to construct a dynamic monitoring method for water distribution. The study focused on overcoming the technical challenges associated with extracting surface water distribution in mountainous undulating areas and produced a 10-m resolution monitoring product for water dynamic changes in the region in 2022. To be more precise, the procedure starts with the incorporating of earlier multi-source data extraction to provide high-confidence water training data. Secondly, a local adaptive XGBoost water extraction model was built to determine the yearly maximum water distribution region using the results of percentile synthesis of Sentinel-2 spectrum data and SDGSAT-1 nighttime thermal infrared data. Ultimately, the HKH's 10 m resolution 2022 temporal water body and the annual stable water body were acquired, as illustrated in Fig. 4.5, after the annual water body variation in the extracted largest water area was monitored using the temporal variation characteristics of Sentinel-2 and SDGSAT-1 nighttime thermal infrared data.

(1) Automatic extraction of training samples based on multi-source data

As seen in Fig. 4.6, SDGSAT-1's nighttime thermal infrared band is resistant to mountain shadows, enabling it to efficiently minimize their effects. Differentiating between the two qualities (water bodies and mountain shadows) is challenging because daytime thermal infrared images, including those from remote sensing and the widely used water body index (NDWI), contain shaded regions that mimic water bodies. On the contrary, the use of nighttime thermal infrared imaging revealed a clear distinction between shades and water bodies. As a result, SDGSAT-1's nighttime thermal infrared loading provides substantial benefits for extracting water bodies in mountainous locations.

The automatic production of high-confidence training data for water bodies and non-water bodies was accomplished by first merging the multi-source data with the SDGSAT-1 nighttime thermal infrared data, based on the distinctive quality of the nighttime thermal infrared data. Using a 5 × 5 spatial sliding window, homogenous

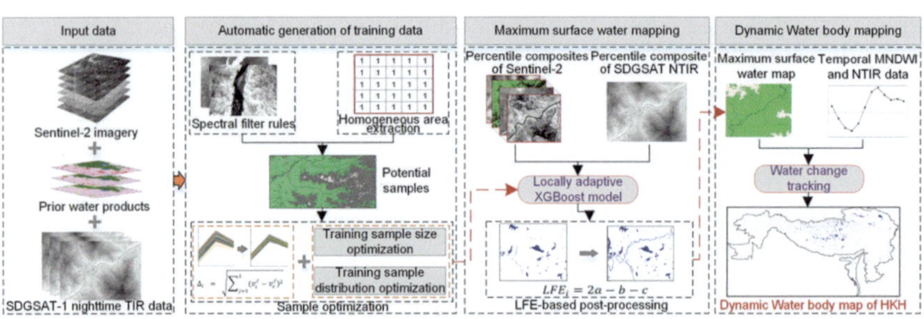

Fig. 4.5 Mapping flow of 10-m-resolution water body dynamics change products with SDGSAT-1 nighttime thermal infrared and Sentinel-2

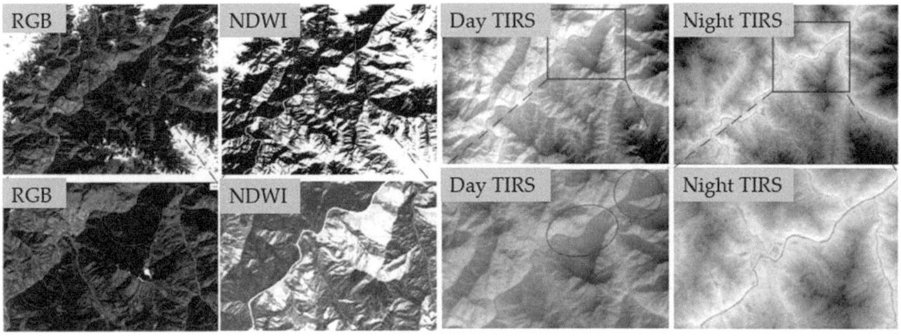

Fig. 4.6 Examples of the sensitivity of remotely sensed RGB (Red, Green, Blue) imagery, NDWI indices, and SDGSAT-1 daytime thermal infrared and nighttime thermal infrared data to mountain shadows in a mountainous undulating region

waters from Joint Research Centre Global Surface Water (JRC-GSW) were identified and designated as possible water bodies (Pekel et al., 2016). In accordance with earlier research by Zhang et al., (2020), if the frequency of water bodies in the 5 × 5 window equaled 1, meaning there were 25 water pixels in the frame, then the center pixel was considered a possible water sample. Secondly, the produced homogenized region was overlaid with the 10-m resolution ESA10 water body surface cover product to increase the accuracy of the water samples. Furthermore, certain bodies of water identified in JRC-GSW and ESA10 were excluded from our input imagery due to the temporal difference between them and our input remote sensing data. Accordingly, only homogeneous area pixels with MNDWI more than 0.3 and short-wave infrared 1 (SWIR1) less than 0.11 (Chen et al., 2020; Yamazaki et al., 2015) (Eq. 4.1) were kept for further sample optimization based on the Sentinel-2 data. The non-water regions identified in JRC-GSW and ESA10 were then intersected, and the homogenous non-water regions were extracted using the same 5 × 5 window. These non-water regions were subsequently utilized as possible non-water samples. Furthermore, given the HKH area's multi-terrain undulation, certain shadows in JRC-GSW and ESA10 could be mistaken for bodies of water. To improve the samples, the nighttime thermal infrared band of SDGSAT-1 was further utilized. Using the Euclidean distance approach shown in Eq. 4.2, outlier samples were eliminated with reference to earlier research by Xie et al. (2019). Considering that the user accuracy of water bodies in ESA10 is 89.4% (Zanaga et al., 2022) and that most non-water bodies generally have higher nighttime temperatures than water bodies in the summer months, we end up removing water samples with Euclidean distances below 11%. We exclude water samples with Euclidean distances below 11% since the user accuracy of water bodies in ESA10 is 89.4% (Zanaga et al., 2022). Additionally, most non-water bodies often have higher nighttime temperatures than water bodies in the summer. In a similar way, we eliminated the possible non-water samples below 18% in terms of absolute distance, based on the overall accuracy of ESA10 in Asia (82.1%) (Zanaga et al., 2022).

$$MNDWI > 0.3 \& SWIR1 < 0.11 \tag{4.1}$$

$$\Delta_i = \sqrt{\sum_{j=1}^{3} (v_i^j - v_c^j)^2} \tag{4.2}$$

where v_i^j and v_c^j refer to the pixel value and centroid value of TIR band j in the water/non-water categories, respectively.

(2) XGBoost-based maximum water coverage extraction

Obtaining the maximum coverage of the HKH's surface water bodies requires the utilization of the XGBoost classifier through the generation of training sample features. The generation of training samples requires the use of time-series SDGSAT-1 and Sentinel-2 images and DEM data as training features. Among them, since nighttime SDGSAT-1 TIR data help to identify water bodies in mountainous areas and the maximum coverage of water bodies in the HKH region occurs in summer (June to October) (Wang et al., 2022), cloud-free summer SDGSAT-1 TIR data was collected and used as input features. Because water bodies typically have lower TIR values in summer (Wang et al., 2023), the 15th percentile of the summer SDGSAT-1 nighttime TIR bands was utilized to generate images of the maximum water body coverage area. Furthermore, considering the spatial and temporal heterogeneity of the water bodies and background, information on elevation and terrain slope, the Blue and SWIR bands, the Modified Normalized Difference Water Index (MNDWI) (Xu, 2006), the Land Surface Water Index (LSWI) (Chandrasekar et al., 2010), and the Enhanced Vegetation Index (EVI) (Huete et al., 2002) indices were utilized. Also, to represent waterbody and non-waterbody features under conditions of maximal water cover, the 15th percentile composites of the Blue and SWIR bands and the 85th percentile composites of the MNDWI and LSWI were utilized. Using the 15th and 85th percentile syntheses of EVI, it was possible to differentiate between vegetation and waterbodies (DeVries et al., 2015). Table 4.1 lists the specifics of the training features.

(3) Tracking dynamic changes in water bodies

In accordance with Zhang et al. (2019) and Yamazaki et al. (2015), we used MNDWI as an indicator for every Sentinel-2 image to record the dynamic changes of water

Table 4.1 Training features for maximum surface water mapping

Data	Features
SDGSAT-1 TIR	TIR: minimum percentile of three night-time TIR bands in summer season
Sentinel-2 MSI	Reflectance: 15th percentile of SWIR and Blue
	Normalized indices: 85th percentile of MNDWI and LSWI, and 15th and 85th percentiles of EVI
DEM data	Elevation and slope

bodies. Stable water bodies have a higher MNDWI rating because they are more stable. Temporary water bodies, on the other hand, exhibit seasonal changes in their MNDWI values (high MNDWI values during the rainy season and low MNDWI values during the dry season) because of transitions between water body and non-water body states. Consequently, a water pixel is probably a temporary water body if its MNDWI value during the dry season is less than a certain threshold. As a result, the 15th percentile composite of MNDWI was used to reflect the lowest water cover during the dry season. Nevertheless, the nighttime TIR imagery from SDGSAT-1 was concurrently used to mitigate the impacts of mountain shadows, as some water bodies may be obscured and appear to have water body-like properties during the dry season. Because of their high heat capacity, aquatic bodies often maintain higher temperatures than the surrounding land during winter and early spring (Adeniran et al., 2022). Hence, we utilized the minimal synthesis of the nighttime TIR bands collected during the dry season to reflect the minimum water cover during this time. By using a training sample of water bodies located within the JRC-GSW stable watershed region to calculate the water body characteristics (mean ($\overline{x_T}, \overline{x_M}$) and standard deviation (σ_T, σ_M) of TIR and MNDWI) during this phenological period, we were able to construct two Self-Normalized Water Index SNWI_S: the SNWI_T and the SNWI_M, to reflect the likelihood of a pixel as water body during this period (Eq. 4.3). Specifically, SNWI_M was used to differentiate between water bodies and other ground cover types, while SNWIT was used primarily to differentiate between water bodies and terrain shading. As shown in Fig. 4.7a and b, SNWI_M can distinguish water bodies from most ground covers, but there is still shadow interference around water bodies. In contrast, SNWI_T can effectively mitigate the effect of terrain shading around water bodies (Fig. 4.7c). Therefore, the two indices can effectively complement each other. Considering that our SNWI was constructed based on samples of stable waterbodies and that the characteristics of large stable waterbodies usually conform to a normal distribution (Feyisa et al., 2014; Yang et al., 2015), according to the statistical characteristics of a normal distribution (Chen et al., 2020; Huang et al., 2010), 99% of waterbodies should have SNWI values below 3. Therefore, if the values of SNWI_M and SNWI_T are less than 3, the pixel is considered a stable water body pixel, otherwise the pixel is identified as a temporary water body.

$$SNWI_M = \sqrt{\left(\frac{x_N - \overline{x_M}}{\sigma_M}\right)^2}$$

$$SNWI_T = \sqrt{\frac{1}{m}\sum_{j=1}^{m}((x_{T_j} - \overline{x_{T_j}})/\sigma_{T_j})^2}, m = 3 \tag{4.3}$$

where x_M refers to the MNDWI value of a pixel; x_{T_j} refers to the value of nighttime TIR band j for specific pixel and m is the number of nighttime TIR bands in SDGSAT-1; $\overline{x_M}, \sigma_N, \overline{x_{T_j}}$, and σ_{T_j} are the mean and standard deviation of MNDWI, and the mean and standard deviation of the TIR band j, respectively.

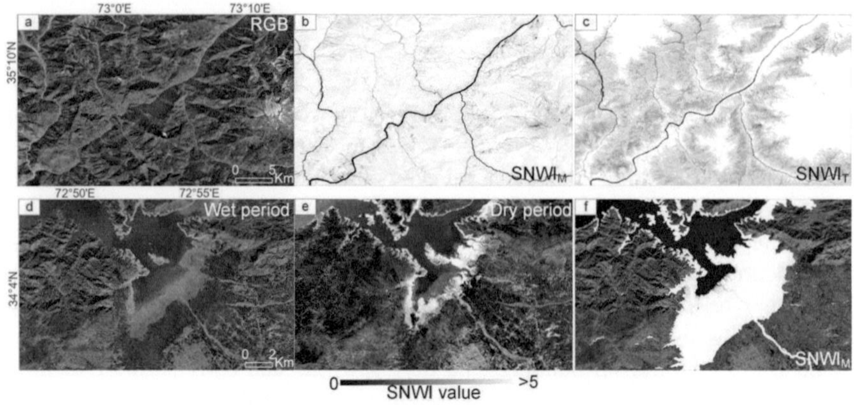

Fig. 4.7 The performance of SNWI$_M$ and SNWI$_T$. **a–c** The functions of SNWI$_M$ and SNWI$_T$; **d–f** SNWI$_M$'s ability to capture changes in water bodies within the maximum water area. The smaller the value of SNWI, the higher the probability that the pixel represents a water body

(2) ***Monitoring of HKH dynamic distribution of surface water distribution changes based on Landsat long time-series data in the HKH region***

To effectively monitor the long-term changes in water distribution in the Hindu Kush-Himalaya (HKH) region, this study employed a comprehensive approach using Landsat satellite data, DEM data, JRC-GSW time-series surface water distribution products, and previously produced 10-m resolution water dynamic distribution map for the HKH region (DWH10) (Sect. 3.2.1). Additionally, other prior datasets were integrated to create a long-term water dynamic change monitoring dataset with a resolution of 30 m for the years 2020, 2010, and 2000. The methodology centered around the XGBoost classification model, which was pivotal in achieving accurate water body extraction. Initially, high-confidence water training data were meticulously extracted by combining multi-source prior data. Subsequently, an XGBoost water extraction model was constructed using Landsat spectral data to identify temporal water bodies and annual stable water bodies with a 30-m resolution. This robust approach allowed for a detailed and dynamic analysis of water distribution changes over two decades, providing valuable insights into water resource management and environmental monitoring in the HKH region. The overall process, illustrated in Fig. 4.8, highlights the integration of advanced remote sensing techniques and machine learning models to enhance the accuracy and reliability of long-term water distribution monitoring.

(1) Automatic extraction of training samples based on multi-source data

The accuracy of prior datasets, such as the Joint Research Centre Global Surface Water (JRC-GSW), in the mountainous regions of the Hindu Kush-Himalayan (HKH) area is uncertain due to the influence of mountain shadows. To address this issue, our study utilized the DWH10 data, which was specifically processed to account

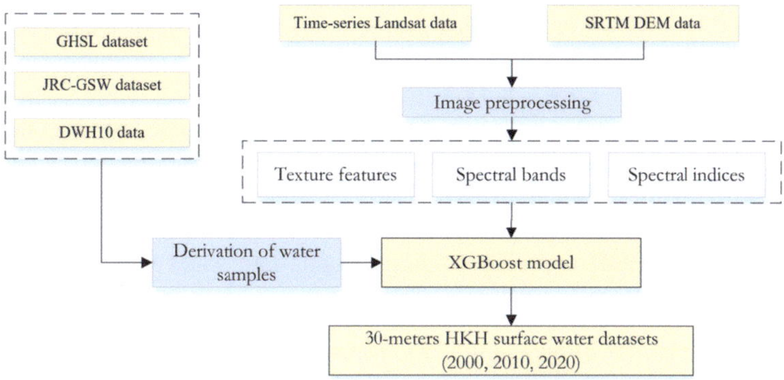

Fig. 4.8 Processing of mapping the dynamic distribution of water bodies in HKH

for mountain shadows, providing better accuracy in the HKH region. We combined DWH10 and JRC-GSW to automatically extract high-confidence time-series surface water training data. Initially, a 5 × 5 pixel spatial sliding window was employed to identify homogeneous water areas in the 2020 surface water mapping results from both JRC-GSW and DWH10. The homogeneous areas obtained from these two products were intersected to identify stable water areas and homogeneous temporal water bodies. To further refine the accuracy of these homogeneous areas, we calculated the 15th and 85th percentile composite values of the Modified Normalized Difference Water Index (MNDWI) for each pixel. Pixels with both 15th and 85th percentile MNDWI values greater than 0.1 were considered high-confidence stable water samples. Conversely, pixels with a 15th percentile MNDWI value less than 0.1 and an 85th percentile MNDWI value greater than 0.1 were categorized as high-confidence temporal water samples. Additionally, Global Human Settlement Layer (GHSL) data was utilized to exclude urban areas from the analysis.

Upon constructing the 2020 water body training samples, we extended the methodology to obtain time series training samples for 2010 and 2000. This involved superimposing the JRC-GSW mapping results for 2020, 2010, and 2000 to identify areas of unchanged water bodies across these years. The samples located in these unchanged areas, derived from the 2020 training data, were then used as time series water system mapping training samples for the HKH region. This approach allowed us to accurately map the water systems for each target year, enhancing the reliability of surface water data in the HKH region despite the challenges posed by mountain shadows.

(2) Surface water cover extraction based on XGBoost

To accurately map surface water distribution in the HKH region for the years 2000, 2010, and 2020, we utilized time-series Landsat images and Digital Elevation Model (DEM) data as training features. Drawing on the methodology described in Sect. 4.2.1 for DWH10 mapping, we employed the Blue and Short-Wave Infrared (SWIR) bands of Landsat imagery and calculated the Modified Normalized Difference Water Index (MNDWI), Land Surface Water Index (LSWI), and Enhanced Vegetation Index

(EVI). Additionally, we incorporated elevation and terrain slope data. The inclusion of the Blue and SWIR bands, alongside the 15th and 85th percentile composite results of the MNDWI, LSWI, and EVI indices, allowed us to capture water and non-water characteristics under both minimum and maximum water cover conditions. This approach was instrumental in distinguishing between temporal and stable water bodies. Utilizing these comprehensive features and the generated training samples, we employed the XGBoost classifier to produce detailed surface water distribution maps for the HKH region, providing insights into water body dynamics and supporting environmental monitoring efforts.

(3) ***Monitoring water quality changes in surface water bodies in the HKH region***

According to Chen et al. (2022) and Shen et al. (2020), water body transparency is a measure of visibility and can be used to reflect comprehensive information on the water quality of inland water bodies because of its close correlation with the five core water quality parameters that are relevant to the sixth Sustainable Development Goal (SDGs 6) of the United Nations. To invert the water transparency in the HKH region, we first developed a weather-consistent image synthesis method for the HKH region by combining meteorological and climatological auxiliary knowledge of the HKH region as well as the spatiotemporal characteristics of the remote sensing images to produce weather-consistent and cloud-free synthetic images of the HKH region for the years 2000, 2010, and 2020. Second, using the water transparency inversion method based on watercolor parameters and synthetic weather-consistent cloud-free images of the HKH region, the first set of fine-resolution (30 m) time-series water transparency products (2000, 2010, 2020) of the HKH region were developed using the GEE cloud computing platform, which was followed by an evaluation of the mapping results' accuracy. Figure 4.9 depicts the entire mapping procedure.

(1) Synthesized images for the HKH region with consistent phenology

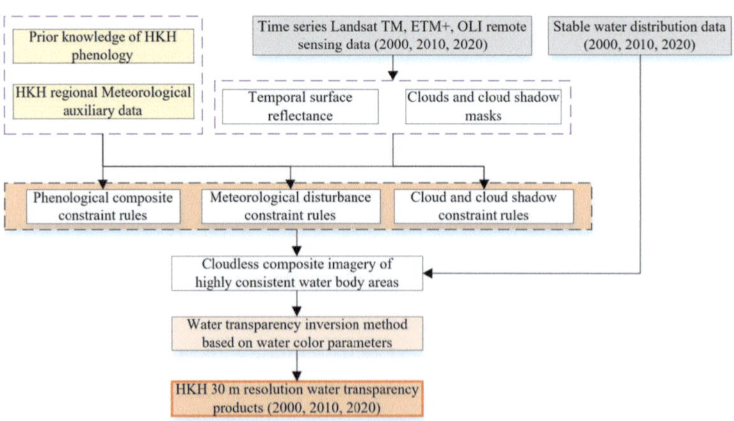

Fig. 4.9 Flow of mapping high consistent water transparency in HKH region based on watercolor parameters

Owing to cloud opacity and shadows, cloud-free synthetic images for each target year must be acquired beforehand to create water body transparency products with full spatial coverage. The study first integrates prior knowledge of the regional phenology of the Hindu Kush-Himalayan (HKH) region and the cloud coverage status of time-series remote sensing data, considering the dynamic change of water body status due to seasonal and rainfall factors. It then proposes a highly coherent image synthesis method with synergistic temporal synthesis constraints, environmental perturbation constraints, as well as cloud and cloud shading constraints to ensure the consistency of synthetic imagery in all the periods, to avoid affecting the results of the temporal analysis of transparency due to variations in water body status (Eq. 4.4). To determine which observations should be synthesized into pixels, the synthesis technique scores the observations according to the synthesis criteria as illustrated in Eq. 4.4 below.

$$\text{Score}_{\text{Total}} = \text{Score}_{\text{DOY}} + \text{Score}_{\text{Env}} + \text{Score}_{\text{Cloud}} \tag{4.4}$$

where $\text{Score}_{\text{Total}}$ is the total score of each observation at a location, and the observation with the highest score is the final synthesized pixel at that location; $\text{Score}_{\text{DOY}}$, $\text{Score}_{\text{Env}}$, and $\text{Score}_{\text{Cloud}}$ are the scores of the phenology, environment, and cloud and cloud shading constraint rules, respectively. Among them, the phenology constraint rule adopts a Gaussian distribution model to maximize the score of the observations during the target monitoring window as the optimal observation to achieve the constraint on the phenological consistency of the data:

$$\text{Score}_{DOY} = \frac{1}{\sigma\sqrt{2\pi}} e^{-\frac{1}{2}\left(\frac{x_j-\mu}{\sigma}\right)^2} \tag{4.5}$$

where μ and σ represent the day of year (DOY) and standard deviation of the target, respectively. Since the water-abundance period of the Hindu Kush is from June to October (Wang et al., 2022), its center date (DOY of 232, i.e., August 15) is used as the DOY of the target synoptic period, the standard deviation of DOY is set to 75 to match the length of the summer period, and x_j is the observation j of the observation time. In addition, considering that environmental factors such as flooding and heavy rainfall during the summer period can increase the amount of suspended particulate matter to be washed into the water body and reduce the transparency of the water body (Liu et al., 2021; Murshed et al., 2014), the Normalized Difference Turbidity Index (NDTI) construction rule was applied to reflect the changes in water body turbidity (Lacaux et al., 2007). Since floods and heavy rains reduce the transparency of water bodies (Liu et al., 2021; Murshed et al., 2014) and NDTI values fluctuate compared to normal conditions (Islam, 2006), the image composition was conducted based solely on the observed values of NDTI at each pixel location between its temporal 30th and 70th quantile composite values, where NDTI_{j}, $\text{NDTI}_{\text{30th}}$, and $\text{NDTI}_{\text{70th}}$ are the NDTI of observation j, as well as the 30th quartile and 70th quartile NDTI synthesized results of the pixel location, respectively.:

Finally, for clouds and cloud shadows, the constraint rules were constructed with reference to existing studies (White et al., 2014):

$$\mathrm{Score_{Env}} = \begin{cases} 1, NDTI_{30th} < NDTI_j < NDTI_{70th} \\ 0, others \end{cases} \tag{4.6}$$

$$\mathrm{Score}_{CloudDisctance} = \frac{1}{1 + e^{(-0.2(\mathrm{D}_{min}(D_i, D_{req}) - \frac{D_{req}}{2}))}} \tag{4.7}$$

where D_i is the distance from the pixel to the cloud/cloud-shadow identifier pixel, D_{req} is the "buffer" minimum distance (default value is 50), D_{min} is the minimum distance between observed pixels (default value is 0), and the cloud and cloud-shadow masks are from the QA bands of the quality control. The settings of all parameters of this rule were derived from studies by Chen et al. (2021) and White et al. (2014).

(2) Water body transparency inversion model based on Landsat and watercolor parameters

The development of inversion models of water body transparency is based on Forel-Ule index (FUI) since research conducted in recent years has shown a strong relationship between the water body color parameter FUI and transparency (Li et al., 2016; Wang et al., 2020). The robustness of transparency inversion was proved by Chen et al. (2022) in their production of a 1985–2020 Chinese regional water body transparency product based on Landsat and FUI watercolor characteristics. Accordingly, this work used a similar water transparency inversion model based on FUI watercolor characteristics to invert and map the water transparency in the HKH region. To perform this water transparency inversion model, it is first necessary to determine the CIE tri-stimulus values X, Y, and Z from the reflectance in the visible band of Landsat images. Various models were utilized to determine the corresponding tri-stimulus values for each sensor image because the loadings varied between Landsat 5, 7, and 8. In this instance, a linearly weighted summation of the band reflectance shown in Eqs. 4.8–4.10 was used to approximate the tri-stimulus values because the Landsat 8 image has four bands in the visible range (Chen et al., 2022):

$$\mathrm{X} = 11.053R(443) + 6.950R(482) + 51.135R(561) + 34.457R(655) \tag{4.8}$$

$$\mathrm{Y} = 1.320R(443) + 21.053R(482) + 66.023R(561) + 18.034R(655) \tag{4.9}$$

$$\mathrm{Z} = 58.038R(443) + 34.931R(482) + 2.606R(561) + 0.016R(655) \tag{4.10}$$

where R denotes the reflectance of the corresponding band. And since Landsat 5 and 7 have only three bands of RGB in the visible band, the tri-stimulus values of Landsat 5 and 7 images were inverted based on an inversion method oriented to sensors with only three bands in the visible range (Wang, 2018):

$$\mathrm{X} = 1.1302R(482) + 1.7517R(561) + 2.7689R(655) \tag{4.11}$$

$$Y = 0.0601R(482) + 4.5907R(561) + 1.0000R(655) \tag{4.12}$$

$$Z = 5.5934R(482) + 0.0565R(561) + 0.0000R(655) \tag{4.13}$$

Secondly, we further inverted the chromaticity angle α by calculating the chromaticity coordinates (x, y) using the inversely performed CIE color tri-stimulus values X, Y, and Z (Chen et al., 2022).

$$x = \frac{X}{X+Y+Z}, y = \frac{Y}{X+Y+Z} \tag{4.14}$$

$$x\prime = y - 1/3, y\prime = x - 1/3 \tag{4.15}$$

$$\alpha = \left(ARCTAN2\frac{y\prime}{x\prime}\right) * 180/\pi \tag{4.16}$$

However, since there is a deviation (Δα) between the color perceived by the human eye and the color acquired by the image (van der Woerd & Wernand, 2015), the derived chromaticity angle α needs to be corrected by basing α + Δα on the required deviation Δα (van der Woerd & Wernand, 2018; Wang, 2018):

$$\Delta\alpha = a_1 \cdot b^5 + a_2 \cdot b^4 + a_3 \cdot b^3 + a_4 \cdot b^2 + a_5 \cdot b + a_6, b = \alpha/100 \tag{4.17}$$

The specific parameters in each sensor are shown in Table 4.2.

Lastly, using the FUI lookup table (Table 4.3) created by Novoa et al. (2013), the FUI may be inverted depending on the corrected angle. According to Wang et al. (2020), an inversion model of the water transparency based on the water colour parameter was created after they discovered that the chromaticity angle α and the water colour parameter FUI demonstrated a clear inverse proportional relationship ($R^2 = 0.95$) with the water transparency SD based on the measured data. Therefore, to facilitate our research, a similar inversion approach was used in the creation of time-series transparency products.

$$FUI < 8, SD = 794630.86 \cdot \alpha^{-1.66} \tag{4.18}$$

Table 4.2 The deviation correction polynomial coefficients of chromaticity angle α for Landsat-TM/ETM+/OLI sensor

Sensors	a_1	a_2	a_3	a_4	a_5	a_6	R^2
Landsat-TM	25.85	−177.40	476.69	−653.30	463.33	−94.41	0.59
Landsat-ETM+	30.47	−203.40	498.80	−570.90	324.73	−56.72	0.52
Landsat-OLI	21.36	−199.29	703.30	−1132.20	801.60	−201.34	0.69

Table 4.3 The 21-class FUI indices and the corresponding range of hue angle α (Chen et al., 2021; Wang, 2018)

FUI	α (°)	FUI	α (°)	FUI	α (°)
1	(35.00, 42.83)	8	(160.97, 175.98)	15	(219.34, 224.87)
2	(42.83, 49.02)	9	(175.98, 186.67)	16	(224.87, 230.23)
3	(49.02, 60.01)	10	(186.67, 195.44)	17	(230.23, 235.09)
4	(60.01, 79.23)	11	(195.44, 202.05)	18	(235.09, 239.56)
5	(79.23 106.94)	12	(202.05, 207.82)	19	(239.56, 243.66)
6	(106.94, 137.03)	13	(207.82, 213.57)	20	(243.66, 247.25)
7	(137.03, 160.97)	14	(213.57, 219.34)	21	(247.25, 252.00)

$$FUI \geq 8, SD = 30380 \cdot FUI^{-2.621} \tag{4.19}$$

4.2.2 *Surface Water Distribution and Transparency Changes*

(1) ***HKH region's 10-m resolution water body distribution product***

Based on our algorithm, we developed a 10-m resolution surface water body distribution product for the HKH region (10-m Dynamic Water body map of the HKH (DWH10)), as shown in Fig. 4.10, which possess the ability to extract the stable and temporary water body types in the HKH region. Overall, surface water bodies were found in the HKH region covering an area of more than 1.45×10^5 km^2, with most of the water bodies found in the central and western parts of the territory. Furthermore, the region's water bodies saw notable seasonal variations due to the existence of several temporary bodies of water. By aggregating our DWH10 to a spatial resolution of 0.1° × 0.1° and calculating the area of waters in both latitudinal and longitudinal directions, the statistical results revealed that stable water bodies were more abundant between 80°E and 95°E longitude and 28°N and 40°N latitude, whereas temporary water bodies were primarily concentrated between 90°E and 95°E longitude and 34°N and 40°N latitude. Moreover, at the regional level, 77.16% of the waters in the HKH region were distributed within China, followed by Myanmar at 7.15% and India at 6.37%. On the contrary, the smallest shares of water bodies in the region were observed in Bhutan (0.25%) and Bangladesh (0.84%), respectively. Furthermore, more stable water bodies than temporary water bodies were detected within China, while temporary water bodies were more abundant in other countries. Specifically, about 6.9×10^4 km^2 of stable water bodies and 4.2×10^4 km^2 of temporary water bodies were detected in China. In Afghanistan, the areas of stable and temporary water bodies were 6.5×10^2 km^2 and 2.5×10^3 km^2, respectively. In Pakistan, these figures were 1.2×10^3 km^2 and 4.2×10^3 km^2, while in India, they measured 2.1×10^3 km^2 and 7.1×10^3 km^2. Nepal had 6.2×10^2

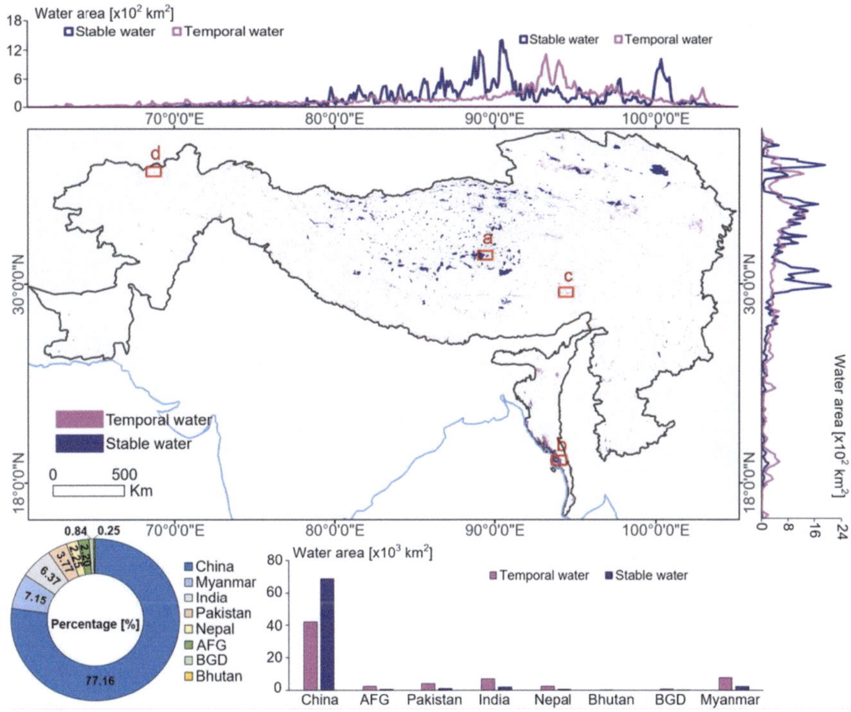

Fig. 4.10 2022 10-m resolution dynamic distribution of water bodies in the HKH region

km² of stable water bodies and 2.6×10^3 km² of temporary water bodies. Similarly, Bhutan recorded 1.1×10^2 km² and 2.5×10^3 km², whereas Bangladesh had 3.8×10^2 km² and 8.2×10^2 km² of stable and temporary water bodies, respectively.

To conduct a comprehensive quantitative assessment of the accuracy of our DWH10 product, we selected 12 validation regions (Fig. 4.11) randomly tagged as l from four HKH countries: Pakistan (a, b, c), China (d, e, f, g, h, and l), Nepal (i), and Myanmar (k and j). Each region generated approximately 1000 validation points. The accuracy of DWH10 in each of these 12 validation locations is presented in Table 4.4. We employed eight assessment metrics to evaluate accuracy: producer and user accuracy for land, producer and user accuracy for temporary and stable water bodies, overall accuracy, and kappa coefficient. Based on the validation data from all 12 locations, DWH10 achieved an average accuracy of 95% for water body monitoring, accompanied by a kappa coefficient of 0.93. Specifically, DWH10 exhibited enhanced accuracy in detecting transient water bodies, with a producer accuracy of 94% and a user accuracy of 88%. In monitoring stable water bodies, DWH10 showed strong performance with producer and user accuracy rates of 92 and 97%, respectively. These results unequivocally demonstrate the effectiveness of our DWH10 in water body monitoring.

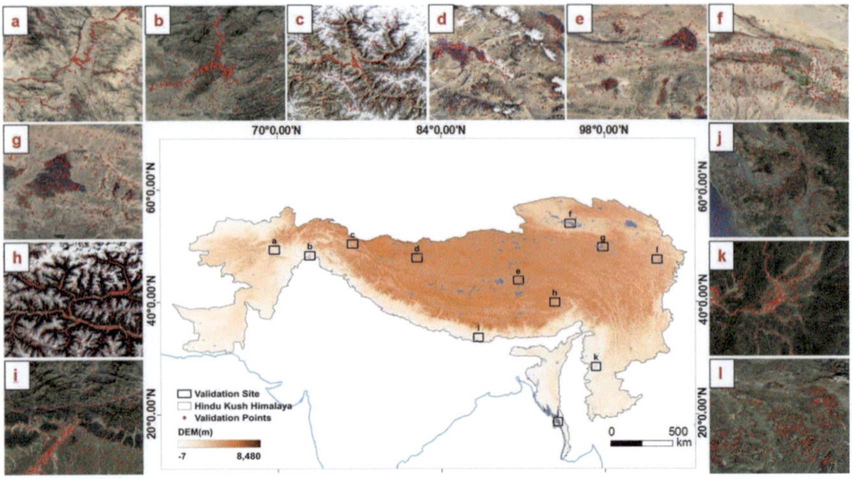

Fig. 4.11 Distribution of validation data (site **a**–**c** in Pakistan, sites **d**–**h** and **l** in China, site **i** in Nepal and sites **k** and **j** in Myanmar)

Table 4.4 The confusion matrix of our DWH10 obtained using validation samples from the 12 sites

		a	b	c	d	e	f	g	h	i	j	k	l	A.A
Npts	T.S	226	240	262	111	86	323	128	146	296	302	207	548	/
	S.S	261	388	137	532	592	321	570	393	221	553	290	109	/
	L.S	423	335	524	319	252	286	230	386	384	113	401	310	/
DWH10	P.T	0.94	0.93	0.92	0.97	0.87	0.95	0.93	0.91	0.93	0.98	0.95	0.99	**0.94**
	U.T	0.85	0.82	0.89	0.84	0.91	0.95	0.86	0.84	0.90	0.94	0.85	0.96	**0.88**
	P.S	0.86	0.93	0.81	0.96	0.99	0.97	0.97	0.96	0.92	0.98	0.86	0.83	**0.92**
	U.S	0.98	0.98	0.82	0.99	0.99	0.95	0.99	0.99	0.94	1.00	0.98	0.98	**0.97**
	P.L	0.99	0.95	0.98	0.97	0.99	0.96	0.96	0.96	0.95	0.95	0.99	0.97	**0.97**
	U. L	0.97	0.97	0.99	0.97	0.96	0.99	0.97	0.96	0.96	0.95	0.97	0.98	**0.97**
	O.A	**0.94**	**0.94**	**0.94**	**0.97**	**0.98**	**0.96**	**0.96**	**0.95**	**0.93**	**0.97**	**0.94**	**0.97**	0.95
	K.C	**0.91**	**0.90**	**0.89**	**0.94**	**0.96**	**0.94**	**0.94**	**0.93**	**0.90**	**0.95**	**0.91**	**0.94**	0.93

Noted T.S., temporal water samples; S.S., stable water samples; L.S., land samples; P.T., producer's accuracy for temporal water; U.T., user's accuracy for temporal water; P.S., producer's accuracy for stable water; U.S., user's accuracy for stable water; P.L., producer's accuracy for land; U.L., user's accuracy for land; O.A., overall accuracy; K.C., kappa coefficient

(2) ***Monitoring results of time-series surface water distribution in the HKH region (2000, 2010, 2020)***

The following analysis presents the results of our time-series dynamic water body dataset of the HKH (TDWH) region for the years 2000, 2010, and 2020 (Fig. 4.12).

The data shows that the distribution of surface water bodies has been abundant since 2000, with significant changes observed by 2020, particularly in the Qinghai-Tibet Plateau region, India, Afghanistan, and Pakistan. In addition, for temporal water bodies, there has been a significant increase in temporal water body coverage in 2020 compared to 2000, which was believed to be prompted by the significant reduction in the permafrost coverage within the HKH. China experienced the largest increase, with an additional 6905.4 km^2, followed by India and Pakistan, with increases of 1100.8 km^2 and 967.6 km^2, respectively. Afghanistan, Myanmar, Nepal, Bangladesh, and Bhutan also saw increases of 642.9 km^2, 440.2 km^2, 435.7 km^2, 71.3 km^2, and 63.5 km^2, respectively. In terms of stable water bodies, the area in Bangladesh and Nepal has decreased by approximately 38.5 km^2 and 5.3 km^2, respectively, compared to 2000. Conversely, China, Pakistan, India, Myanmar, Afghanistan, and Bhutan have all seen increases in stable water body areas by 4300.4 km^2, 151.1 km^2, 94.5 km^2, 60.7 km^2, 51.0 km^2, and 25.8 km^2, respectively. Overall, the data indicates a general increase in water bodies across most regions of the HKH over the past two decades.

Furthermore, the details for some areas in China are illustrated in Fig. 4.13 to demonstrate the effectiveness of our dataset in monitoring changes in water bodies. Notably, the area of stable water bodies has increased significantly from 2000 to 2020, as seen in Fig. 4.13a and b. Additionally, some areas initially classified as temporal water bodies have transformed into stable covered waters, with an extended duration of water coverage. Conversely, in Fig. 4.13c, small and medium-sized stable lakes have become seasonal covered waters, resulting in a decrease in the total coverage area compared to 2000.

We meticulously evaluated the accuracy of our time series products using validation samples from 12 selected areas as described in Sect. 4.2.1. Utilizing high-resolution Google Earth images, we visually interpreted the true categories of all validation samples for the years 2010 and 2020. Unfortunately, due to the limited availability of images in Google Earth for the year 2000, we were unable to complete the category interpretation for that year. Consequently, our accuracy assessment focused solely on the mapping results of 2010 and 2020. To ensure a comprehensive evaluation, we used eight indicators, including producer accuracy and user accuracy for both temporary and stable water bodies, as well as land, along with overall accuracy and kappa coefficient. Additionally, we cross-compared our mapping results with those from the Joint Research Centre Global Surface Water (JRC-GSW) dataset for the years 2000, 2010, and 2020 to further validate our product's reliability.

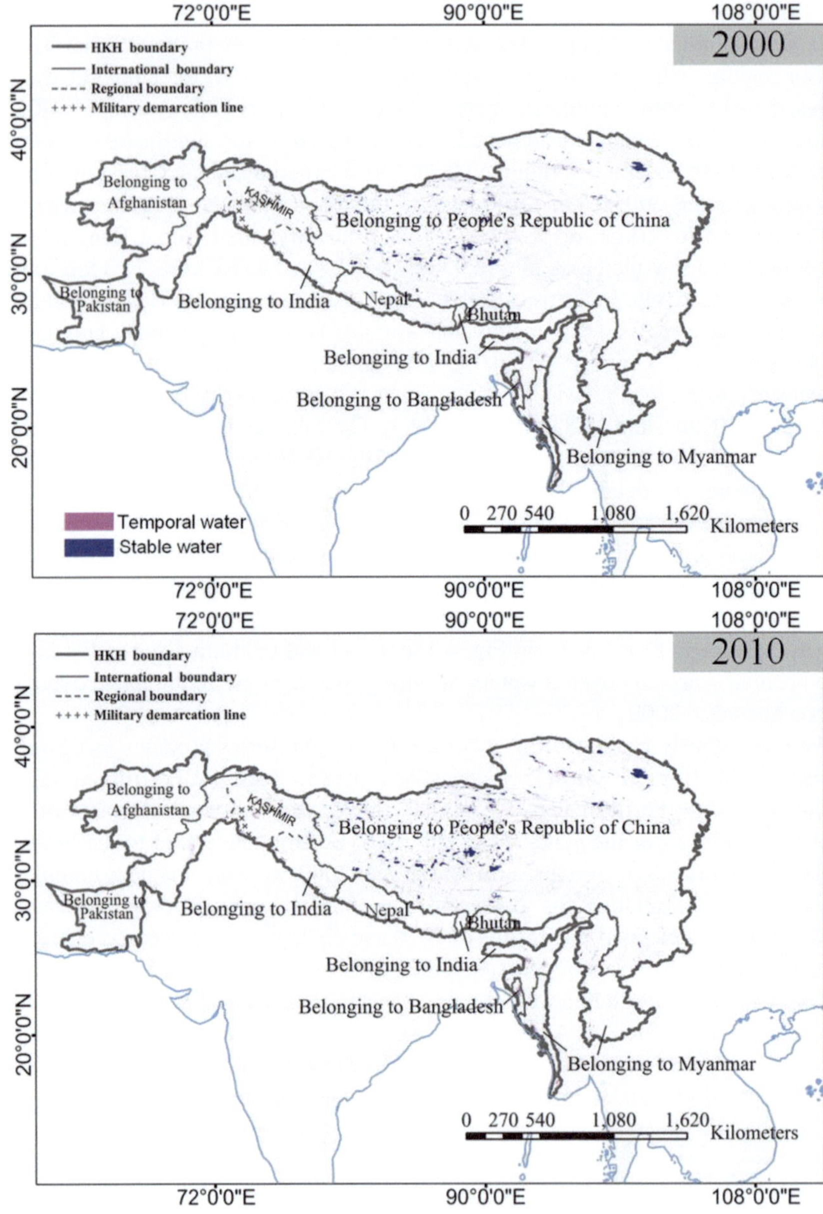

Fig. 4.12 Time series monitoring results of surface water distribution in the HKH region (2000, 2010, and 2020)

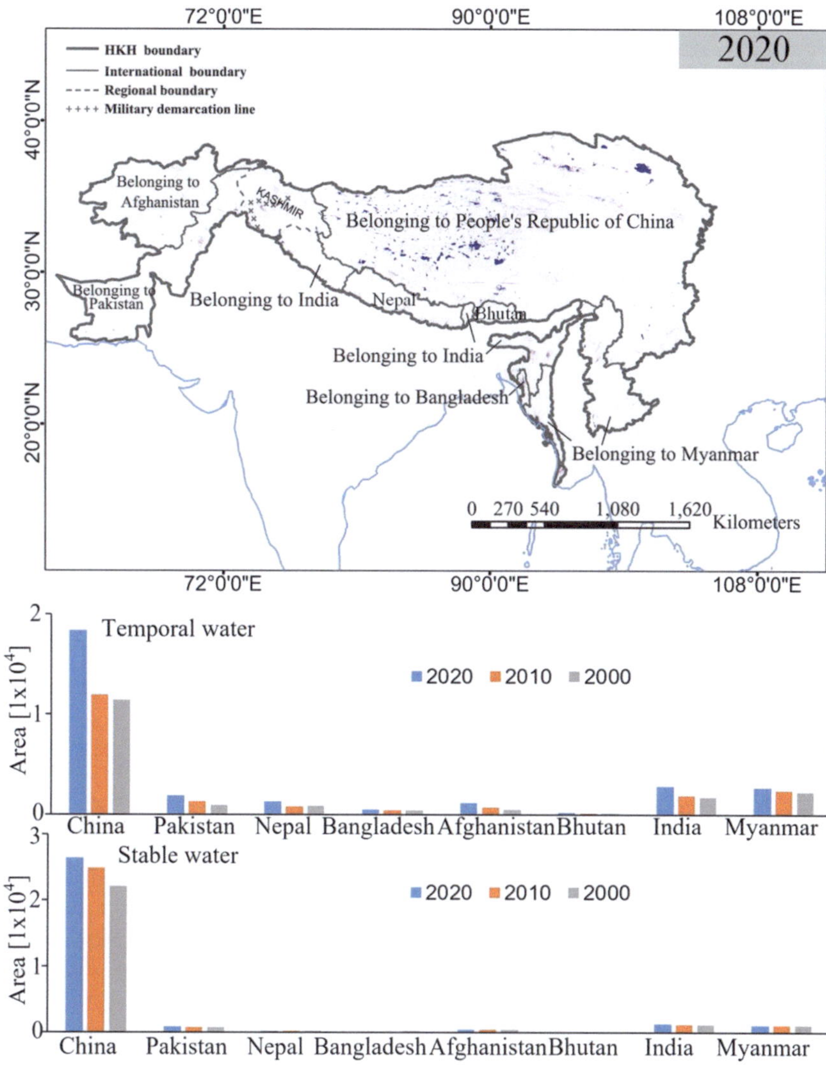

Fig. 4.12 (continued)

The verification results from all 12 validation points revealed that the average accuracy of our Time-series Dynamic Water Habitats (TDWH) product for water body monitoring in 2010 and 2020 reached an impressive 93%, with kappa coefficients of 0.89 and 0.90, respectively. Our TDWH product also demonstrated high accuracy in identifying temporal water bodies, achieving producer accuracies of 91 and 92% for 2010 and 2020, respectively, and user accuracies of 88 and 89%. For stable water bodies, the producer accuracy was consistent at 90% for both years, while the user accuracy was slightly higher, at 93% for 2010 and 95% for 2020

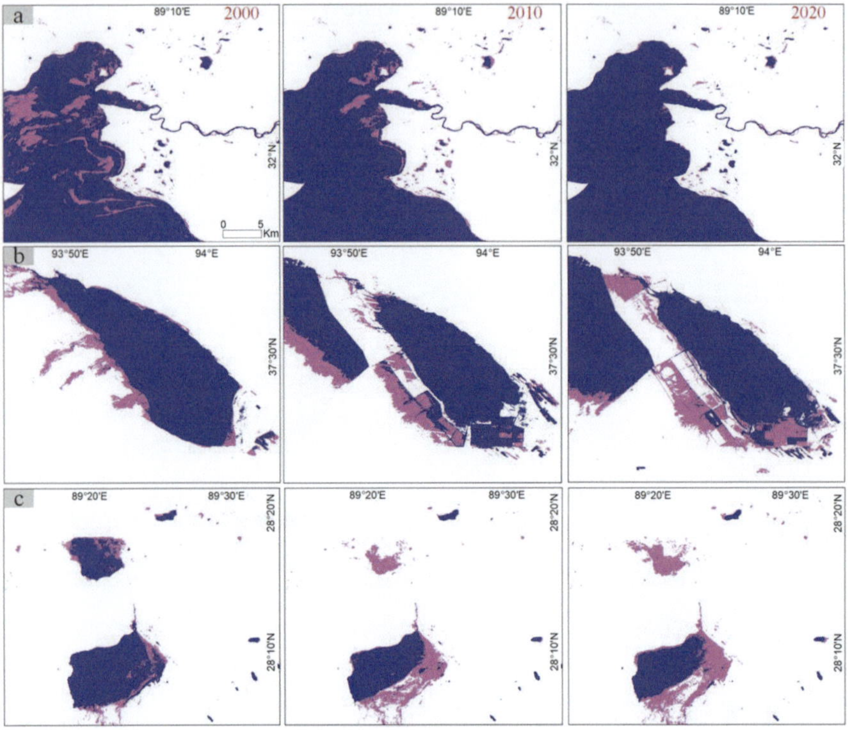

Fig. 4.13 Local details of changes in time series monitoring results of surface water distribution in the HKH region on the Qinghai-Tibet Plateau

(Table 4.5). These high accuracy levels highlight the robustness and reliability of our TDWH product in monitoring various types of water bodies over time.

In addition to individual accuracy metrics, we compared our TDWH results with the JRC-GSW dataset to assess consistency. The consistency (R^2) between the TDWH and JRC-GSW datasets for stable water bodies was remarkably high, exceeding 0.96 for the years 2000, 2010, and 2020, with significant consistency ($P < 0.05$). Similarly, the comparison for temporal water bodies also showed significant consistency across the three years, with the lowest R^2 value being 0.63. This strong agreement between the TDWH and JRC-GSW datasets further confirms the reliability of our time series water body product in the HKH region. Our comprehensive evaluation, incorporating multiple accuracy indicators and cross-dataset comparisons, demonstrates the effectiveness of the TDWH product in accurately monitoring water bodies over an extended temporal scale (Fig. 4.14).

(3) ***Time-series water transparency datasets for the HKH region***

This study developed a fine-resolution long time-series water transparency dataset (2000, 2010, 2020) for the HKH region based on the produced synthetic cloud-free data with consistent phenology for the years 2000–2020 and the constructed

Table 4.5 Accuracy evaluation table of TDWH in 2020 and 2010

		a	b	c	d	e	f	g	h	i	j	k	l	A.A
2020														
Npts	T.S	221	242	245	111	82	315	125	143	294	294	205	540	/
	S.S	253	385	137	532	575	318	568	390	220	550	282	101	/
	L.S	423	331	524	319	246	285	230	386	384	113	401	310	/
TDWH	P.T	0.96	0.88	0.87	0.97	0.86	0.92	0.91	0.91	0.93	0.98	0.86	0.95	**0.92**
	U.T	0.78	0.97	0.86	0.84	0.90	0.91	0.86	0.83	0.90	0.94	0.90	0.97	**0.89**
	P.S	0.84	0.99	0.84	0.86	0.97	0.93	0.91	0.96	0.92	0.95	0.86	0.81	**0.90**
	U.S	0.96	0.97	0.82	0.94	0.99	0.94	0.97	0.99	0.94	0.93	0.93	0.96	**0.95**
	P.L	0.99	0.95	0.98	0.96	0.96	0.96	0.96	0.96	0.95	0.93	0.96	0.95	**0.96**
	U. L	0.98	0.97	0.95	0.92	0.94	0.96	0.91	0.92	0.96	0.95	0.97	0.95	**0.95**
	O.A	0.92	0.93	0.94	**0.87**	**0.96**	**0.93**	**0.91**	**0.94**	**0.93**	**0.96**	**0.92**	**0.98**	**0.93**
	K.C	**0.88**	**0.90**	**0.88**	**0.84**	**0.93**	**0.90**	**0.86**	**0.93**	**0.90**	**0.94**	**0.91**	**0.96**	**0.90**
2010														
Npts	T.S	221	242	232	110	80	310	120	140	287	285	202	532	/
	S.S	243	361	130	520	570	304	545	380	214	547	278	100	/
	L.S	433	355	548	324	238	305	253	401	404	131	426	324	/
TDWH	P.T	0.91	0.86	0.85	0.94	0.86	0.93	0.90	0.92	0.93	0.97	0.86	0.96	**0.91**
	U.T	0.78	0.97	0.82	0.82	0.88	0.92	0.88	0.81	0.91	0.92	0.91	0.97	**0.88**
	P.S	0.84	0.96	0.81	0.83	0.95	0.90	0.94	0.95	0.93	0.94	0.87	0.84	**0.90**
	U.S	0.88	0.97	0.80	0.96	0.97	0.92	0.97	0.99	0.94	0.91	0.93	0.96	**0.93**
	P.L	0.93	0.92	0.95	0.98	0.96	0.96	0.96	0.97	0.97	0.96	0.95	0.97	**0.96**
	U. L	0.95	0.91	0.92	0.99	0.96	0.96	0.96	0.98	0.96	0.95	0.94	0.96	**0.95**
	O.A	0.89	0.90	0.93	**0.88**	**0.94**	**0.92**	**0.97**	**0.95**	**0.94**	**0.96**	**0.91**	**0.98**	**0.93**
	K.C	**0.86**	**0.88**	**0.87**	**0.86**	**0.91**	**0.89**	**0.89**	**0.93**	**0.91**	**0.92**	**0.88**	**0.93**	**0.89**

water transparency inversion model based on watercolor parameters. To extract the stable water body coverage areas in the HKH region for each target year, the study combined the results of the water distribution mapping in Sect. 4.2.2. Then, based on the extracted water body areas, the study completed the inversion and mapping of water transparency products. A clearer water body is represented by increased transparency in Fig. 4.15, which shows the outcomes of the water body transparency products developed for the three target years in the Hindu Kush-Himalayan region: 2000, 2010, and 2020. A closer look at the two water body regions shown in Fig. 4.15b and c shows that the water body transparency products generated by our research can accurately depict the spatial transparency status of the water body, with a smooth transparency excess extending from the lake's center to the shore. Furthermore, it is evident from the results of 2000, 2010, and 2020 that the target waterways' water transparency varies over time. From 2000 onwards, there was an increase in

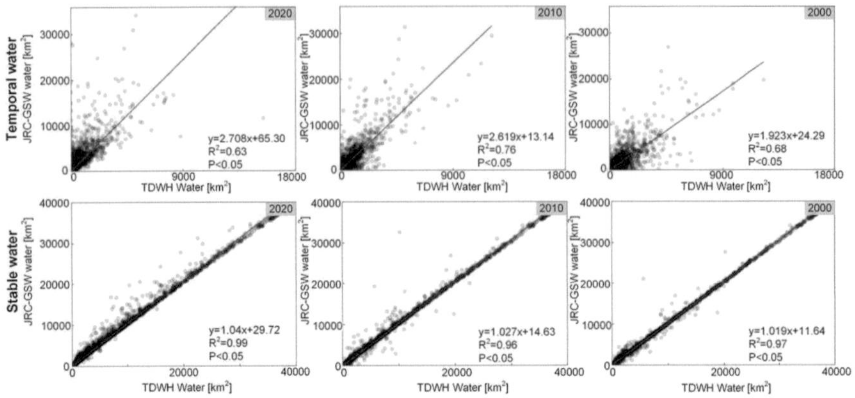

Fig. 4.14 Scatter plot of the research mapping results (time-series DWH, TDWH) and the time-series cross-comparison of JRC-GSW in the HKH region from 2000 to 2020

transparency in both selected waterways, with the water bodies getting clearer over time.

4.2.3 Summary

Utilizing a sophisticated algorithm, we developed a high-resolution (10-m) dynamic water body distribution map (DWH10) for the Hindu Kush-Himalayan (HKH) region. This map distinguishes between stable and temporary water bodies. The analysis revealed that the HKH region hosts over 1.45×10^5 km^2 of surface water bodies, predominantly in its central and western parts. Seasonal variations were significant due to the presence of numerous temporary water bodies. The aggregated data showed that stable water bodies are more prevalent between 80°E and 95°E longitude and 28°N and 40°N latitude, whereas temporary water bodies are mainly found between 90°E and 95°E longitude and 34°N and 40°N latitude. Notably, China encompasses 77.16% of the HKH region's water bodies, with Myanmar and India following at 7.15 and 6.37%, respectively Nepal and Pakistan held a coverage of 2.25 and 3.77 respectively. The smallest shares were observed in Bhutan (0.25%) and Bangladesh (0.84%). Validation of the DWH10 product using data from 12 randomly selected regions demonstrated an average accuracy of 95% for water body monitoring, with a kappa coefficient of 0.93, highlighting the product's reliability and precision.

A time-series analysis from 2000 to 2020 showed substantial changes in the distribution of surface water bodies in the HKH region, particularly in the Qinghai-Tibet Plateau, India, Afghanistan, and Pakistan. The area of stable water bodies increased significantly during this period. Temporal water bodies transformed into stable ones in some regions, extending the duration of water coverage, while some stable lakes

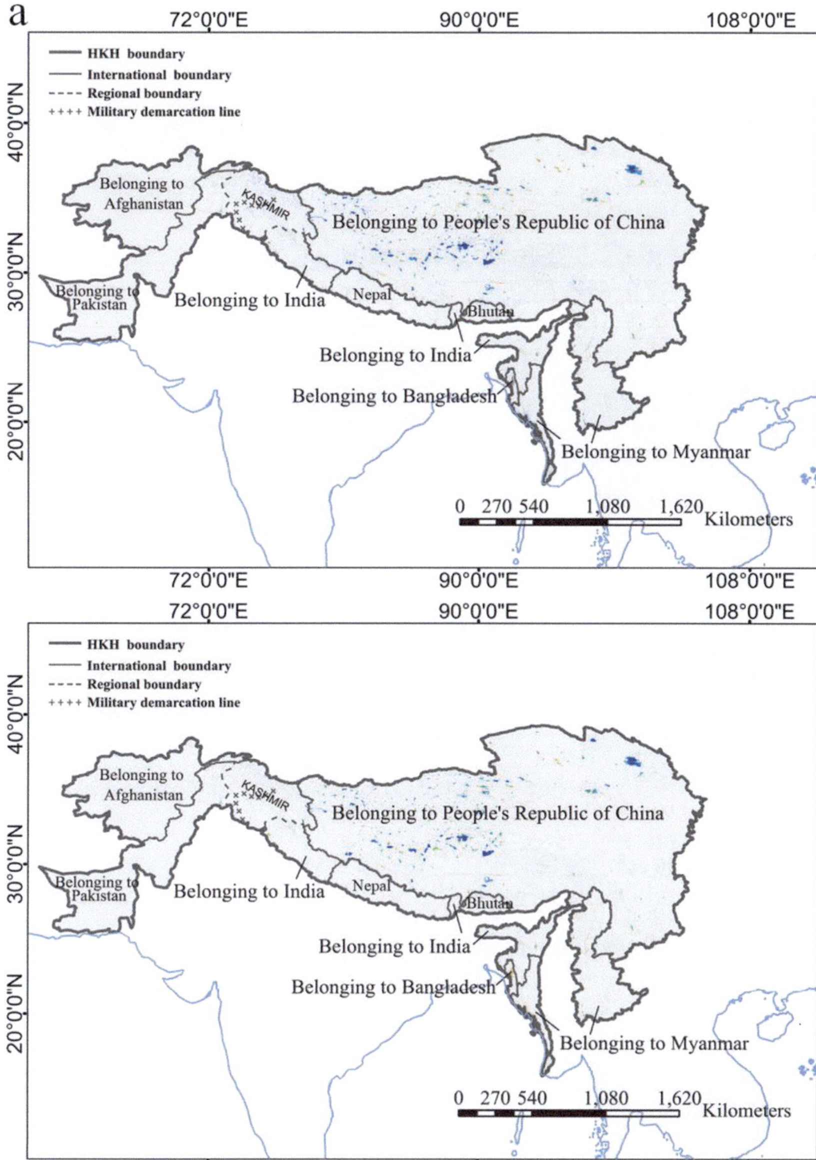

Fig. 4.15 30-m resolution water transparency products for the HKH region in 2000, 2010 and 2020

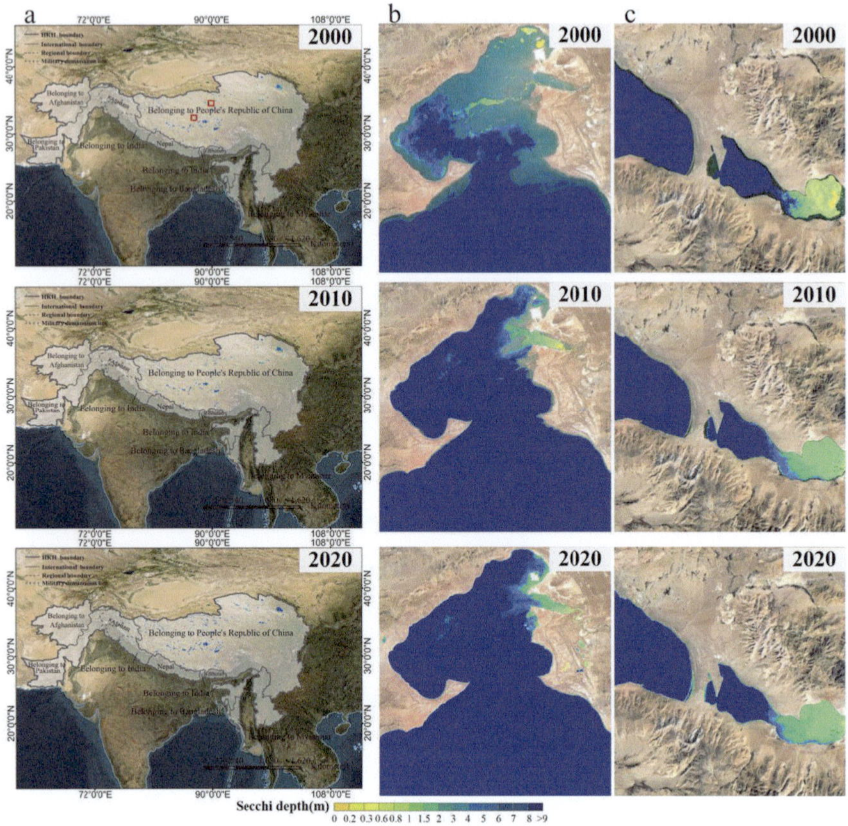

Fig. 4.15 (continued)

became seasonal. The accuracy of our time-series products was validated using high-resolution images and cross-comparisons with the Joint Research Centre Global Surface Water dataset, revealing high consistency and reliability. The comprehensive evaluation demonstrated that our time-series dynamic water body dataset of the HKH (TDWH) effectively monitors water bodies over time, achieving an average accuracy of 93% with a kappa coefficient of 0.90. Furthermore, a 30-m resolution time-series water transparency dataset (2000, 2010, and 2020) for the HKH region was also developed. This dataset was generated using our TDWH dataset, along with the produced synthetic cloud-free data featuring consistent phenology for the years 2000–2020, and watercolor parameters. Our water transparency dataset results from the years 2000, 2010, and 2020 indicate that the HKH water bodies have become clearer, with increased transparency.

4.3 Evaluation of Climate Change Impacts on Regional Water Resources Changes

With the intensification of global climate change, alterations in regional water resources have garnered increasing attention. Climate change has profoundly impacted various aspects of water resources, including ice, snow, permafrost, water body areas, and water quality. These transformations impact ecosystem stability and human societal development, presenting substantial challenges for the sustainable use of water resources (Wang et al., 2022; Zhang et al., 2022). Rising global temperatures have accelerated the reduction of glaciers, ice, and snow, which significantly influences river runoff, the formation of glacial lakes, and the associated disaster risks (Yang et al., 2019). The degradation and thinning of these cryosphere elements also lead to increased surface water evaporation, thereby affecting water supply in adjacent areas (Yao, 2019). Furthermore, changes in precipitation patterns and the frequency of extreme climate events impact the availability of water resources in rivers, lakes, and groundwater systems (Wang et al., 2022).

Climate change also exacerbates water quality issues, including eutrophication, acidification, and pollution, further diminishing the availability of clean water resources (Chen et al., 2022). These multifaceted impacts of climate change on regional water resources present unprecedented challenges for both the ecological environment and human societies. Despite these recognized challenges, the specific effects have not been comprehensively and quantitatively assessed in the HKH region. This region is particularly sensitive to climate change due to its significant cryospheric components and its crucial role in providing water resources to surrounding areas.

Understanding the impact of key climate factors on water resource changes in the HKH region is critical for ensuring sustainable water use and developing effective water resource management strategies. Comprehensive and quantitative assessments of these climate impacts can guide policymakers and stakeholders in implementing measures to mitigate adverse effects and adapt to the evolving climate conditions. Hence, this study aims to explore the detailed impact of climate change on water resources in the HKH region, focusing on identifying key climate factors and their specific effects on water availability and quality. By addressing these issues, we strive to support the development of resilient and adaptive water resource management practices that safeguard both ecological integrity and human well-being in the HKH region.

4.3.1 Methodology for Assessing Climate Factors' Impact on Water Resource Changes in the HKH Region

To analyze the potential impact of climate factors on water resource changes in the HKH region, we collected data on eight typical climate factors: wind speed (WS), humidity (HU), evapotranspiration (ET), pressure (PR), precipitation (PRE), downward shortwave radiation (DSR), downward longwave radiation (DLR), and temperature (TEM). These factors are generally believed to affect water resource changes in the HKH region (McNicoll, 2019). Data for WS, HU, DSR, DLR, and TEM were obtained from NASA's Global Land Data Assimilation System (GLDAS) product (Rodell et al., 2004). GLDAS provides daily data with a spatial resolution of 0.25 degrees, covering the period from 2000 to 2020. We calculated the annual average WS, HU, DSR, DLR, and TEM for analysis. Data for PR and PRE were sourced from the European Centre for Medium-Range Weather Forecasts Reanalysis v5 (ERA5) product (Hersbach et al., 2023). ERA5 is a reanalysis product covering the period from 1950 to the present, with hourly temporal resolution and 0.1-degree spatial resolution. We obtained the annual average surface pressure and precipitation based on ERA5 data. ET data were derived from the MOD16A2 evapotranspiration product (Running et al., 2017). MOD16A2 provides total surface evapotranspiration on an 8-day scale at a 500-m resolution. We calculated the annual average evapotranspiration for analysis. Using these processed climate factors, we analyzed their relationship with changes in ice and snow, permafrost, surface water bodies, and water quality. Water transparency, a commonly used comprehensive water quality monitoring parameter, was used as an indicator of water quality. We employed the non-parametric Spearman correlation analysis method to evaluate the correlation between various climate factors and water resource changes r.

$$\mathrm{r} = \frac{\sum(R_X - \overline{R_X})(R_Y - \overline{R_Y})}{\sqrt{\sum(R_X - \overline{R_X})^2 \sum(R_Y - \overline{R_Y})^2}} \tag{4.20}$$

The variables X and Y have ranks denoted by R_X and R_Y, respectively, whereas the average ranks are denoted by $(\overline{R_X})$ and $(\overline{R_Y})$. Additionally, we regard the correlation between the variables as significant when the *p*-value is less than 0.05.

4.3.2 *Potential Relationship Between Climate Change and Regional Water Resources Change*

The comprehensive analysis depicted in Fig. 4.16 provides an overview of the diverse climatic trends observed across the Hindu Kush Himalayan (HKH) region from 2000 to 2020. In China, significant increases were noted in evapotranspiration, air pressure, longwave radiation, and temperature over the two-decade period. Conversely, precipitation, humidity, and downward shortwave radiation remained relatively steady, while wind speed exhibited a decreasing trend. Pakistan experienced similar upward trends in humidity, evapotranspiration, air pressure, downward longwave radiation, and temperature, with stable conditions in air pressure, precipitation, and downward shortwave radiation, while wind speed decreased. Nepal recorded increases in evapotranspiration, temperature, and precipitation, alongside declining wind speed and stable air pressure, humidity, and both downward longwave and shortwave radiation. In Bangladesh, declining trends were observed in wind speed, humidity, downward longwave radiation, and temperature, while evapotranspiration, precipitation, downward shortwave radiation, and air pressure remained stable. Afghanistan saw declining wind speed alongside increases in evapotranspiration and downward longwave radiation, with other climate factors remaining stable. Bhutan and Myanmar both showed decreasing wind speeds, with Bhutan indicating stability across the other climate factors and Myanmar also showing declines in humidity and longwave radiation, with stability observed in the remaining factors. In India, declining wind speeds were noted, while evapotranspiration and temperature increased, with stability across the other climate factors throughout the analyzed period from 2000 to 2020.

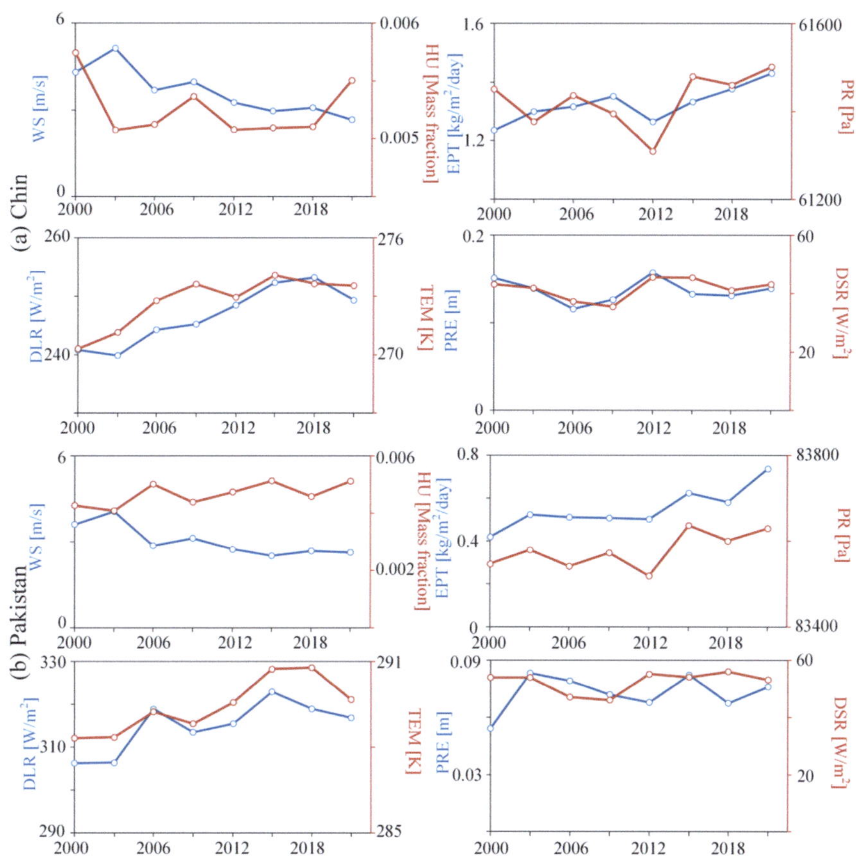

Fig. 4.16 Wind speed (WS), humidity (HU), evapotranspiration (ET), pressure (PR), precipitation (PRE), and downward shortwave radiation (downward) in the HKH region from 2000 to 2020 Changes in eight climate factors: shortwave radiation (DSR), downward longwave radiation (DLR) and temperature (Temperature, TEM)

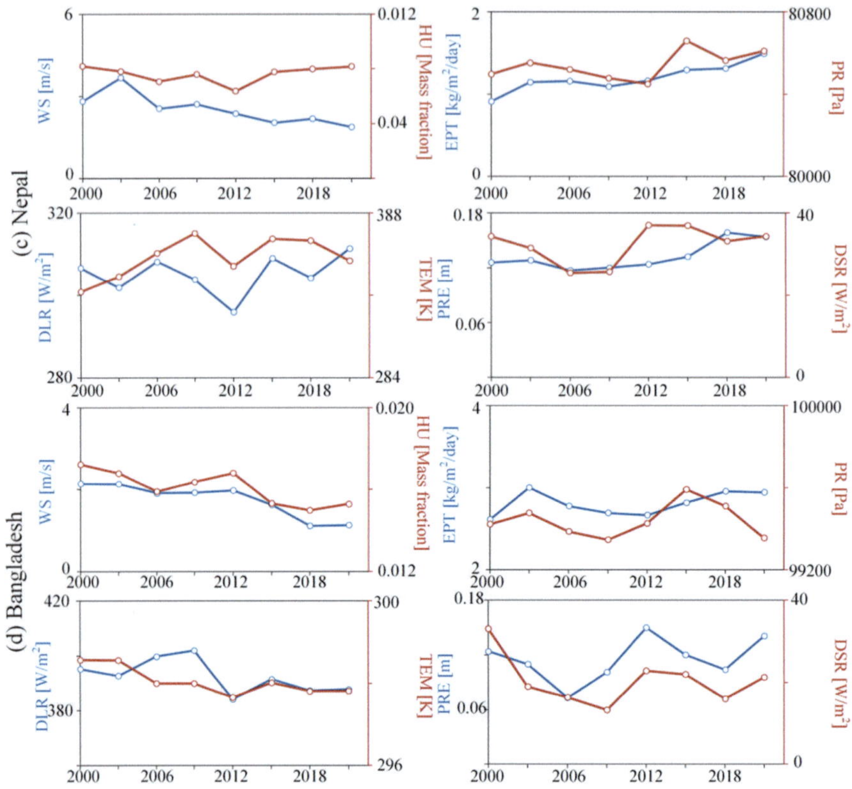

Fig. 4.16 (continued)

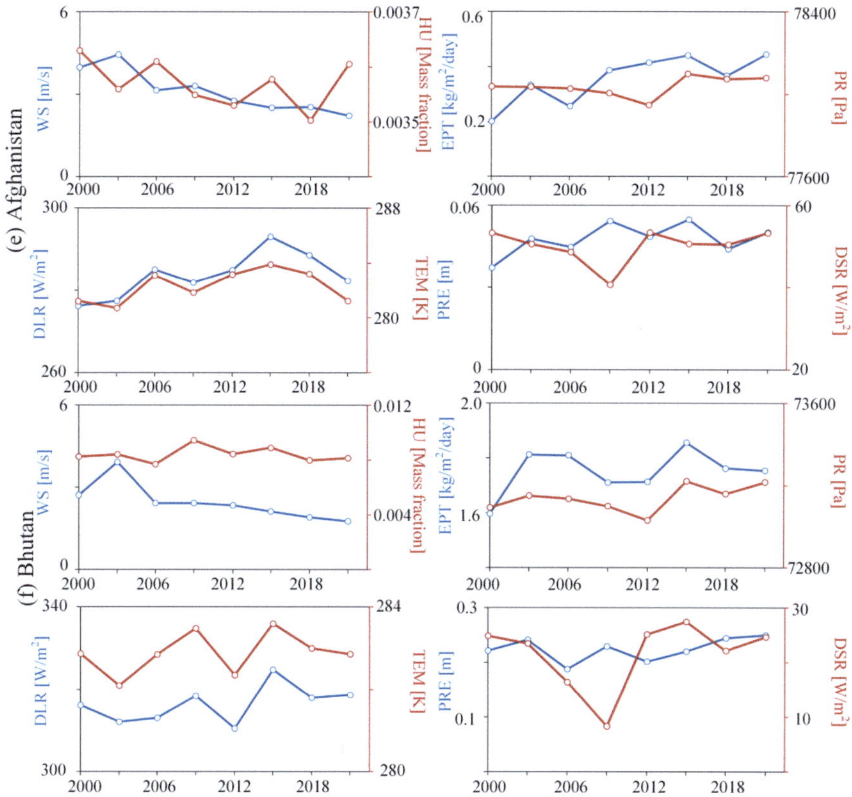

Fig. 4.16 (continued)

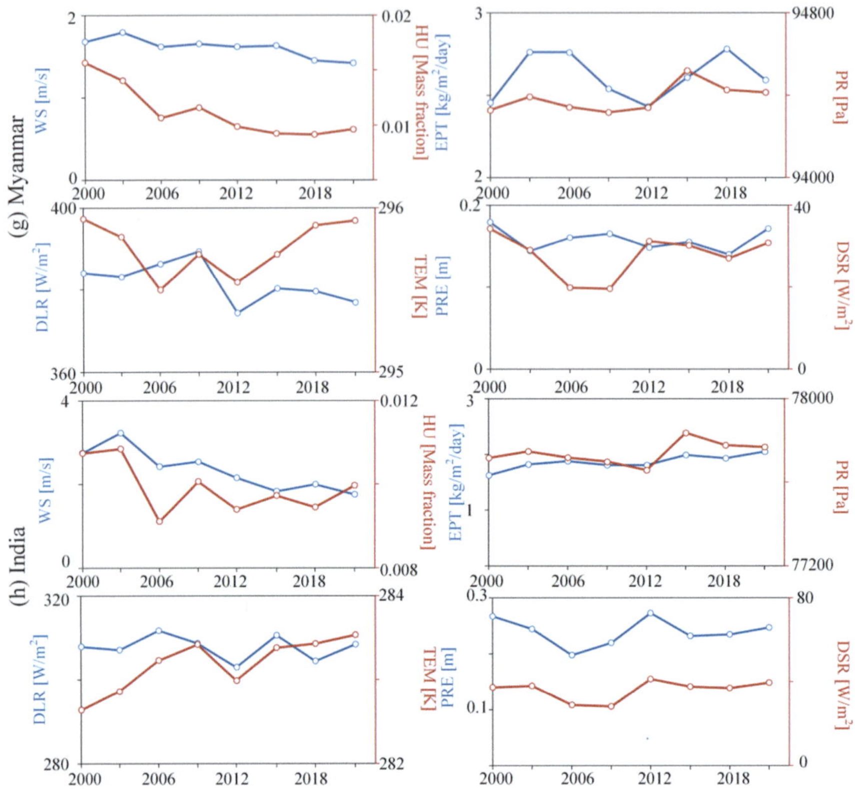

Fig. 4.16 (continued)

As shown in Fig. 4.17, it is evident that the relationship between environmental variables such as permafrost, wind speed, downwelling long-wave radiation, temperature, and water body area varies across different countries in the HKH region. For instance, the changes in permafrost and wind speed are positively correlated in most countries, except Myanmar. Specifically, China, Nepal, and Afghanistan exhibit significant negative correlations between these two variables. This implies that as the permafrost decreases, wind speed tends to increase in these regions. The relatively small distribution of the HKH region in Myanmar and the insignificant change in wind speed may explain the weak negative correlation observed there.

Additionally, downwelling long-wave radiation is negatively correlated with changes in permafrost across the HKH region, with significant negative correlations noted in China, Nepal, and Afghanistan. This means that higher levels of downwelling long-wave radiation are associated with a reduction in permafrost in these countries. The temperature in Myanmar, on the other hand, shows a significant positive correlation with its permafrost area, suggesting that higher temperatures are linked to a decrease in permafrost. Furthermore, changes in the permafrost area and water bodies show a significant negative correlation in China and Nepal. This indicates

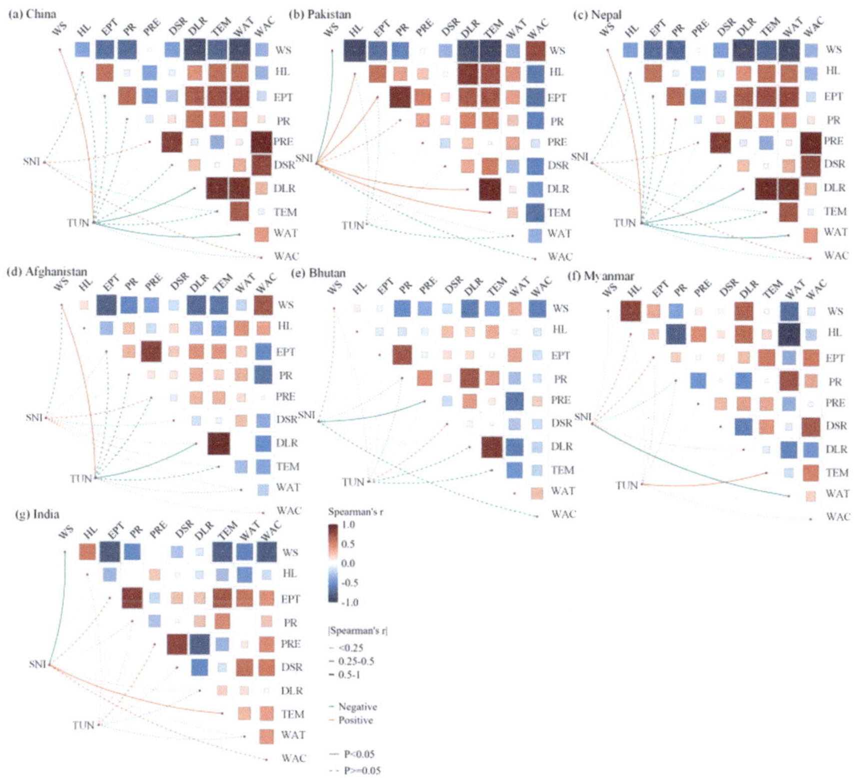

Fig. 4.17 Correlation between changes in snow and ice (SNI) and permafrost (Tundra, TUN) in the HKH region from 2000 to 2020 and various typical climate factors. *Noted* There is no permafrost or snow in Bangladesh

that a reduction in permafrost area is associated with an increase in water body area, likely due to the melting of permafrost. Non-significant negative correlations between these variables are also observed in Pakistan, India, and Afghanistan, hinting similar but less pronounced trends. When considering ice and snow, wind speed changes in China, Nepal, Afghanistan, Pakistan, and India are negatively correlated with changes in ice and snow. India and Pakistan show significant negative correlations, suggesting that reduced wind speeds may lead to an increase in snow cover in these countries. Given that over 99% of the ice and snow in the HKH region is in these five countries, the observed negative correlation is crucial in understanding regional climate dynamics.

Moreover, temperature shows a significant positive correlation with ice and snow in Pakistan and India, indicating that higher temperatures are associated with increased ice and snow cover in these regions. Changes in humidity, evapotranspiration, and downwelling long-wave radiation also exhibit significant positive correlations with ice and snow changes in Pakistan, further emphasizing the complex interactions between these environmental factors. Lastly, there is a significant negative correlation between changes in ice and snow and changes in water area in Myanmar. This suggests that as the ice and snow decrease, the water area increases, likely due to melting ice contributing to surface water bodies, however, on the long run this could lead to a more devastating effect on the hydrological system. These findings collectively highlight the intricate and region-specific responses of the HKH region's environmental variables to climatic changes, underscoring the importance of localized studies for effective environmental management and policymaking.

The relationship between climate factors and changes in surface water bodies across different regions, particularly in China, Bangladesh, Nepal, Myanmar, Pakistan, and India, reveals significant correlations that highlight the complex interactions between climate variables and water body dynamics (Fig. 4.18). In China, changes in surface water bodies exhibit a noteworthy positive correlation with increases in downward long-wave radiation and evapotranspiration, while showing a negative correlation with wind speed. The intensified greenhouse effect due to increased downward long-wave radiation accelerates the melting of ice, snow, and permafrost, potentially influencing water body expansion. Moreover, elevated radiation enhances evapotranspiration rates, further affecting water body areas positively. Conversely, higher wind speeds increase water body evaporation rates, which explains the negative correlation observed with water body area changes.

In Bangladesh, changes in water bodies correlate positively with temperature variations, suggesting a relationship where warmer temperatures may contribute to water body expansion. In Nepal, changes in water bodies are positively correlated with permafrost performance, indicating that alterations in permafrost conditions influence water body dynamics. Conversely, in Myanmar, changes in water bodies show negative correlations with humidity and ice/snow variations, indicating that drier conditions and reduced ice/snow cover may lead to water body contraction. Turning to water body transparency, which serves as an indicator of water quality, significant correlations with climate factors are observed across different countries. In China and Nepal, water body transparency shows a positive correlation with precipitation, indicating that increased precipitation may lead to clearer water bodies and improved water quality. Additionally, in China, downwelling shortwave radiation also correlates positively with water transparency, further suggesting that increased solar radiation enhances water clarity. Conversely, in Pakistan, higher wind speeds positively correlate with water transparency, suggesting that increased wind may clear turbid waters. However, in India, higher wind speeds negatively correlate with water transparency, indicating that increased wind may stir up sediments and reduce water clarity.

Overall, these findings underscore the diverse impacts of climate change on water environments across different regions. While some regions may experience clearer water and improved quality due to increased precipitation and solar radiation, others may face turbidity or changes in water body extent due to variations in temperature,

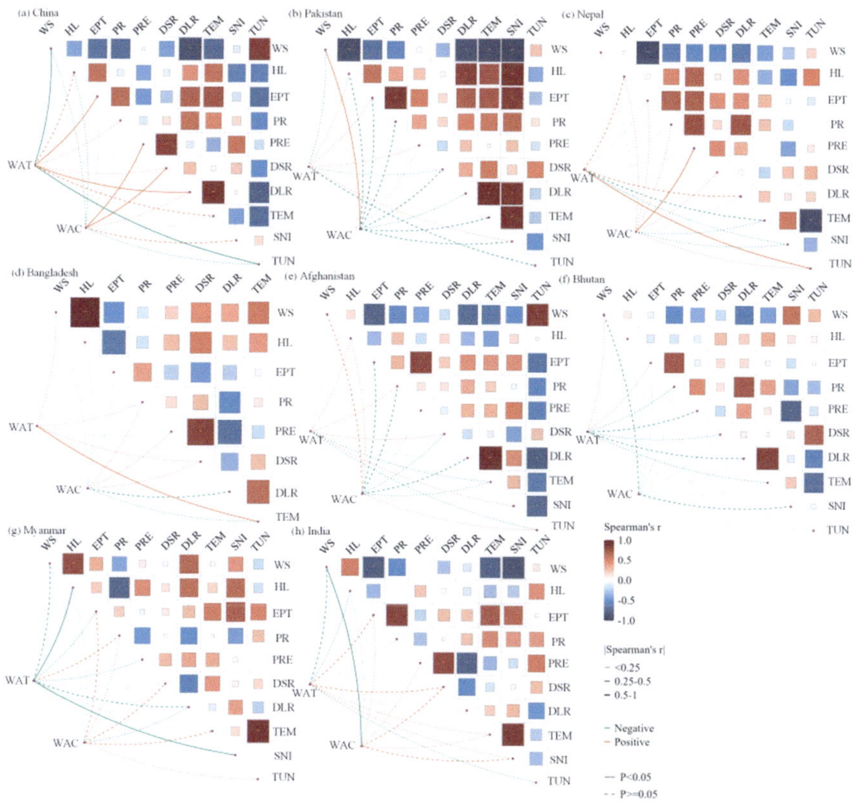

Fig. 4.18 Correlation between changes in surface water area (WAT) and water clarity (water clarity (WAC)) in the HKH region from 2000 to 2020 and various typical climate factors

humidity, and wind patterns. Understanding these complex relationships is crucial for predicting and mitigating the impacts of climate change on water resources and ecosystems globally.

4.3.3 Summary

The summary drawn from the comparison of climatic factors in the HKH region reveals several significant correlations and implications for environmental dynamics. Wind speed and downward longwave radiation exhibit positive correlations with changes in permafrost, suggesting that these climatic variables contribute to alterations in the extent of permafrost across the region. Conversely, there is a negative correlation between changes in permafrost and the area of water bodies, indicating that as permafrost diminishes, there is a concurrent expansion of surface water bodies in most HKH regions, likely due to the melting processes involved. Furthermore, variations in ice and snow cover generally correlate negatively with changes in wind speed, indicating that faster winds may contribute to the reduction of ice and snow.

Moreover, the study highlights the diverse impacts of temperature, humidity, evapotranspiration, and other factors on permafrost, ice and snow cover, and surface water bodies across different regions within the HKH area. These findings underscore the complexity and variability in how regional climate factors influence surface water area and water quality. Therefore, effective protection and management strategies for the HKH region must consider the unique characteristics and sensitivities of each area to climate change, ensuring adaptive measures are tailored to mitigate the specific environmental impacts observed.

4.4 Evaluation of the Impacts of Human Activities on Regional Water Quality Changes

The water quality in the HKH is increasingly threatened by various human activities. One major source of pollution is agricultural runoff from the Indo-Gangetic Plains, which brings sewage, chemical fertilizers, and other contaminants into the HKH's water systems. Deforestation and land-use changes further exacerbate water quality issues by increasing sedimentation and turbidity in rivers and streams.

In addition to local pollution, global climate change intensifies water quality problems in the HKH. The accelerated melting of glaciers due to rising temperatures releases accumulated pollutants and debris into water sources. Climate anomalies, such as extreme precipitation events, have also been linked to outbreaks of waterborne diseases like Dengue and Chikungunya in the region (ICIMOD, 2023). Another significant anthropogenic threat is the overexploitation and declining recharge of groundwater resources. The drying up of springs, crucial for drinking water in many mountain communities, is closely tied to unsustainable groundwater usage and changing precipitation patterns, leading to further water quality deterioration as contaminants become more concentrated (Kamboj et al., 2020; Sharma et al., 2023).

Protecting the water quality of the HKH is vital for the well-being of both human and natural ecosystems, necessitating concerted and collaborative action to ensure a sustainable water future for this critical region. Hence to tackle these complex challenges, we will adopt a multi-faceted approach to assess the anthropogenic impact on the HKH water quality.

4.4.1 Methodology

The methodology employed in the study involves utilizing remote sensing and demographic data to analyze changes in land cover and human activity factors across the Hindu Kush Himalayan (HKH) region. Specifically, MODIS land cover product

MCD12Q1 was utilized to extract temporal changes in impervious surfaces (IMP), crops (CRO), and various vegetation types (VEG) such as grassland, woodland, and shrubs. This dataset, known for its spatial and temporal coverage, enabled the researchers to track land cover dynamics from 2000 to 2020.

In addition to land cover data, WorldPop population distribution data at a resolution of 100 m were incorporated, spanning the years from 2000 to 2021. This demographic data provided insights into population trends and distributions across the HKH region, facilitating the assessment of human impacts on the environment. Moreover, Gridded global GDP and Electricity Consumption (GGEC) data were employed, offering grid-level information on Gross Domestic Product (GDP) and electricity consumption (EC) worldwide from 1992 to 2019. For the year 2020, where GGEC data was unavailable, linear regression was applied to extrapolate GDP and EC values based on trends observed from 1992 to 2019.

To evaluate the potential influence of human activities on HKH water resources, the study utilized the non-parametric Spearman correlation coefficient. This statistical method assessed relationships between variables such as land cover changes (IMP, CRO, VEG), demographic shifts (POP), and economic indicators (GDP, EC). Significance of relationships was determined using a p-value threshold of 0.05, indicating statistical significance where applicable. By integrating these datasets and methodologies, the study aimed to comprehensively understand the interplay between human activities and environmental changes in the sensitive HKH region, crucial for sustainable resource management and conservation strategies.

4.4.2 Potential Relationship Between Human Activities and Regional Water Resources Change

The analysis in Fig. 4.19 reveals notable trends across various indicators in the HKH region over the past two decades. Population (POP), Gross Domestic Product (GDP), and Energy Consumption (EC) have exhibited consistent upward trajectories across the region. With respect to vegetation cover, most countries in the HKH region have experienced an expansion, except for Myanmar where a decrease is observed. Bangladesh and Bhutan stand out for their stable impervious area trends amidst general regional increases. Conversely, China, Nepal, Bangladesh, Bhutan, and India have witnessed declines in cultivated land area, while Pakistan and Afghanistan have remained relatively stable. Myanmar notably demonstrates a significant increase in cultivated land area, marking substantial agricultural changes over the study period.

The results from Fig. 4.20 reveal several key correlations between human activities and environmental changes across different regions within the HKH region. Firstly, changes in vegetation area across all regions exhibit a consistent negative correlation with changes in permafrost. This negative relationship is particularly significant in China and Afghanistan, suggesting that as permafrost diminishes, vegetation area tends to decrease. Moreover, changes in permafrost (permafrost) in most regions

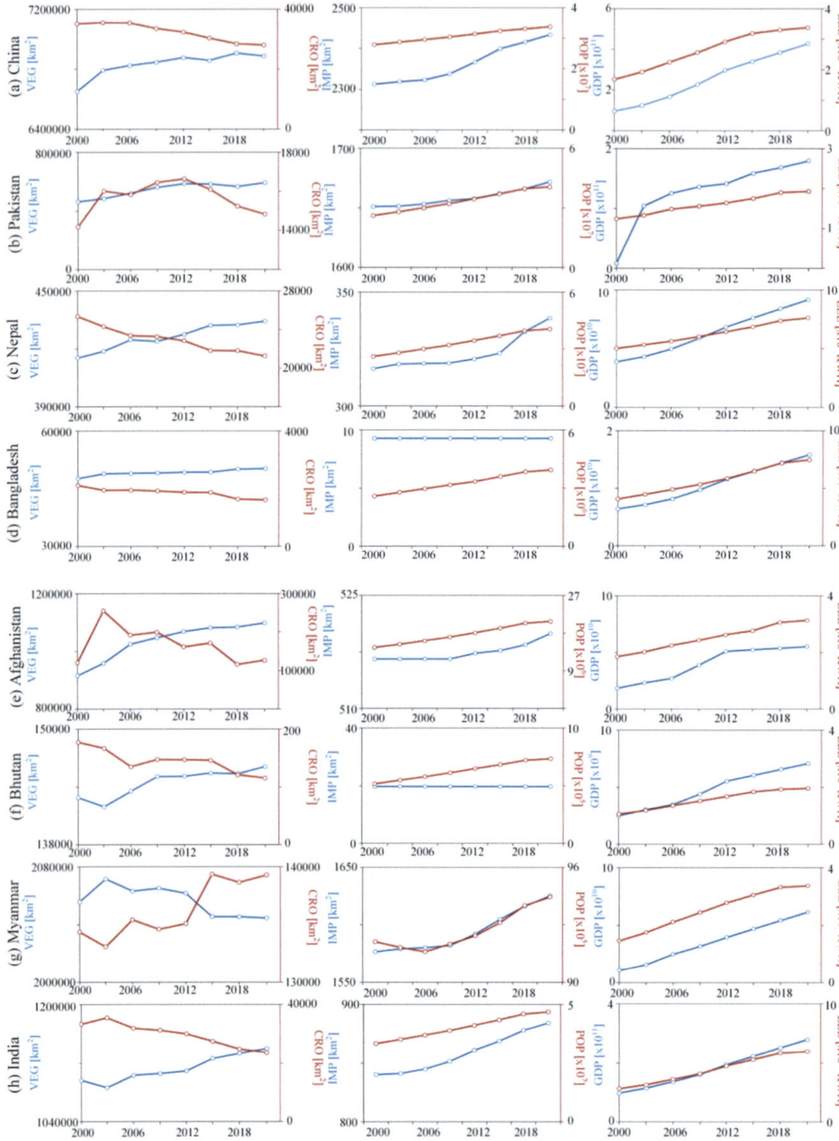

Fig. 4.19 2000–2020 changes in impervious surface (IMP), crops (cropland, CRO), vegetation (VEG), population (population, POP), Gross Domestic Product (GDP) and electricity consumption (EC) in the HKH area

show a negative correlation with impervious surface, population, GDP, and electricity consumption, indicating that increased human activity in terms of urbanization, population growth, economic development, and energy consumption tends to coincide with declining permafrost extent. Notably, China demonstrates a strong negative correlation in this regard, highlighting the profound impact of human-induced factors on permafrost stability.

Conversely, changes in cultivated land generally show a positive correlation with changes in permafrost across the HKH region, with China exhibiting a particularly significant positive relationship. This suggests that expansion of cultivated areas often corresponds with increased permafrost extent, except in Pakistan where the correlation is weakly negative. Regarding ice and snow dynamics, except for Myanmar and Bhutan, changes in these elements positively correlate with impervious surface, population, GDP, and electricity consumption in the region. Significant positive correlations are observed in countries like Pakistan and India, indicating that human activities linked to urban development, population growth, economic output, and

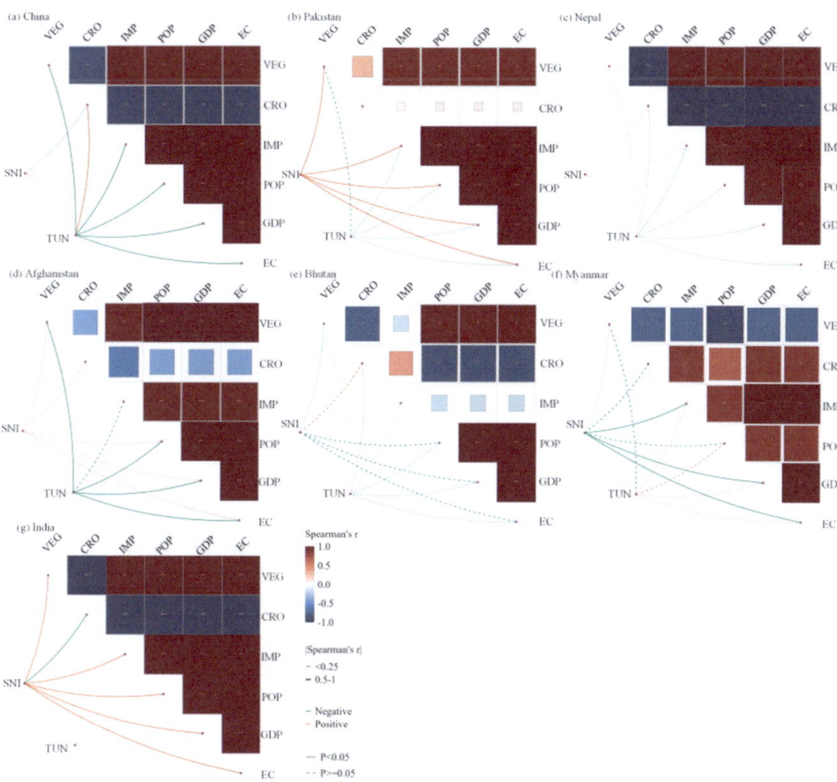

Fig. 4.20 Correlation between changes in snow and ice (SNI) and permafrost (Tundra, TUN) in the HKH region from 2000 to 2020 and various typical human social factors. *Noted* There is no permafrost or snow in Bangladesh

energy use tend to coincide with changes in ice and snow cover. However, Myanmar shows a notable negative correlation between ice and snow changes and impervious surface, cultivated land, and electricity consumption, suggesting unique environmental dynamics in that context. Overall, these findings underscore the varied impacts of human activities on environmental elements within each country in the HKH region, emphasizing the complex interplay between socio-economic factors and natural systems.

The analysis of surface water bodies across different regions reveals diverse correlations between human activities and environmental changes. In both China and Myanmar, increases in impervious surface area, population density, GDP, and electricity consumption positively correlate with changes in surface water bodies. Conversely, cultivated land in China negatively impacts surface water bodies, while in Myanmar, it has a positive effect. Furthermore, vegetation changes exhibit a positive correlation with surface water bodies in China. Notably, other countries studied did not show significant correlations between human activity factors and changes in water bodies (Fig. 4.21).

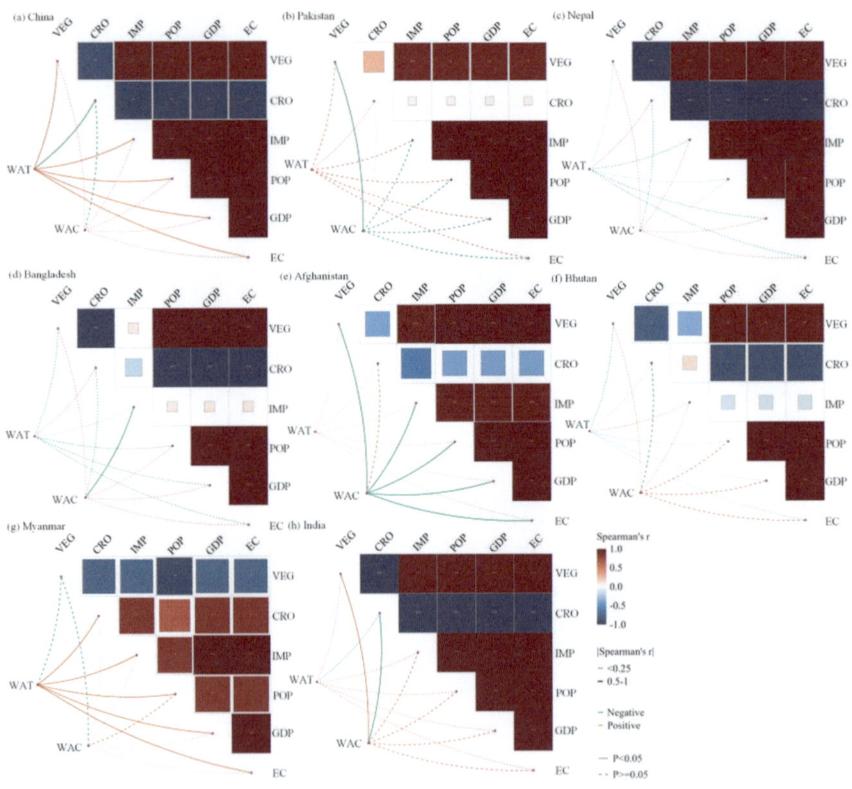

Fig. 4.21 Correlation between changes in surface water area (WAT) and water clarity (WAC) in the HKH from 2000 to 2020 and typical human social factors

Turning to water body transparency, the findings vary across regions. In Pakistan and Afghanistan, vegetation changes negatively correlate with water body transparency, whereas in India, vegetation changes positively affect transparency. Conversely, *changes in cultivated land in India negatively impact transparency.* Impervious surface coverage negatively affects water transparency in Bangladesh and Afghanistan. Afghanistan further demonstrates that population density, GDP, and electricity consumption are inversely related to water transparency (Fig. 4.21). These results underscore the regional variability in how human activities influence both surface water dynamics and water quality, highlighting the complex interplay between environmental factors and socio-economic activities across different contexts.

4.4.3 Summary

The comparison of typical indicators reflecting human social and economic activities in the HKH region reveals a notable trend: most of these indicators, such as impervious surface area, crop area, vegetation area, population, GDP, and electricity consumption, exhibit a negative correlation with changes in permafrost. This suggests that as human activities intensify, there is a corresponding reduction in the area covered by permafrost in the region. This phenomenon underscores the potential reverse impact of human economic development on permafrost dynamics, further contributing to its decline across the HKH region.

Furthermore, the study highlights diverse relationships between human economic activities and environmental factors like ice and snow cover, surface water content, and water quality in different regions. These findings indicate that the influence of human activities on water resources varies significantly depending on local conditions. Unlike climate factors, which may have more consistent impacts across regions, the effects of human activities on water resources exhibit considerable variability. Therefore, the summary emphasizes the necessity for countries to adopt tailored measures based on their specific socio-economic and environmental contexts. Such targeted strategies are crucial for mitigating adverse impacts on water resources and promoting sustainable development in the HKH region and beyond.

4.5 Summary

In summary, this comprehensive study on the Hindu Kush Himalayan (HKH) region has provided significant insights into the spatial distribution and temporal dynamics of key environmental indicators, namely ice, snow, permafrost, and surface water bodies, alongside their interactions with human activities. Spatially, the study identified concentrated areas of ice and snow primarily in the Himalayas, parts of the Tibetan Plateau, and northern Myanmar, while permafrost was predominantly located in the Qinghai-Tibet Plateau and northern Afghanistan and Myanmar. Over the past two decades, temporal changes in ice, snow, and permafrost coverage exhibited diverse patterns across different countries within the HKH. Countries like Pakistan, Afghanistan, and India saw increases in ice and snow coverage, whereas Nepal experienced initial increases followed by declines. Similarly, permafrost areas in China, Pakistan, and Afghanistan experienced significant declines, whereas Nepal, Bhutan, Myanmar, and India exhibited stable or fluctuating trends.

The development of a time-series dynamic water body dataset with a 30-m resolution for the HKH region (TDWH) (years 2000, 2010, and 2020) and a 10-m resolution water body dynamics product (DWH10) (year 2022) was another pivotal achievement of this study, enabling the identification and monitoring of both stable and temporary water bodies across the HKH region. The time-series TDWH dataset with a 30-m resolution has an average monitoring accuracy of 93% and a Kappa coefficient of 0.90 for detecting water bodies. Additionally, the 10-m DWH10 product achieves an accuracy of 95% and a Kappa coefficient of 0.93, significantly outperforming the widely used JRC-GSW. Moreover, based on the generated water distribution dataset, a time-series water transparency dataset with a 30-m resolution for the HKH region (years 2000, 2010, and 2020) was also produced. This dataset effectively demonstrates the spatiotemporal patterns of water transparency changes in the HKH region. It reveals that the water transparency in the HKH region has been steadily increasing since the year 2000, with water bodies becoming progressively clearer. The above water datasets fill the gap in high-resolution time-series monitoring datasets for the corresponding features.

Furthermore, the study highlighted the complex interplay between human socio-economic activities and environmental indicators like permafrost dynamics and water resources. It revealed a negative correlation between indicators of human economic development (e.g., impervious surface area, crop area, population, GDP) and permafrost coverage, indicating potential adverse impacts of human activities on environmental stability in the HKH region. These findings emphasize the urgency of adopting context-specific strategies for sustainable development that consider both environmental resilience and socio-economic progress.

Looking ahead, future research should focus on refining predictive models for ice, snow, and permafrost dynamics, incorporating advanced remote sensing techniques and climate modeling. Improving the resolution and accuracy of surface water body monitoring tools like DWH10 will also be crucial, along with expanding the temporal scope of analysis to capture longer-term trends and climate impacts. Moreover, fostering international collaboration and data sharing initiatives will enhance the robustness of regional assessments and support the formulation of adaptive policies for sustainable environmental management across the diverse landscapes of the HKH region. By continuing to integrate scientific advancements with practical policy recommendations, stakeholders can effectively address the complex challenges posed by climate change and human development in this ecologically sensitive area.

References

Adeniran, I. A., Zhu, R., Yang, J., Zhu, X., & Wong, M. S. (2022). Cross-comparison between sun-synchronized and geostationary satellite-derived land surface temperature: A case study in Hong Kong. *Remote Sensing, 14*(18), 4444. https://doi.org/10.3390/rs14184444

Chandrasekar, K., Sesha Sai, M. V. R., Roy, P. S., & Dwevedi, R. S. (2010). Land Surface Water Index (LSWI) response to rainfall and NDVI using the MODIS vegetation index product. *International Journal of Remote Sensing, 31*(15), 3987–4005. https://doi.org/10.1080/01431160802575653

Chen, X., Liu, L., Zhang, X., Li, J., Wang, S., Gao, Y., & Mi, J. (2022). Long-term water clarity patterns of lakes across China using Landsat series imagery from 1985 to 2020. *Hydrology and Earth System Sciences, 26*(13), 3517–3536. https://doi.org/10.5194/hess-26-3517-2022

Chen, X., Liu, L., Zhang, X., Li, J., Wang, S., Liu, D., Duan, H., & Song, K. (2021). An assessment of watercolor for inland water in china using a Landsat 8-derived Forel-Ule index and the google earth engine platform. *IEEE Journal of Selected Topics in Applied Earth Observations and Remote Sensing, 14*, 5773–5785. https://doi.org/10.1109/JSTARS.2021.3085411

Chen, X., Liu, L., Zhang, X., Xie, S., & Lei, L. (2020). A novel water change tracking algorithm for dynamic mapping of inland water using time-series remote sensing imagery. *IEEE Journal of Selected Topics in Applied Earth Observations and Remote Sensing, 13*, 1661–1674. https://doi.org/10.1109/JSTARS.2020.2983729

DeVries, B., Verbesselt, J., Kooistra, L., & Herold, M. (2015). Robust monitoring of small-scale forest disturbances in a tropical montane forest using Landsat time series. *Remote Sensing of Environment, 161*, 107–121. https://doi.org/10.1016/j.rse.2015.02.012

Feyisa, G. L., Meilby, H., Fensholt, R., & Proud, S. R. (2014). Automated water extraction index: A new technique for surface water mapping using Landsat imagery. *Remote Sensing of Environment, 140*, 23–35. https://doi.org/10.1016/j.rse.2013.08.029

Friedl, M., & Sulla-Menashe, D. (2019). MCD12Q1 MODIS/terra+aqua land cover type yearly L3 global 500m SIN grid V006. https://doi.org/10.5067/MODIS/MCD12Q1.006

Gurung, D. R., Maharjan, S. B., Shrestha, A. B., Shrestha, M. S., Bajracharya, S. R., & Murthy, M. S. R. (2017). Climate and topographic controls on snow cover dynamics in the Hindu Kush Himalaya. *International Journal of Climatology, 37*(10), 3873–3882. https://doi.org/10.1002/joc.4961

Hersbach, H., Bell, B., Berrisford, P., Biavati, G., Horányi, A., Muñoz Sabater, J., Nicolas, J., Peubey, C., Radu, R., Rozum, I., Schepers, D., Simmons, A., Soci, C., Dee, D., & Thépaut, J.-N. (2023). ERA5 hourly data on single levels from 1940 to present. Copernicus Climate Change Service (C3S) Climate Data Store (CDS).

Huang, C., Goward, S. N., Masek, J. G., Thomas, N., Zhu, Z., & Vogelmann, J. E. (2010). An automated approach for reconstructing recent forest disturbance history using dense Landsat time series stacks. *Remote Sensing of Environment, 114*(1), 183–198. https://doi.org/10.1016/j.rse.2009.08.017

Huete, A., Didan, K., Miura, T., Rodriguez, E. P., Gao, X., & Ferreira, L. G. (2002). Overview of the radiometric and biophysical performance of the MODIS vegetation indices. *Remote Sensing of Environment, 83*(1–2), 195–213. https://doi.org/10.1016/S0034-4257(02)00096-2

ICIMOD. (2019). *The Hindu Kush Himalaya Assessment* (P. Wester, A. Mishra, A. Mukherji, & A. B. Shrestha, Eds.). Springer International Publishing. https://doi.org/10.1007/978-3-319-92288-1

ICIMOD. (2023, June 26). Water ice society ecosystems in the Hindu Kush Himalaya.

Islam, M. M. (2006). Analyses of ASTER and spectroradiometer data with in situ measurements for turbidity and transparency study of Lake Abashiri. *International Journal of Geoinformatics, 87*, 76–77.

Kamboj, V., Kamboj, N., & Sharma, N. (2020). A review on general characteristics, classification and degradation of river systems. In *Environmental Degradation: Causes and Remediation Strategies* (pp. 47–62). Agro Environ Media—Agriculture and Ennvironmental Science Academy, Haridwar, India. https://doi.org/10.26832/aesa-2020-edcrs-04

Klein, I., Gessner, U., Dietz, A. J., & Kuenzer, C. (2017). Global WaterPack—A 250 m resolution dataset revealing the daily dynamics of global inland water bodies. *Remote Sensing of Environment, 198*, 345–362. https://doi.org/10.1016/j.rse.2017.06.045

Lacaux, J. P., Tourre, Y. M., Vignolles, C., Ndione, J. A., & Lafaye, M. (2007). Classification of ponds from high-spatial resolution remote sensing: Application to Rift Valley Fever epidemics in Senegal. *Remote Sensing of Environment, 106*(1), 66–74. https://doi.org/10.1016/j.rse.2006.07.012

Li, J., Wang, S., Wu, Y., Zhang, B., Chen, X., Zhang, F., Shen, Q., Peng, D., & Tian, L. (2016). MODIS observations of watercolor of the largest 10 lakes in China between 2000 and 2012. *International Journal of Digital Earth, 9*(8), 788–805. https://doi.org/10.1080/17538947.2016.1139637

Liu, C., Zhu, L., Li, J., Wang, J., Ju, J., Qiao, B., Ma, Q., & Wang, S. (2021). The increasing water clarity of Tibetan lakes over last 20 years according to MODIS data. *Remote Sensing of Environment, 253*, 112199. https://doi.org/10.1016/j.rse.2020.112199

Lu, S., Jia, L., Jiang, Y., Wang, Z., Duan, H., Shen, M., Tian, Y., & Lu, J. (2021). Progress and prospect on monitoring and evaluation of United Nations SDG 6 (clean water and sanitation) target. *Bulletin of Chinese Academy of Sciences (Chinese Version), 36*(8).

McNicoll, G. (2019). The Hindu Kush Himalaya assessment: Mountains, climate change, sustainability and people. *Population and Development Review, 45*, 442–443.

Murshed, M. F., Aslam, Z., Lewis, R., Chow, C., Wang, D., Drikas, M., & van Leeuwen, J. (2014). Changes in the quality of river water before, during and after a major flood event associated with a La Niña cycle and treatment for drinking purposes. *Journal of Environmental Sciences, 26*(10), 1985–1993. https://doi.org/10.1016/j.jes.2014.08.001

Novoa, S., Wernand, M. R., & van der Woerd, H. J. (2013). The Forel-Ule scale revisited spectrally: Preparation protocol, transmission measurements and chromaticity. *Journal of the European Optical Society: Rapid Publications, 8*, 13057. https://doi.org/10.2971/jeos.2013.13057

Palmer, S. C. J., Kutser, T., & Hunter, P. D. (2015). Remote sensing of inland waters: Challenges, progress and future directions. *Remote Sensing of Environment, 157*, 1–8. https://doi.org/10.1016/j.rse.2014.09.021

Pekel, J.-F., Cottam, A., Gorelick, N., & Belward, A. S. (2016). High-resolution mapping of global surface water and its long-term changes. *Nature, 540*(7633), 418–422. https://doi.org/10.1038/nature20584

Pickens, A. H., Hansen, M. C., Hancher, M., Stehman, S. V., Tyukavina, A., Potapov, P., Marroquin, B., & Sherani, Z. (2020). Mapping and sampling to characterize global inland water dynamics from 1999 to 2018 with full Landsat time-series. *Remote Sensing of Environment, 243*, 111792. https://doi.org/10.1016/j.rse.2020.111792

Rodell, M., Houser, P. R., Jambor, U., Gottschalck, J., Mitchell, K., Meng, C.-J., Arsenault, K., Cosgrove, B., Radakovich, J., Bosilovich, M., Entin, J. K., Walker, J. P., Lohmann, D., & Toll, D. (2004). The global land data assimilation system. *Bulletin of the American Meteorological Society, 85*(3), 381–394. https://doi.org/10.1175/BAMS-85-3-381

Running, S., Mu, Q., & Zhao, M. (2017). MOD16A2 MODIS/terra net evapotranspiration 8-day L4 global 500m SIN grid V006.

Sharma, A. K., Sharma, M., Sharma, A. K., Sharma, M., & Sharma, M. (2023). Mapping the impact of environmental pollutants on human health and environment: A systematic review and meta-analysis. *Journal of Geochemical Exploration, 255*, 107325. https://doi.org/10.1016/j.gexplo.2023.107325

Shen, M., Duan, H., Cao, Z., Xue, K., Qi, T., Ma, J., Liu, D., Song, K., Huang, C., & Song, X. (2020). Sentinel-3 OLCI observations of water clarity in large lakes in eastern China: Implications for SDG 6.3.2 evaluation. *Remote Sensing of Environment, 247*, 111950. https://doi.org/10.1016/j.rse.2020.111950

van der Woerd, H. J., & Wernand, M. R. (2015). True colour classification of natural waters with medium-spectral resolution satellites: SeaWiFS, MODIS, MERIS and OLCI. *Sensors, 15*, 25663–25680. https://doi.org/10.3390/s151025663

van der Woerd, H., & Wernand, M. (2018). Hue-angle product for low to medium spatial resolution optical satellite sensors. *Remote Sensing, 10*(2), 180. https://doi.org/10.3390/rs10020180

Wang, L., Cuo, L., Luo, D., Su, F., Ye, Q., Yao, T., Zhou, J., Li, X., Li, N., Sun, H., Liu, L., Wang, Y., Zeng, T., Hu, Z., Liu, R., Chai, C., Wang, G., Zhong, X., Guo, X., … Yang, W. (2022). Observing multisphere hydrological changes in the largest river basin of the Tibetan Plateau. *Bulletin of the American Meteorological Society, 103*(6), E1595–E1620. https://doi.org/10.1175/BAMS-D-21-0217.1

Wang, S. (2018). Large-scale and long-time water quality remote sensing monitoring over lakes based on water color index.

Wang, S., Li, J., Zhang, B., Lee, Z., Spyrakos, E., Feng, L., Liu, C., Zhao, H., Wu, Y., Zhu, L., Jia, L., Wan, W., Zhang, F., Shen, Q., Tyler, A. N., & Zhang, X. (2020). Changes of water clarity in large lakes and reservoirs across China observed from long-term MODIS. *Remote Sensing of Environment, 247*, 111949. https://doi.org/10.1016/j.rse.2020.111949

Wang, Z., Ishida, Y., & Mochida, A. (2023). Effective factors for reducing land surface temperature in each local climate zone built type in Tokyo and Shanghai. *Remote Sensing, 15*(15), 3840. https://doi.org/10.3390/rs15153840

White, J. C., Wulder, M. A., Hobart, G. W., Luther, J. E., Hermosilla, T., Griffiths, P., Coops, N. C., Hall, R. J., Hostert, P., Dyk, A., & Guindon, L. (2014). Pixel-based image compositing for large-area dense time series applications and science. *Canadian Journal of Remote Sensing, 40*(3), 192–212. https://doi.org/10.1080/07038992.2014.945827

Xie, S., Liu, L., Zhang, X., Yang, J., Chen, X., & Gao, Y. (2019). Automatic land-cover mapping using Landsat time-series data based on google earth engine. *Remote Sensing, 11*(24), 3023. https://doi.org/10.3390/rs11243023

Xu, H. (2006). Modification of Normalised Difference Water Index (NDWI) to enhance open water features in remotely sensed imagery. *International Journal of Remote Sensing, 27*(14), 3025–3033. https://doi.org/10.1080/01431160600589179

Yamazaki, D., Trigg, M. A., & Ikeshima, D. (2015). Development of a global ~90m water body map using multi-temporal Landsat images. *Remote Sensing of Environment, 171*, 337–351. https://doi.org/10.1016/j.rse.2015.10.014

Yang, X., Lu, X., Park, E., & Tarolli, P. (2019). Impacts of climate change on lake fluctuations in the Hindu Kush-Himalaya-Tibetan Plateau. *Remote Sensing, 11*(9), 1082. https://doi.org/10.3390/rs11091082

Yang, Y., Liu, Y., Zhou, M., Zhang, S., Zhan, W., Sun, C., & Duan, Y. (2015). Landsat 8 OLI image based terrestrial water extraction from heterogeneous backgrounds using a reflectance homogenization approach. *Remote Sensing of Environment, 171*, 14–32. https://doi.org/10.1016/j.rse.2015.10.005

Yao, T. (2019). Tackling on environmental changes in Tibetan Plateau with focus on water, ecosystem and adaptation. *Science Bulletin, 64*(7), 417. https://doi.org/10.1016/j.scib.2019.03.033

Zanaga, D., Van De Kerchove, R., Daems, D., De Keersmaecker, W., Brockmann, C., Kirches, G., Wevers, J., Cartus, O., Santoro, M., Fritz, S., Lesiv, M., Herold, M., Tsendbazar, N. E., Xu, P., Ramoino, F., & Arino, O. (2022). ESA WorldCover 10 m 2021 v200 (version v200) . Zenodo

Zhang, G., Yao, T., Chen, W., Zheng, G., Shum, C. K., Yang, K., Piao, S., Sheng, Y., Yi, S., Li, J., O'Reilly, C. M., Qi, S., Shen, S. S. P., Zhang, H., & Jia, Y. (2019). Regional differences of lake evolution across China during 1960s–2015 and its natural and anthropogenic causes. *Remote Sensing of Environment, 221*, 386–404. https://doi.org/10.1016/j.rse.2018.11.038

Zhang, X., Liu, L., Wu, C., Chen, X., Gao, Y., Xie, S., & Zhang, B. (2020). Development of a global 30 m impervious surface map using multisource and multitemporal remote sensing datasets with the Google Earth Engine platform. *Earth System Science Data, 12*(3), 1625–1648. https://doi.org/10.5194/essd-12-1625-2020

Zhang, Y., Gu, J., Liu, S., Wang, X., Jiang, Z., Wei, J., & Zheng, Y. (2022). Spatial pattern of the debris-cover effect and its role in the Hindu Kush-Pamir-Karakoram-Himalaya glaciers. *Journal of Hydrology, 615*, 128613. https://doi.org/10.1016/j.jhydrol.2022.128613

Chapter 5
Remote Sensing Monitoring and Evaluation of Forests and Grasslands

The Hindu Kush Himalayan (HKH) region, one of the world's largest and most ecologically significant mountain systems, is experiencing rapid changes due to both climate change and human activities (Wester et al., 2019). Vegetation dynamics in the region, which are profoundly influenced by these environmental shifts, have a major impact on the structure and function of terrestrial ecosystems (Franklin et al., 2016). Mountain vegetation is not only crucial to the soil, water, and atmospheric conditions, but also serves as a fundamental resource for the region's sustainable development (Bayle et al., 2019; Jiang et al., 2022).

By playing a key role in regulating atmospheric CO_2 levels through processes such as photosynthesis, respiration, and transpiration, vegetation acts as a large carbon sink, thereby regulating both regional and global temperatures. This process helps preserve biodiversity and mitigates the impacts of climate change. Additionally, mountain vegetation supplies raw minerals and wood—vital resources for human society. The well-developed root systems of mountain plants also stabilize the soil and retain water, which significantly reduces the risk of geological disasters like landslides, rockfalls, debris flows, and floods, thereby safeguarding both human lives and property. The stability and proper functioning of mountain ecosystems are deeply reliant on maintaining healthy plant cover.

Vegetation dynamics are highly sensitive to climate variability, which is a major driver of changes in the formation, growth, and distribution of vegetation (Ahlström et al., 2015; Piao et al., 2017; Wang et al., 2017). Changes in the biophysical properties of the land surface, such as albedo and roughness, in turn influence local and global climate systems (Perugini et al., 2017). Climate models predict that ecosystems in the HKH, which are currently constrained by temperature, may experience shifts over the next 50–100 years, potentially opening up new areas for plant growth (Keenan & Riley, 2018). Warming trends are associated with ecosystem changes, such as longer growing seasons, shifts in vegetation biomes, expansion of tundra shrubs, and overall higher productivity in northern regions (Ougahi et al., 2022; Perugini et al., 2017; Xu et al., 2016). At the same time, there is evidence of a global greening trend,

S. Lu et al., *Remote Sensing Assessment of Sustainable Development in the Hindu Kush-Himalayan (HKH) Region*, https://doi.org/10.1007/978-981-96-8675-9_5

which may be linked to increased CO_2 fertilization, leading to changes in vegetation productivity and eco-hydrological responses (Ougahi et al., 2022).

In addition to the effects of climate change, human activities such as overgrazing, land reclamation, and deforestation have dramatically altered the structure and function of mountain vegetation ecosystems in recent decades (Sharma et al., 2016). These anthropogenic pressures are significantly challenging the sustainable development of both the HKH region and its surrounding areas. The impact on the global biological health of mountain vegetation has become a growing concern, especially since the retreat of glaciers and the decreasing snow cover, which have pushed treelines in mountain regions to higher altitudes (Anderson et al., 2020). Between 2001 and 2018, approximately 78 million hectares of mountain forest were lost globally, representing 7.1% of the total global forest loss. Asia, particularly the HKH region, has experienced the greatest loss, due to high population density and the expansion of agriculture and livestock farming into high-altitude areas.

The HKH region, which encompasses more than 60% forest and grassland, and is home to alpine meadows that make up around 80% of its total vegetation cover, depends heavily on these ecosystems for ecological services and livelihoods, particularly for local herders. However, climate warming and excessive grazing have led to significant reductions in alpine and swamp meadows, underlining the urgent need for effective ecosystem protection. Remote sensing technology has proven invaluable in addressing these challenges by providing long-term monitoring, real-time data, and broad coverage. This technology is an indispensable tool for land cover analysis, environmental protection, and ecological monitoring.

Through the use of satellite resources, remote sensing can extract detailed information about forest and grassland cover in the HKH region, enabling continuous monitoring of ecosystem changes. This approach helps researchers and policymakers assess the dynamic shifts in vegetation, evaluate human impacts, and design effective conservation policies. In alignment with the SDG 15.4, this study utilizes machine learning techniques and the Google Earth Engine (GEE) platform to analyze forest and grassland cover at a 30-m resolution, using Landsat images from 2000, 2010, to 2020. By examining these changes over the past two decades, the study provides vital data for sustainable development planning and ecological assessment in the HKH region.

5.1 Forest and Grass Extraction and Analysis of Spatiotemporal Changes

Extracting land cover types such as forests and grasslands is crucial for land use, environmental monitoring, and resource management. Utilizing long-term series and high spatial resolution satellite data allows for effective extraction and analysis of these land cover types and their changes over time. Numerous scholars and institutions have developed various methods for land cover inversion, resulting in valuable land cover products that provide critical data support and scientific foundations for various fields, including land use, environmental monitoring, and resource management.

For instance, the MODIS land cover type product (MCD12Q1) offers long-term global land cover information, making it suitable for studying land cover change trends on a global scale. However, its low spatial resolution (500 m) limits its ability to capture fine-grained land cover details, particularly in the complex terrain of the HKH region, where it may result in significant errors. In contrast, the ESA WorldCover product, which uses data from Sentinel-1 and Sentinel-2, provides land cover products with a higher spatial resolution (10 m), offering more detailed and accurate land cover information. Despite its advantages in fine-grained extraction and monitoring, its short time coverage does not meet the needs of HKH regional research that requires long-span data.

To address these limitations, we used the Landsat series of images and advanced machine learning algorithms to extract high-precision data on forests and grasslands in the HKH area. By training and optimizing our models, we accurately identified forest and grassland coverage areas across different years and generated corresponding vegetation coverage inversion maps. These maps not only detail the forest and grass cover but also reveal their spatial distribution characteristics, providing a comprehensive understanding of land cover changes in the HKH region.

5.1.1 Methodology

Figure 5.1 illustrates the process of extracting forest and grass coverage in the HKH region. The key data sources used in this process included Landsat and DEM images. For the year 2000, data from Landsat 5 and Landsat 7 were utilized, while in 2010, only Landsat 5 data were employed. By 2020, the extraction process relied on data from Landsat 8. To enhance the accuracy of extracting mountain forest and grassland information, the study incorporated elevation data from NASADEM to better understand the topography of the study area.

(1) **Data preprocessing**

Google Earth Engine's (GEE) Landsat data includes a band called "pixel_qa," which provides quality information for each pixel, such as the presence of clouds, shadows, and water bodies. In this study, pixels containing clouds were marked as "NAN" (invalid values) based on the "pixel_qa" band. Subsequently, the normalized difference vegetation index (NDVI) was calculated using the reflectance values of the non-cloud pixels following Eq. 5.1 below:

$$NDVI = \frac{NIR - R}{NIR + R} \tag{5.1}$$

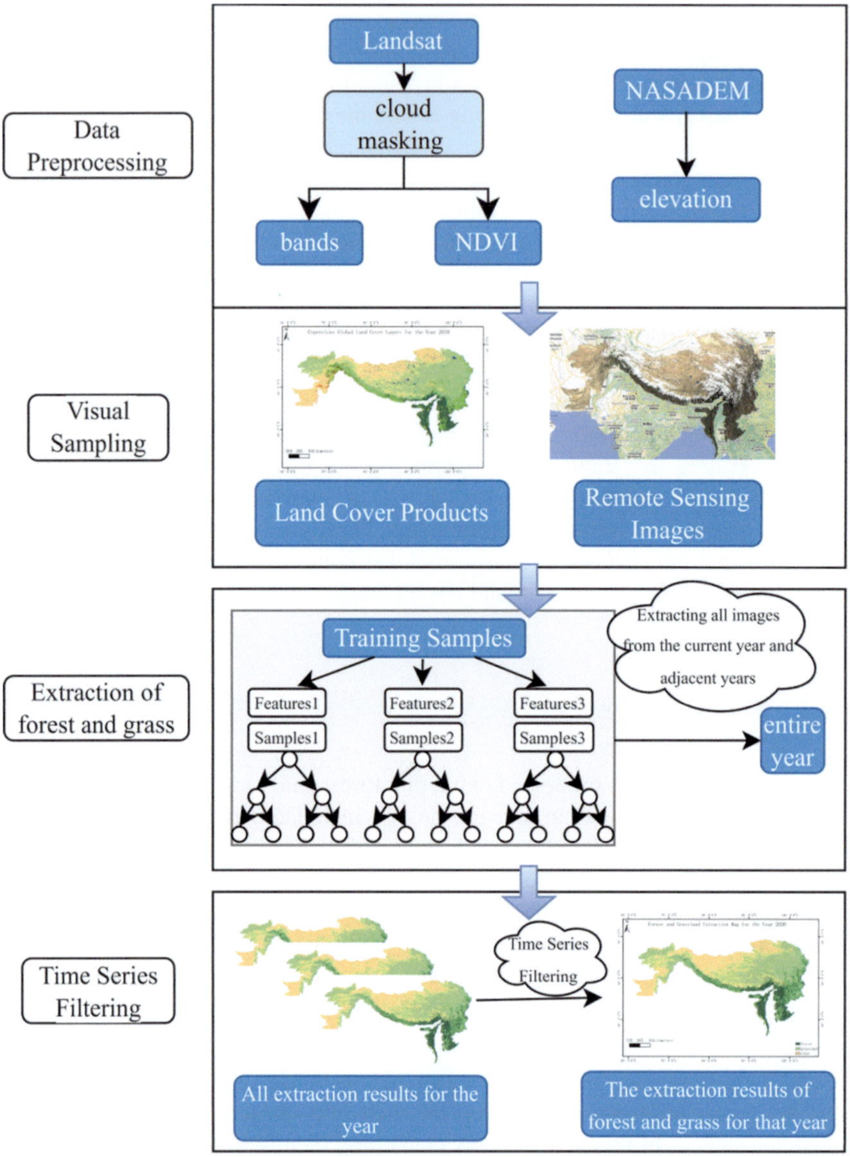

Fig. 5.1 The technical process of forest and grass extraction

where NIR is the near-infrared band and R is the red band. The NDVI is the ratio of the difference to the sum of the near-infrared band (NIR) and the red band (RED) (Rouse et al., 1974). This is based on the simple principle that as the more visible sunlight a plant absorbs, the greater its photosynthesis activity and production and

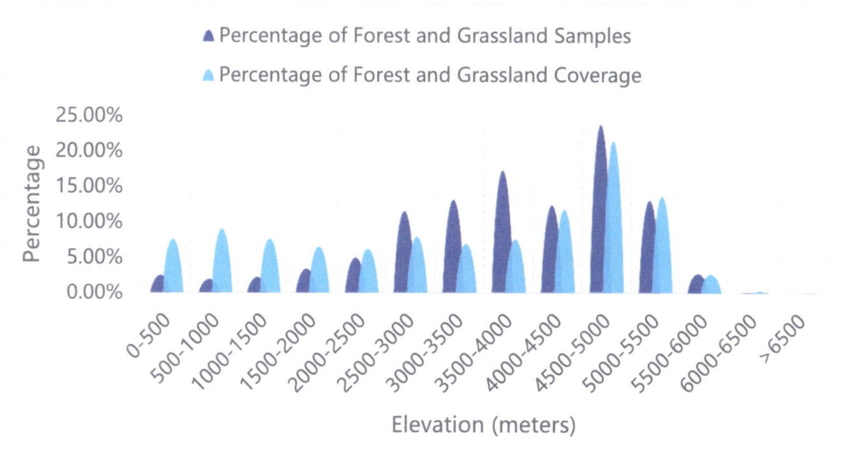

Fig. 5.2 Distribution of samples at different altitudes in the HKH

vice versa. The NDVI for a given pixel always results in a number between −1 and +1.

(2) **Sample generation**

To generate the samples needed for this study, we first selected sample points within the study area using a random function. These points were then overlaid with land cover classification products and remote sensing images, including true and false color images. The categories of the sample points were determined through visual discrimination, and the sample labels were classified into three categories: forest, grassland, and other. Figure 5.2 illustrates the distribution proportion of the selected samples at different altitudes.

Based on the geographical location information of the sample points, combined with reflectance, normalized difference vegetation index (NDVI), and elevation image data, we obtained the reflectance NDVI value and elevation information for the pixels at these locations. Reflectance NDVI and elevation were used as attribute features of the samples, while the different categories, such as forest and grassland, served as labels for the samples.

(3) **Extraction of forest and grassland**

This study utilized the Random Forest method for the extraction of forest and grass areas. Random Forest, an ensemble learning algorithm used for classification and regression tasks, was proposed by Breiman (2001). It has since become one of the most esteemed and widely used algorithms in machine learning. The core concept of Random Forest is to enhance the model's accuracy and generalization capability by constructing multiple decision trees and combining their prediction results. The study aimed to map the distribution of forests and grasslands for the years 2000, 2010, and 2020. To minimize extraction and evaluation errors, as well as uncertainties caused by factors such as image quality, the study referenced extraction results from neighboring years before and after each target year. This comprehensive evaluation, which considers neighboring years, will enhance the accuracy of the final outcomes. Table 5.1 details the satellite data types, and time ranges used in the study.

(4) **Time series filtering**

Due to various factors such as weather conditions, cloud coverage, and sensor limitations, single-scene images often contain missing values or outliers. These abnormal pixel values can significantly affect the accuracy of the model's results. Therefore, it is essential to aggregate long-term image sequences to ensure more reliable and representative results. By evaluating the entire years' worth of data, the impact of potential outliers at specific moments or under certain conditions can be minimized, resulting in more stable outcomes. Following the approach outlined in Sect. 5.1.1. Section 5.3, classification results for all phases within a year can be generated.

The process involves coding the results initially, and labeling cells as forest, grassland, or other categories. Subsequently, as illustrated in Fig. 5.3, the forest and grass results are extracted for each of the four seasons in a year. In cases where multiple results occur in a specific area during a season (i.e., multiple images within one season), a voting mechanism is employed. Essentially, the mode of all results within that season is considered as the outcome for that period. This approach helps to ensure consistency and reliability in the quarterly results.

To address the issue of vegetation variations across seasons and obtain accurate forest and grass coverage throughout the year, it is necessary to develop a method that eliminates seasonal influences. This study adopts a strategy to comprehensively assess the results of the four seasons (Winter, Autumn, Spring and Summer) extracted in the previous step. Specifically, for a given pixel, if forest appears in one of the four seasons, it is classified as forest; if there is no forest but grassland appears, it is classified as grassland; and if neither forest nor grassland appears in the four seasons, it is classified as other.

(5) **Extraction results and accuracy evaluation**

Forest and grass conditions for the years 2000, 2010, and 2020 were extracted using random forest and time series screening methods, and related accuracy evaluation was conducted. The sample points generated in Sects. 5.1.1 and 5.2 were divided into training and validation samples using a random method. The accuracy results based on validation samples are shown in Table 5.2. It is observed from the table that the kappa coefficients of the extracted models are all above 0.94, indicating that the models have achieved a certain level of accuracy on the training data.

Table 5.1 Satellite image dataset from 2000 to 2020

The time point to be extracted	2000	2010	2020
Time range for image usage	1998–2002	2009–2011	2019–2021
Satellites used	Landsat5/Landsat7	Landsat5	Landsat8

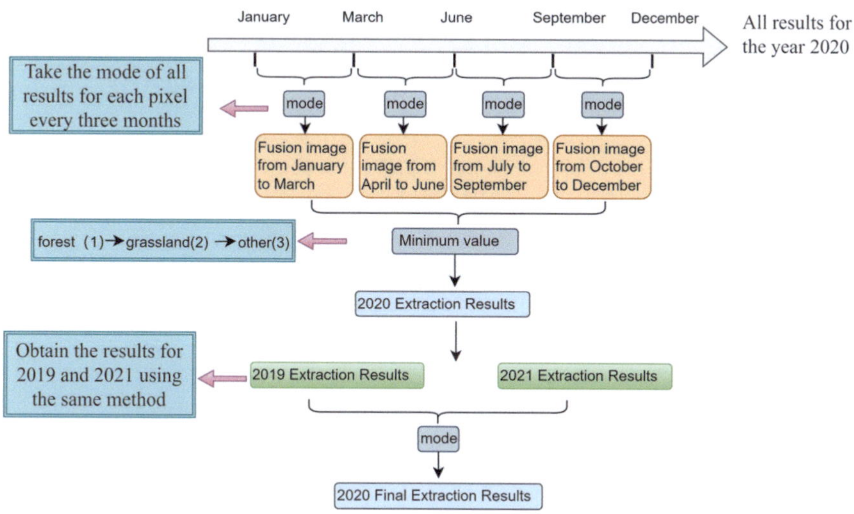

Fig. 5.3 Time series screening process (taking 2020 as an example)

Table 5.2 Accuracy evaluation of forest and grassland products

Year	Kappa	Accuracy
2000	0.945	0.991
2010	0.944	0.991
2020	0.968	0.993

Furthermore, this study compared the extraction results with the classification results of several medium and high-resolution land cover products covering the HKH area. This comparison aimed to evaluate the accuracy and consistency of the extracted forest and grass products, further verifying the reliability of the research findings. Among these products, the Regional Land Cover Monitoring System (RLCMS) stands out as an operational service system designed specifically for the HKH region. It provides annual land cover mapping and change analysis (Tables 5.3, and 5.4).

Overall, the correlation between this product and the RLCMS product is stronger for forest but weaker for grassland. The cross-validation accuracy for grassland in 2020 is only 0.66. Examining the confusion matrix reveals 235 samples were categorized differently in the RLCMS product but identified as grassland in this study. This discrepancy may be attributed to the time series processing undertaken in this product to determine maximum forestland and grassland coverage. Grassland experiences seasonal changes, resulting in varied outcomes depending on the timeframe used, highlighting the strengths of this research method.

The ESA WorldCover 10 m 2020 v100 product, produced by the European Space Agency (ESA), is a global land cover product for 2020 based on Sentinel-1 and Sentinel-2 data (Zanaga et al., 2021). This dataset offers 10 m high resolution and encompasses 11 land categories. However, since this product only provides results

Table 5.3 Cross-validation accuracy with RLCMS products

	2000				2010				2020			
	Precision	Recall	F1-score	Support	Precision	Recall	F1-score	Support	Precision	Recall	F1-score	Support
Forest	0.85	0.94	0.89	370	0.86	0.91	0.88	364	0.87	0.87	0.87	362
Grassland	0.76	0.75	0.76	733	0.74	0.79	0.76	781	0.66	0.76	0.72	695
Other	0.83	0.81	0.82	867	0.83	0.76	0.79	825	0.85	0.73	0.79	913
Weighted avg	0.81	0.81	0.81	1970	0.80	0.80	0.80	1970	0.79	0.77	0.78	1970

Table 5.4 Cross-validation confusion matrix with RLCMS products

	2000			2010			2020		
	Forest	Grassland	Other	Forest	Grassland	Other	Forest	Grassland	Other
Forest (RLCMS)	346	24	0	331	33	0	314	48	0
Grassland (RLCMS)	41	550	142	38	615	128	33	547	115
Other (RLCMS)	20	148	699	17	180	628	13	235	665

after 2020, cross-validation with this product is limited to that year. The validation results are presented in Tables 5.5 and 5.6.

The results show that both products achieved a high degree of fit in classifying forests, but the accuracy of grassland classification was low. The main reason for the inconsistent results was identified to be the seasonal changes in grassland. This finding was further supported by cross-validation with the previous product.

The CGLS-LC100 is a global land cover map with a 100 m spatial resolution. This land cover map (v3.0.1) provides global classification results for 2015–2019, derived from the PROBA-V 100 m time series, a database of high-quality land cover sites, and multiple ancillary datasets. The accuracy of Level 1 reaches 0.8. As the latest data available is only for products in 2019, this study cross-validated its products for 2020 with the 2019 CGLS products. The validation results are presented in Tables 5.7 and 5.8.

Upon comparison, it was found that the forest and grassland demonstrate a certain degree of overall compatibility, with forest exhibiting higher accuracy and grassland displaying poorer accuracy. The primary reason for the substantial difference between the two lies in their temporal inconsistency, yet the overall situation can still achieve a certain level of accuracy.

Table 5.5 Cross-validation accuracy with ESA WorldCover 10 m 2020 v100 product

	Precision	Recall	F1-score	Support
Forest	0.90	0.75	0.82	435
Grassland	0.60	0.86	0.71	586
Other	0.92	0.76	0.83	949
Weighted avg	0.82	0.79	0.79	1970

Table 5.6 Cross-validation confusion matrix with ESA WorldCover 10 m 2020 v100 product

	Forest	Grassland	Other
Forest (ESA WorldCover)	325	106	4
Grassland (ESA WorldCover)	29	502	55
Other (ESA WorldCover)	6	222	721

Table 5.7 Cross-validation accuracy with CGLS products

	Precision	Recall	F1-score	Support
Forest	0.96	0.72	0.83	481
Grassland	0.69	0.86	0.76	663
Other	0.88	0.84	0.86	826
Weighted avg	0.84	0.82	0.82	1970

Table 5.8 Cross-validation confusion matrix with CGLS product

	Forest	Grassland	Other
Forest	347	130	4
Grassland	8	569	86
Other	5	131	690

In general, this product aligns well with other products, indicating that the results are reliable to some extent. Additionally, this product possesses several advantages: it utilizes the Landsat dataset with higher resolution to extract more detailed forest and grass coverage, and it employs the time series screening method to address extraction errors caused by seasonal changes in vegetation to a certain extent.

5.1.2 Analysis of Forest and Grass Distribution at Different Altitudes

The HKH area, known as the "Third Pole" of the Earth, has a unique ecological environment. Within this region, the distribution of vegetation exhibits remarkable vertical gradient characteristics. Through a analysis of the proportions of forest and grassland across different altitude ranges, we explore the distribution of forest and grassland at various elevations over a span of two decades.

As depicted in Fig. 5.4, for grasslands in areas below 3000 m in altitude, the 2000–2500 m range had the highest proportion of grassland in 2000 and 2010. However, in 2020, this high-proportion zone shifted to the 1500–2000 m range. In areas above 3000 m, the 4000–4500 m region consistently maintained the highest proportion of grassland throughout the years of 2000, 2010, and 2020. In regions below 3000 m, the proportion of grassland in each gradient initially demonstrated an upward trend from 2000 to 2010, but then underwent a decline from 2010 to 2020. Conversely, in areas above 3000 m, the proportion of grassland in each gradient exhibited a upward trend from 2000 to 2020. When it comes to forestland, as shown in Fig. 5.5, the proportion of forest across the three years generally displayed a decreasing trend with increasing altitude.

This study also explored the spatial change characteristics of forests and grasslands during the periods of 2000–2010, 2010–2020, and 2000–2020.

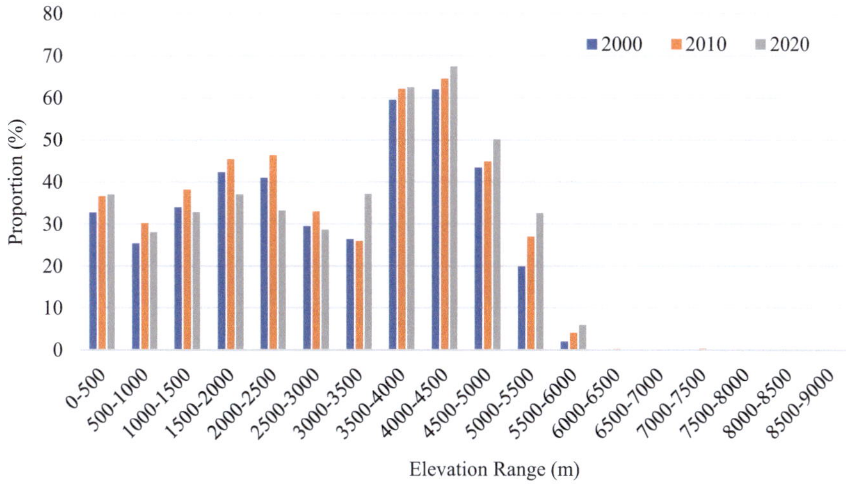

Fig. 5.4 Proportion of grassland in different altitude ranges

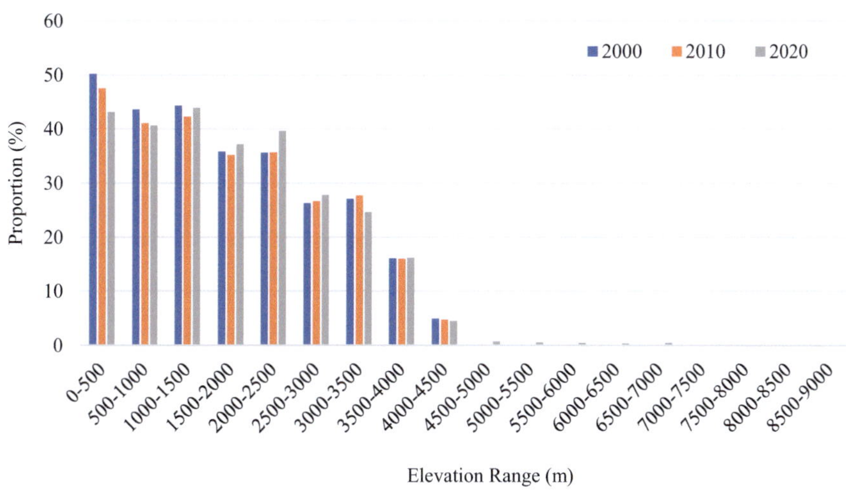

Fig. 5.5 Proportion of forest land in different altitude ranges

Figures 5.6 illustrate the changes in forestland and grassland in the HKH area from 2000 to 2010. During this period, there has been an increasing trend in grassland coverage, while forestland has shown a decreasing trend (Lu et al., 2017). In Southeast Asia, deforestation activities are particularly severe (You Hui, 2013), resulting in the conversion of most forestland to grasslands, explaining the increased grassland area and decreased forestland area in Myanmar. In the southeastern Qinghai-Tibet Plateau and western Sichuan, China, there is a trend of grassland replacing forestland. The central region of the HKH has experienced severe degradation of grasslands, aligning

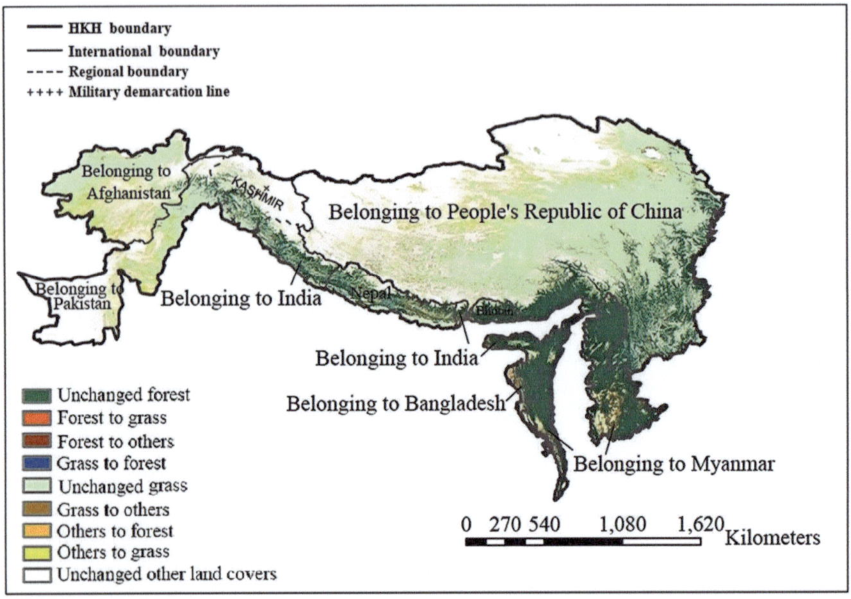

Fig. 5.6 Changes in forest and grass coverage in the HKH area from 2000 to 2010

with most research findings (Lu et al., 2017).In the northeastern Tibetan Plateau, grassland coverage has significantly increased, with partial increases observed in the eastern and western parts of the plateau. Moreover, there has been a significant increase in grassland cover in Afghanistan and Pakistan (Panday & Ghimire, 2012). This trend reflects shifts in land use and environmental conditions across the HKH region.

The observed conversion from forest to grassland, particularly in regions like Southeast Asia and parts of China, underscores the complex interplay of human activities and environmental factors in shaping land cover patterns. Overgrazing, as seen in the central HKH, has led to severe degradation of grasslands, highlighting the need for sustainable land management practices. In areas like the northeastern Tibetan Plateau, increased grassland coverage signifies positive environmental change. Key contributors to the improvement of grasslands include increased precipitation and rising temperatures (Li et al., 2023). However, precipitation and temperature alone cannot account for all observed changes; various factors, including land management strategies, soil conditions, altitude, and topography, also play significant roles in altering land cover (Chen, 2020).

Understanding these shifts in land cover is essential for managing natural resources and ecosystems effectively in the HKH region. By considering the multifaceted influences on land cover change, from climate to human activities, policymakers and conservationists can develop strategies that promote sustainable land use and biodiversity conservation. Additionally, continued monitoring and research into the drivers of land cover change will be crucial for maintaining the ecological integrity and resilience of the HKH landscape in the face of ongoing environmental challenges.

As illustrated in Fig. 5.7, the state of forestland in parts of Southeast Asia shows some improvement; however, most areas continue to experience a decline. Forest loss is particularly severe in Myanmar (Behera et al., 2018). In Afghanistan and Pakistan, there has been a significant reduction in grassland. Conversely, in western Sichuan and southeastern Tibet in China, forestland coverage has increased due to conservation policies, such as converting farmland back to forests. While some grasslands in the central region of the Qinghai-Tibet Plateau are showing signs of recovery, grassland degradation persists in central and eastern Tibet (Chen & Kuang, 2019).

As shown in Fig. 5.8, from 2000 to 2020, the grassland area in the region exhibited an increasing trend, consistent with previous research findings (Meng et al., 2022). Specifically, grasslands in the northeastern part of the Tibetan Plateau show a significant increase, with growth also observed in the western region. However, some grasslands in western Sichuan and southeastern Tibet were converted into forest. Overall, the changes in grasslands in Afghanistan and Pakistan are also very significant. In terms of forestland, there was an increase in southeastern Tibet and western Sichuan, China. In Southeast Asia, particularly in Myanmar, excessive deforestation and over-exploitation led to a significant loss of forestland.

To better understand these changes, the forest-grass area ratio was introduced, reflecting the distribution shifts between forestland and grasslands in the HKH region. The calculation method for this variable is as follows:

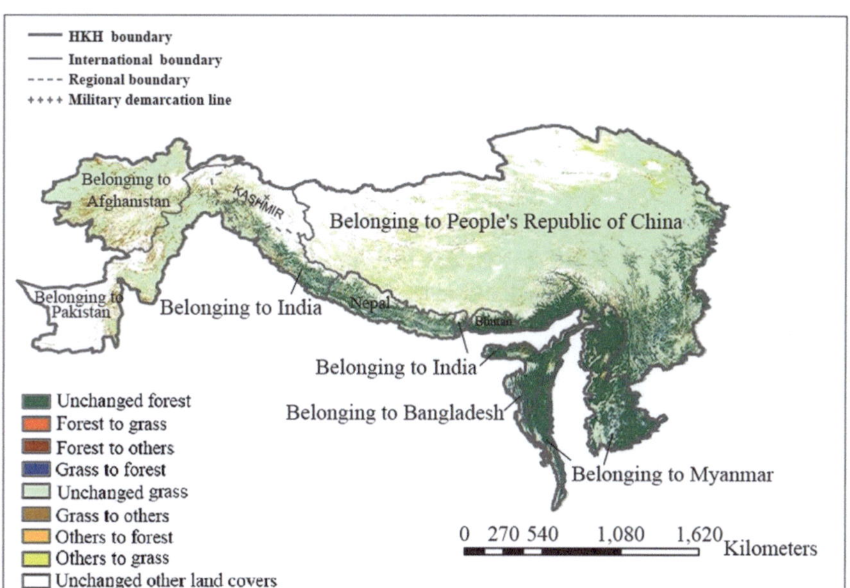

Fig. 5.7 Changes in forest and grass coverage in the HKH area from 2010 to 2020

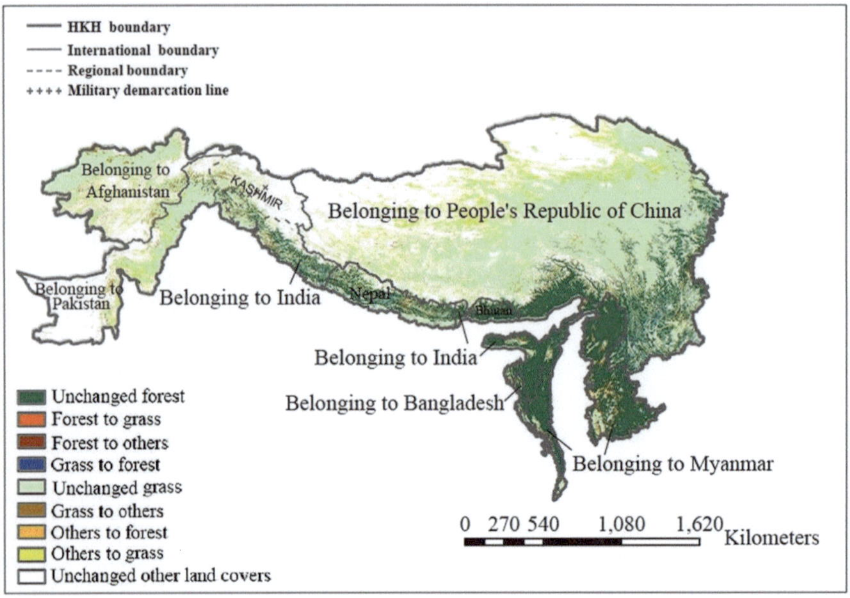

Fig. 5.8 Changes in forest and grass coverage in the HKH area from 2000 to 2020

$$Forest\ and\ grass\ area\ ratio = \frac{S_{forest/grassland}}{S_{region}} \tag{5.2}$$

where $S_{forest/grassland}$ is the forest and grassland area of the HKH region, and S_{region} is the area of the entire HKH region.

Figure 5.9 illustrates the proportion of forest and grassland coverage in the HKH region across three time periods: 2000, 2010, and 2020. The vegetation coverage ratio for forest and grassland was 56.7, 59.47, and 60.47%, respectively, demonstrating a steady overall increase over the past two decades. During this time, the grassland coverage showed a continuous increase, while forest coverage showed an overall decline. The reduction in forest areas has been most prominent in regions such as Southeast Asia, where extensive logging activities have significantly contributed to forest loss. On the other hand, changes in grassland coverage are concentrated in the Qinghai-Tibet Plateau, Afghanistan and Pakistan.

5.1.3 Summary

The analysis of forestland and grassland distribution across different altitudes in the HKH region from 2000 to 2020 reveals significant spatial and temporal variations. Below 3000 m, the highest proportion of grassland was in the 2000–2500-m range in 2000 and 2010, shifting to 1500–2000 m in 2020. Above 3000 m, the 4000–4500-m

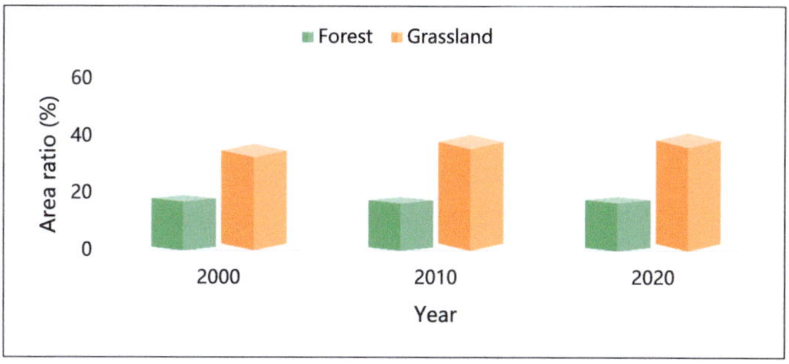

Fig. 5.9 Area ratio of forest and grassland in the HKH area in 2000, 2010 and 2020

region consistently had the highest proportion. Grassland below 3000 m increased from 2000 to 2010 but declined from 2010 to 2020, while areas above 3000 m showed a consistent upward trend during the same period. Regarding forestland, the proportion of forest generally decreased with increasing altitude over the three years. Spatial analysis showed that regions like Southeast Asia and parts of China experienced severe deforestation, converting forests to grasslands, while other areas like the northeastern Tibetan Plateau saw increases in grassland coverage driven by climatic improvements and conservation efforts.

Overall, the HKH region demonstrated complex patterns of land cover change over the past two decades, highlighting the interplay between climate change, human activities, and natural factors. Grasslands, more sensitive to temperature changes than forests, showed varied responses to climatic and anthropogenic pressures, with significant degradation observed in overgrazed central regions and notable recovery in areas with improved precipitation and conservation measures. Forest areas, particularly in Southeast Asia, faced extensive losses due to logging, while regions like western Sichuan and southeastern Tibet saw forest regrowth due to policy interventions. Understanding these shifts is crucial for managing natural resources and promoting sustainable land use in the HKH region. Policymakers and conservationists must consider these multifaceted influences to develop effective strategies for preserving the ecological integrity and resilience of this critical landscape.

5.2 Retrieval of Forest and Grass Fractional Vegetation Cover and Analysis of Spatiotemporal Changes

Fractional Vegetation Cover (FVC), which is defined as the vertical projection area ratio of green vegetation within a specific landscape (Carlson & Ripley, 1997), represents an important biophysical parameter for characterizing vegetation coverage. This metric has been widely applied in quantifying vegetation dynamics across various spatial scales, from regional ecosystems to global environmental assessments. In the face of global climate change and the escalating intensity of human activities, forest and grassland ecosystems are under immense pressure. Consequently, dynamically monitoring the forest and grass coverage in mountainous regions has emerged as a highly crucial long-term concern that demands consistent and dedicated attention.

To explore the spatiotemporal variations of vegetation coverage in the HKH region, we utilized remote sensing images from 2000, 2010, to 2020 to obtain the fractional vegetation cover (FVC) for three phases, and then analyzed the spatiotemporal changes patterns.

5.2.1 *Methodology*

Considering the gradient characteristics of the mountainous topography and the heterogeneity of the surface in the HKH region, we built a FVC inversion methodology suitable for mountainous area. This approach incorporates altitude gradient and surface parameters to enhance inversion accuracy in mountainous area. The overall process of inverting the FVC of forest and grass areas in the HKH area is shown in Fig. 5.10.

The main steps are as follows: First, generate an FVC sample library in the HKHarea, taking into account the land cover type and the characteristics of the altitude gradient. Second, construct spatiotemporal transfer learning models to obtain the distribution of FVC.

(1) **Development of the FVC sample library in the HKH region**

When applying deep learning methods to extract FVC, a crucial prerequisite is to construct a sufficient number of highly accurate and representative samples. In

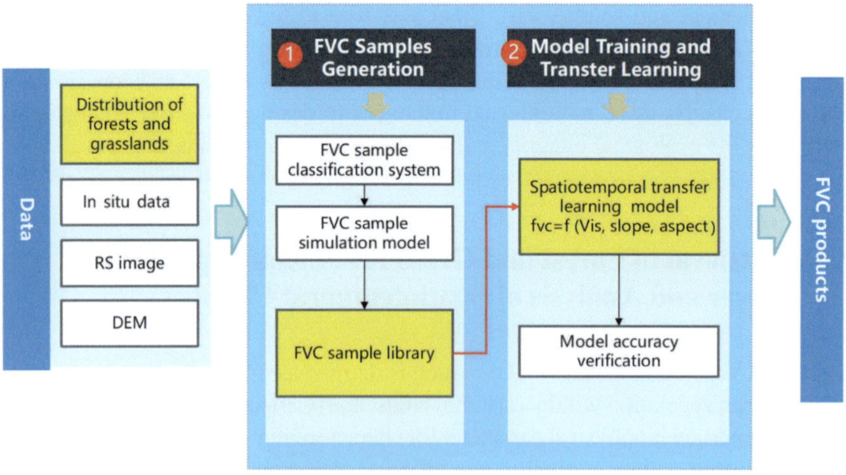

Fig. 5.10 FVC inversion process in situ

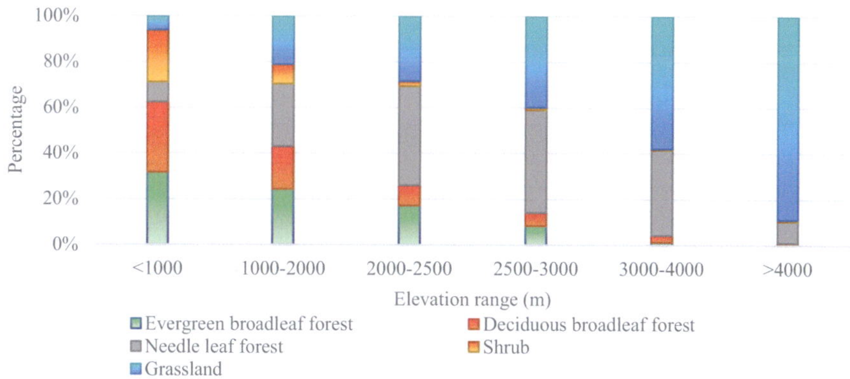

Fig. 5.11 Distribution proportions of vegetation types across each altitude gradient in the HKH experimental area

plateau mountainous areas, due to their unique topographic and climatic conditions (Anderson et al., 2020), collecting FVC sample data on site has proven to be a challenging task. Additionally, the currently available FVC sample data has issues such as insufficient accuracy and a lack of sufficient representativeness of the unique characteristics of mountainous regions.

To ensure the samples' regional representativeness, we must consider the characteristics of regional surface vegetation cover and altitude gradients when constructing a sample classification system. In this study, we analyze the gradient characteristics of forest and grassland distribution based on existing land cover products and DEM data that cover the target area. Such an analysis provides strong support for the establishment of the sample classification system. Figure 5.11 shows the proportional distribution of vegetation types in the HKH mountains of according to the altitude gradient.

Considering the spectral differences of different vegetation types under the existing remote sensing image scale and based on the vertical distribution characteristics of vegetation in the HKH region published by ICIMOD, this study constructs a sample system from two perspectives: vegetation type and altitude gradient. Vegetation types are divided into two main categories: forest and grassland. The altitude gradient is classified into six categories: (1) < 2000 m, (2) 2000–2500 m, (3) 2500–3000 m, (4) 3000–3500 m, (5) 3500–4000 m, and (6) >4000 m. Due to the nearly complete absence of forest distribution above 4000 m, the altitude classification for forests is divided into five categories. Based on the above basis, the mountain FVC sample system for this study was utilized as illustrated in Fig. 5.12.

After establishing the sample system for the study area, the next step is to construct high precision FVC samples. These samples will serve as the foundation for accurately extracting FVC using deep learning methods. This study proposed a method to simulate FVC samples based on the PROSAIL model (Jacquemoud et al., 2009).

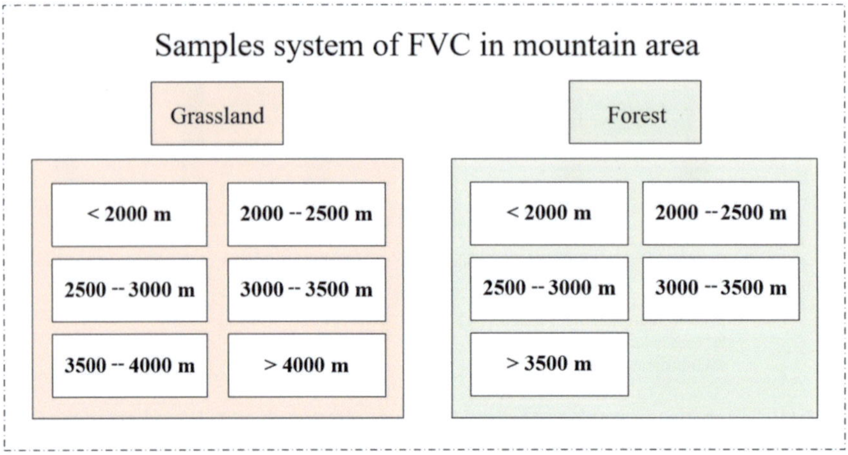

Fig. 5.12 Vegetation coverage sample system in mountainous areas considering topographic characteristics

Compared with other radiative transfer models, PROSAIL takes into account comprehensive parameters and has high computational efficiency. It can simulate the radiative transfer processes of leaves and canopies of various vegetation types at different scales, and is one of the most widely used radiative transfer models currently. The PROSAIL model has now been widely applied to the inversion of various vegetation remote sensing parameters such as vegetation chlorophyll content, leaf area index, and fractional vegetation cover in various scenarios such as farmland, hilly areas, and mountainous regions. Figure 5.13 is a schematic diagram of the construction of simulated FVC samples based on the PROSAIL model.

Firstly, the PROSAIL model is employed to acquire simulated spectra and FVC samples. By inputting the physical and chemical parameters of the leaves, structural parameters, and parameters such as light and soil, the spectral reflectance of the vegetation canopy in the range of 400–2500 nm is simulated. The simulated FVC data can be obtained using the following equations:

$$P_0(\theta) = e^{-\lambda_0 \frac{G(\theta,\theta_1)}{cos\theta} \cdot LAI} \tag{5.3}$$

$$FVC = 1 - P(0^{\circ}) \tag{5.4}$$

where, $P_0(\theta)$ is the function of gap fraction, θ is the sun zenith angle, θ_1 is defined as the average leaf angle, and $G(\theta, \theta_1)$ is the function of solar zenith angle and mean leaf inclination, expressed as an orthographic projection of the unit leaf area. λ_0 is the leaf dispersion or clumping, and it can be considered that the canopy leaves are randomly distributed in this study, that is, λ_0=1.

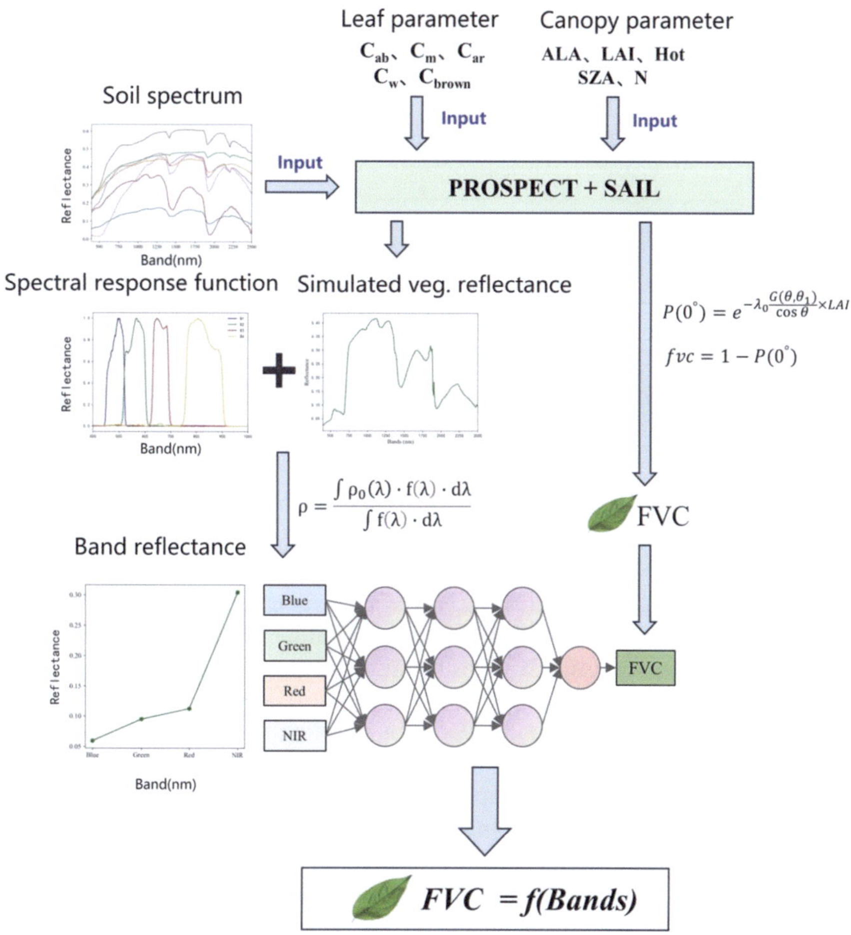

Fig. 5.13 Flow chart of the construction of the simulated FVC samples based on the PROSAIL model

Subsequently, an artificial neural network is utilized to establish the corresponding relationship between FVC and spectral characteristics. This relationship functions as a generator for high-precision FVC samples (Huang et al., 2023; Huang, 2024). Based on this generator, by inputting relevant satellite spectral parameters, the FVC samples are obtained. By integrating the vegetation indices and characteristic variable parameters such as slope and aspect acquired from remote sensing data, an FVC sample library is constructed. The specific process to construct the sample library is shown in Fig. 5.14.

(2) **Construction of the FVC inversion model based on spatio-temporal transfer learning**

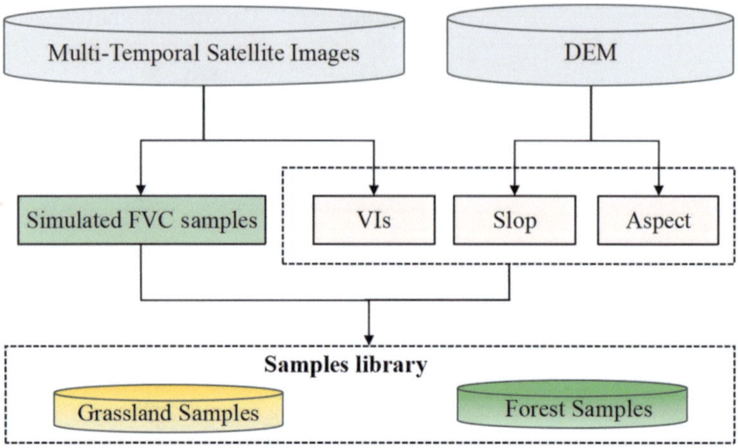

Fig. 5.14 Flow chart of the construction of the sample library

Although the sample library can provide sufficient samples for the inversion of FVC within a specific mountainous area, for FVC with obvious seasonal variation characteristics, spatial discontinuities are likely to occur during large-scale inversion. This is mainly due to data loss caused by the climatic conditions in mountainous areas and the differences in image acquisition times. In recent years, spatiotemporal transfer learning has emerged as a new development direction in machine learning. This method enables the neural network structure to learn the characteristics of the time and space scales of the existing samples. Then, it applies these characteristics to the data of other regions or time series. In this way, it compensates for the spatiotemporal discontinuity of data in the target research area.

Figure 5.15 illustrates the pre-training and fine-tuning process of the FVC inversion model in mountainous areas using transfer learning. In the pre-training stage, through the constructed network model, the temporal and spatial characteristics are extracted from FVC samples in the source domain, and the initial model weights are obtained. In the fine-tuning process, the neural network model is used to extract important features from the domain data, and a small amount of data in the target domain is used to optimize the model. Finally, based on the remote sensing images of the target domain, the fine-tuned inversion model is used to achieve the spatiotemporal transfer inversion of the FVC in the mountainous area.

In this study, as shown in Fig. 5.16, two models, the Convolutional Neural Network (CNN) and the Long Short-Term Memory (LSTM) network, were selected to build a CNN—LSTM coupled model. The CNN structure extracts spatial scale features from the sample library, while the LSTM structure captures time series features, enabling the simultaneous transfer of data across different spatiotemporal scales in mountainous areas. The vegetation index for each altitude gradient in the sample library changes seasonally and dynamically with vegetation growth, whereas terrain

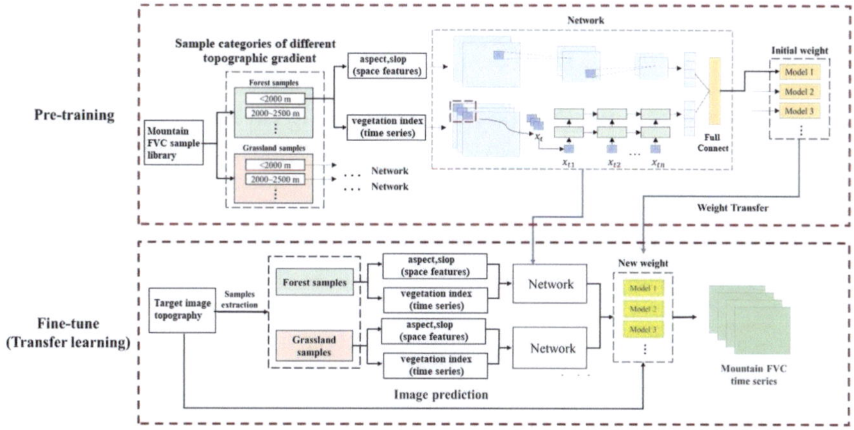

Fig. 5.15 Flow chart of mountainous FVC inversion model based on transfer learning

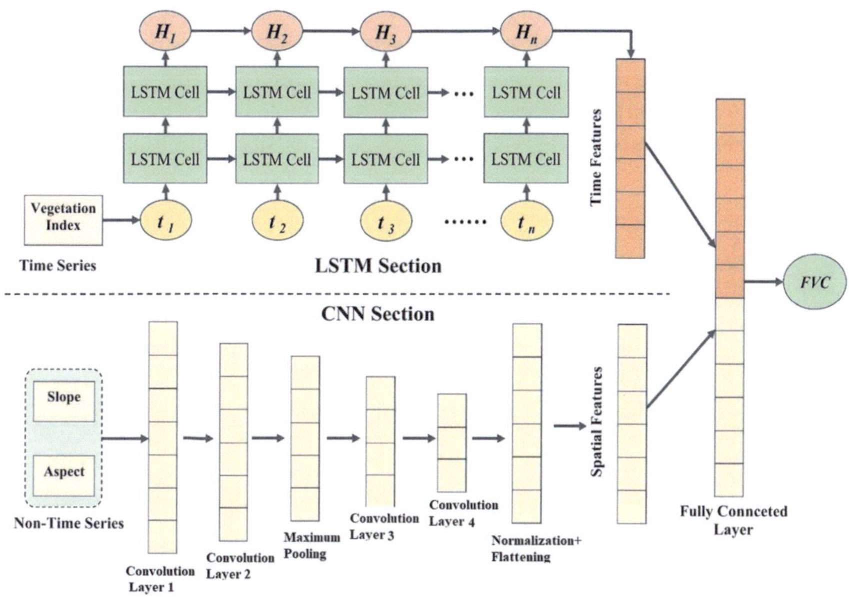

Fig. 5.16 Network structure

features such as slope and aspect remain constant over time. Consequently, the vegetation index features of the time series data are fed into the LSTM network, which excels at processing time series information. Meanwhile, the terrain features are input into the CNN to extract the spatial features of remote sensing images. This combined approach allows for a comprehensive analysis of FVC in mountainous regions by leveraging both temporal and spatial data.

As shown in Fig. 5.16, the CNN structure of the model comprises of four convolutional layers and one max pooling layer. The convolutional layers contain 64, 128, 256, and 128 neurons, respectively. Each convolutional layer uses a 3 × 3 convolution kernel, which moves along the side length of the image to extract spatial features. A max pooling layer is inserted between the second and third convolutional layers to downscale the data, thereby eliminating redundant parameters generated during training and reducing the data volume. After the fourth convolutional layer extracts feature values, these values are regularized and flattened, resulting in the output of the spatial scale features of the mountainous area (Fig. 5.17).

The network structure of the LSTM model is illustrated as follows in Fig. 5.18: it comprises four LSTM layers designed to extract the time series characteristics of vegetation index variables. The number of LSTM units in each layer is configured as 64, 128, 128, and 64, respectively.

(3) **Results and accuracy evaluation**

The implementation of the spatiotemporal transfer learning model in the HKH area is as follows. First, an FVC sample library for the HKH region, which takes into account land cover types and topographic gradients, was constructed based on the time-series Sentinel-2 data, DEM data and in-situ datasets. As the source domain data for the pre-training of the network model, the sample library contains not only the FVC values but also topographic features such as slope and aspect extracted

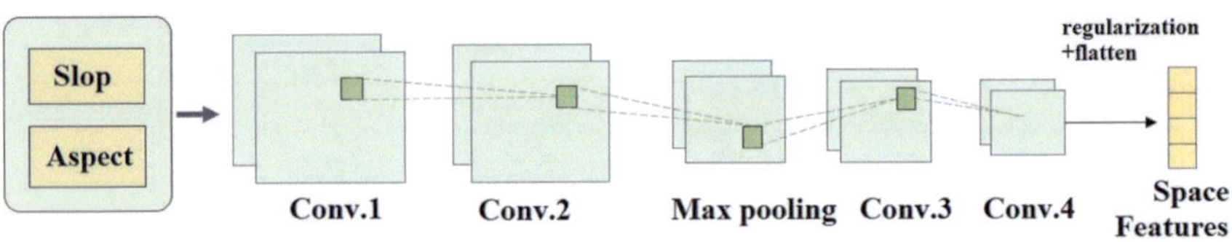

Fig. 5.17 CNN part of the vegetation coverage inversion model

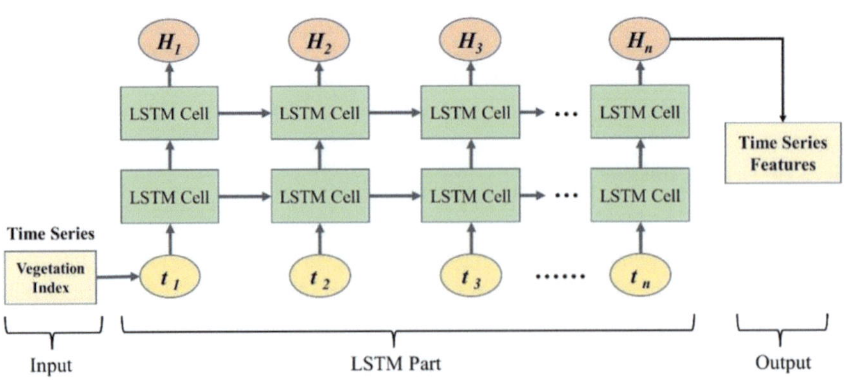

Fig. 5.18 Vegetation coverage inversion model LSTM part

from the DEM data, along with the time-series vegetation indices extracted from the time-series Sentinel-2 remote sensing images.

To validate the accuracy of the FVC samples generated in this study, we collected a set of ground data from three regions: Huzhu in Qinghai, China (where data was collected between September 10 and 12, 2022), Lijiang in Yunnan, China (with data collection spanning from November 24 to 28, 2023), and Kathmandu in Nepal (where data was gathered from September 24 to 29, 2023). In addition, this study also incorporates a set of FVC data based on Unmanned Aerial Vehicle (UAV) imagery published by Geng et al. (2022) to supplement the validation samples. Figure 5.19 shows the spatial distribution of the ground points used in this study.

Figure 5.20 displays a scatter plot that compares the simulated FVC samples with the ground collected data. The accuracy of samples is quantified by two statistical metrics: An R^2 value of 0.945, which indicates a very high degree of correlation between the predicted and actual values, and a Root Mean Square Error (RMSE) of 0.0675.

Subsequently, the data from the sample library is input into the network model for pre-training. During this process, the spatiotemporal features of the training samples are extracted to determine the initial model weights. In the model transfer process, Landsat 8 images are used as the target domain dataset. A small number of samples from the target domain are used to fine-tune the pre-trained model. Based on the fine-tuned model, the FVC product for the HKH area can be acquired (see Fig. 5.21).

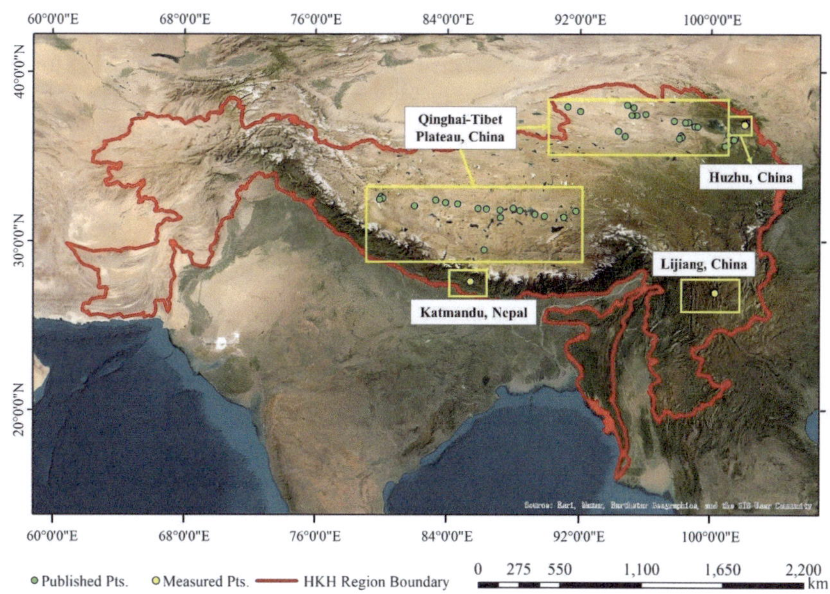

Fig. 5.19 Spatial distribution of ground collected points

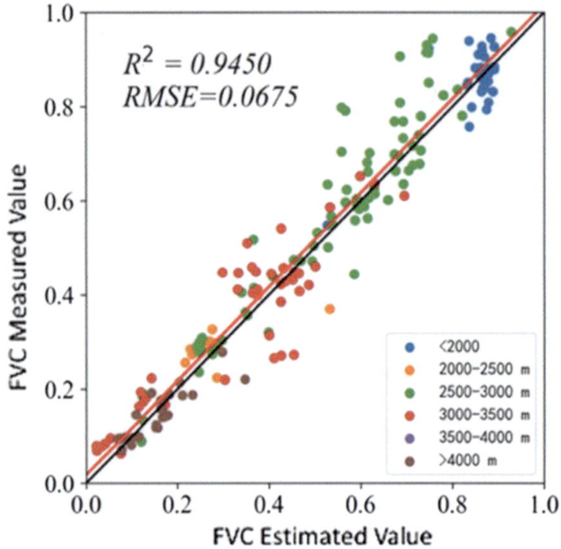

Fig. 5.20 Accuracy assessment of samples based on ground collected data

5.2.2 Analysis of Spatiotemporal Variations in FVC

Figures 5.22 and 5.23 depict statistical analyses of forest and grassland's FVC distribution at various altitude ranges within the HKH area. The findings reveal that forest vegetation coverage remains consistently high at each altitude in the HKH area, gradually decreasing with higher elevations. Conversely, grassland demonstrates a pattern of initially increasing and then decreasing as altitude rises.

Figure 5.24 display the variations in FVC in the HKH area from 2000 to 2010. From it we can see that the FVC in the forest and grassland areas exhibited an overall downward trend, with the most prominent decline occurring in the southern and eastern parts of the HKH regions. This decline can be attributed to changes in vegetation types, which are the consequences of human activities and climate change, ultimately resulting in a reduction of FVC. For instance, deforestation activities in Southeast Asia and the Himalayas have transformed many forests into grasslands. Additionally, grassland areas in Tibetan Plateau, Qinghai, and Sichuan in China have undergone degradation. However, in some areas, vegetation coverage has increased due to strengthened ecological protection measures or other factors such as climate change, as observed in Afghanistan and Pakistan.

Figure 5.25 indicates changes in vegetation coverage in the HKH region, showing a gradual recovery overall from 2010 to 2020. This improvement is attributed to enhanced vegetation ecological protection efforts across various regions. However, there is still a obviously decline in grassland vegetation coverage in the southern Qinghai province in China.

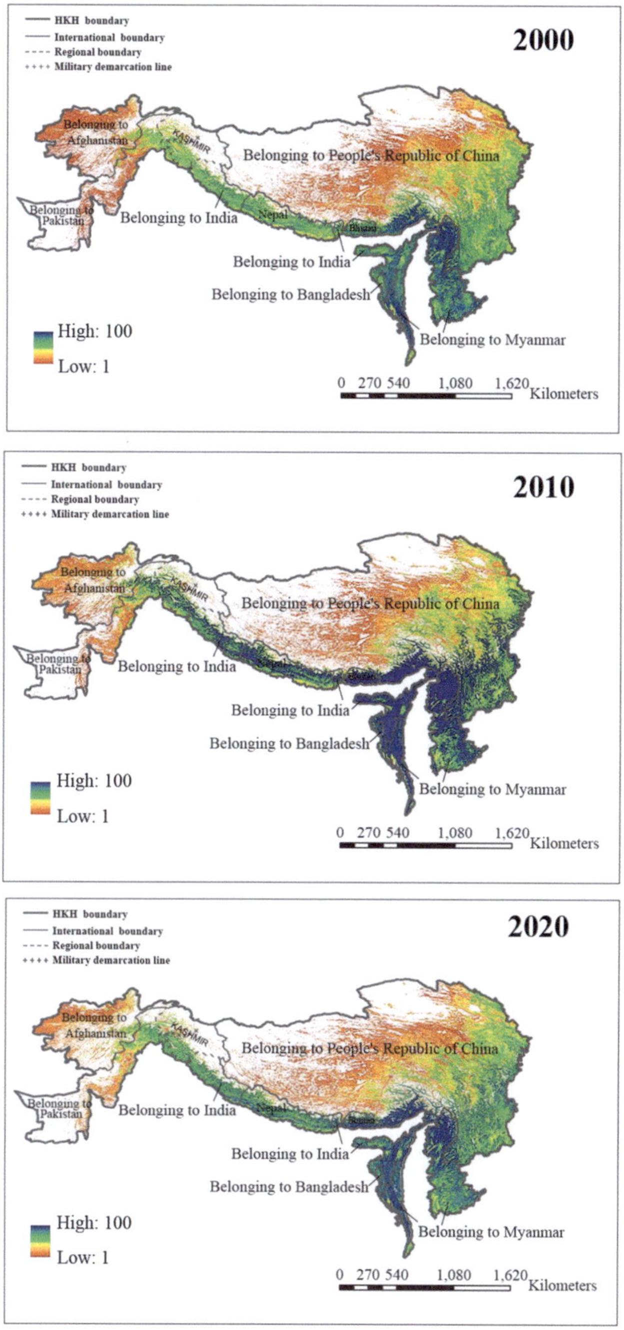

Fig. 5.21 Spatial distribution map of FVC in the HKH area in 2000, 2010 and 2020

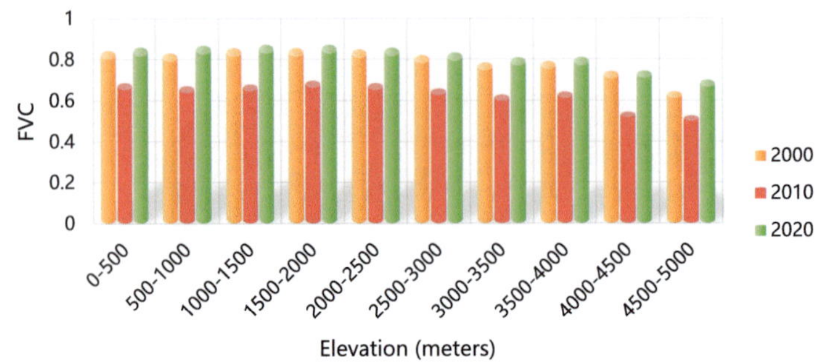

Fig. 5.22 Distribution of FVC across varying altitude ranges in the forest areas of the HKH region

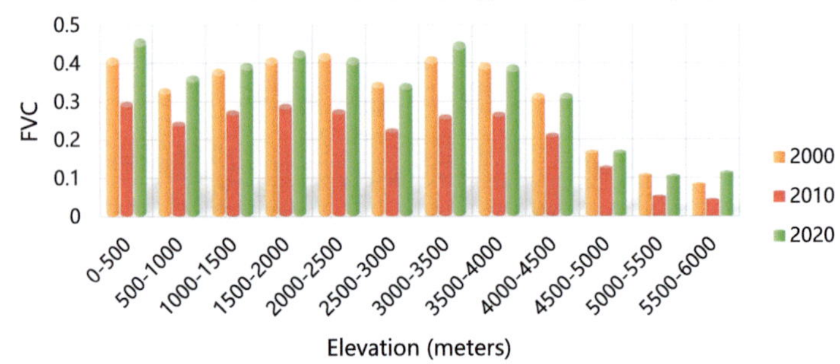

Fig. 5.23 Distribution of FVC across varying altitude ranges in the grassland areas of the HKH region

Figure 5.26 displays the overall situation of FVC from 2000 to 2020. Approximately 75% of the forest and grassland vegetation coverage in the HKH area demonstrates an increasing trend. This growth is largely attributed to the strengthened protection of mountainous vegetation ecosystems. The restoration of forest and grassland in the northeastern Tibetan Plateau and the Himalayas stands out as the primary area for this increase in vegetation coverage within the HKH region. However, in Southeast Asia, particularly in countries like Myanmar and northeastern India, vegetation is declining due to excessive logging and urban development. Additionally, factors such as climate warming have led to the expansion of forest and grass vegetation coverage towards higher altitude areas.

In this study, we also constructed an index to evaluate the average greening status across the entire HKH area. The calculation method is as follows in **Eq. 5.5**:

$$\text{Regional vegetation coverage} = \frac{\sum_{k=0}^{n} \text{vegetation coverage}_k}{n} \tag{5.5}$$

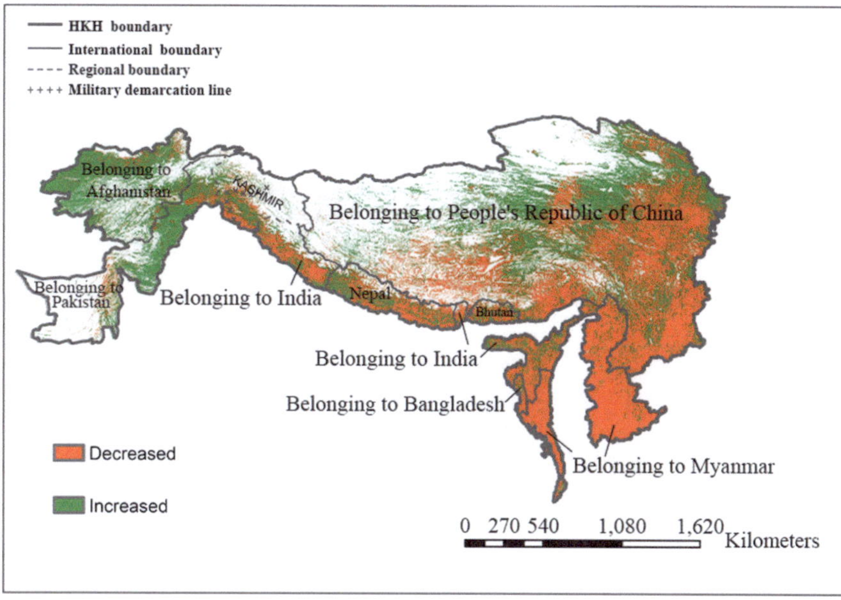

Fig. 5.24 Variations in FVC in the HKH area from 2000 to 2010

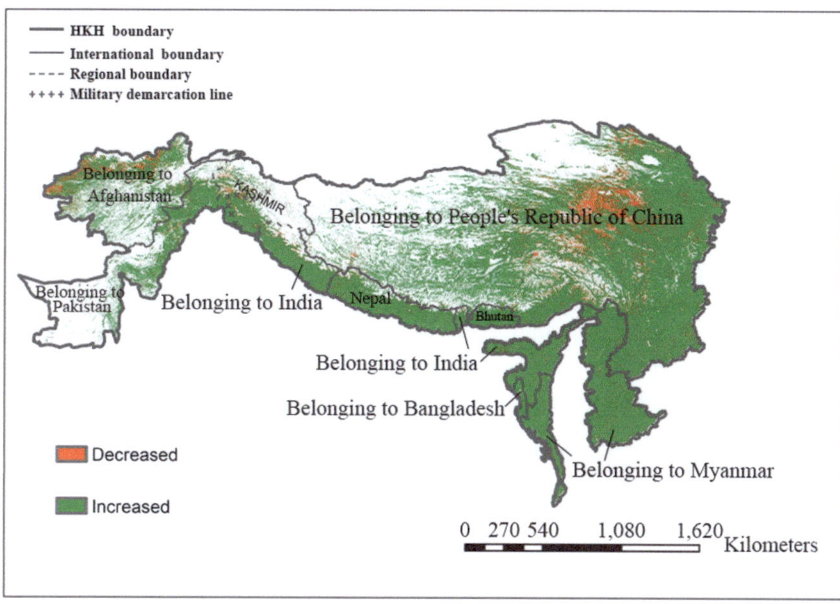

Fig. 5.25 Variations in FVC in the HKH area from 2010 to 2020

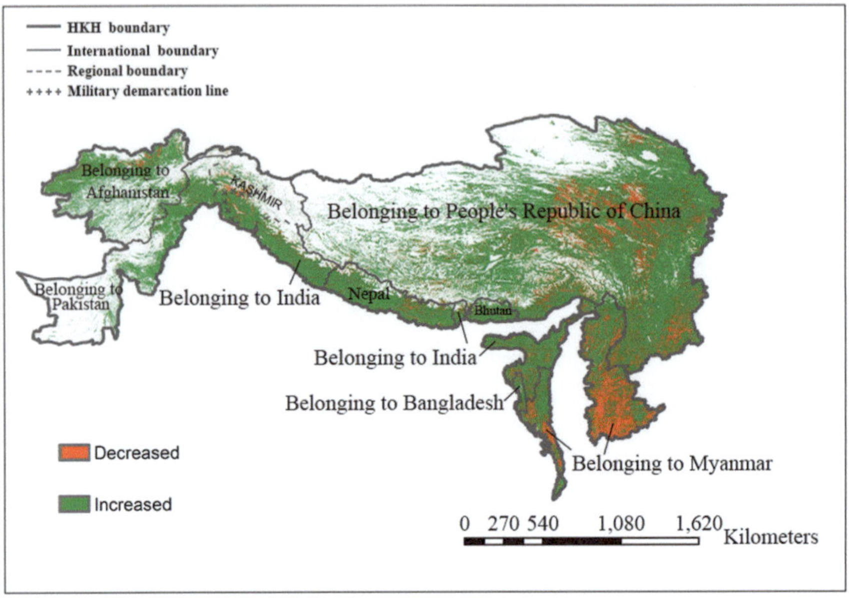

Fig. 5.26 Variations in FVC in the HKH area from 2000 to 2020

The formula uses 'n' to represent the total number of pixels in the remote sensing image of the HKH area. Based on the estimation results of the vegetation coverage inversion method proposed in this study, we calculate the average vegetation coverage of the entire HKH for the years 2000, 2010, and 2020, as shown in Table 5.9.

By calculating the vegetation coverage across the entire HKH over three periods, the results indicate a trend of initial decrease followed by an increase from 2000 to 2020. During this period, the overall greening situation in the HKH has seen improvement. In the first decade, factors such as climate change and human development led to the degradation of mountain vegetation ecology. However, with increased efforts in protecting mountain vegetation, the greening status of most regions has improved in the subsequent decade. This restoration indicates progress in the detection and protection of vegetation ecology in the HKH in recent years.

Table 5.9 Average FVC of forest and grassland in different phases in the HKH area

	2000	2010	2020
FVC	0.401	0.335	0.452

5.2.3 *Summary*

The spatiotemporal analysis of FVC in the HKH region from 2000 to 2020 reveals significant trends and patterns shaped by varying geographical conditions, human activities, and climate change. The study found that the FVC of forest remained consistently high across different altitudes, gradually decreasing at higher elevations, while the FVC of grassland initially increased before declining with altitude. Overall, the HKH region experienced a positive trend in FVC. Notably, the period from 2000 to 2010 saw a decline in FVC due to factors such as deforestation, overgrazing, and climate impacts, particularly in regions like the Himalayas, the Tibetan Plateau, western Sichuan, and Southeast Asia. However, from 2010 to 2020, enhanced ecological protection efforts and climate factors contributed to a recovery in FVC in many areas, especially in Afghanistan and Pakistan.

Looking at the overall situation from 2000 to 2020, approximately 75% of the HKH region showed an increasing trend in forest and grassland FVC, underscoring the positive impact of strengthened protection measures. The northeastern Tibetan Plateau and the Himalayas emerged as primary areas of vegetation restoration, while Southeast Asia, particularly Myanmar and northeastern India, faced challenges due to excessive logging and urban development. Climate warming also contributed to the expansion of vegetation coverage towards higher altitudes. The region's average FVC for 2000, 2010, and 2020 reflect an initial decline followed by an increase, indicating overall improvement in the greening status of the HKH. This trend highlights the effectiveness of recent efforts to protect and restore mountain vegetation ecosystems, demonstrating significant progress in the ecological conservation of the HKH region.

5.3 Evaluation of the Impact of Climate Change on Forest and Grass Vegetation Ecosystems

Climate change poses significant challenges to forest and grassland ecosystems in the HKH region, with diverse impacts on vegetation patterns and ecological dynamics. These ecosystems, including mountain forests and grasslands, are particularly sensitive to changes in temperature, precipitation, and other climate variables. Shifts in temperature and precipitation regimes influence vegetation distribution, phenology, and productivity, altering the structure and functioning of forest and grass ecosystems. Additionally, the HKH's complex topography leads to microclimatic variations along altitude gradients, further exacerbating the vulnerability of these ecosystems to climate change impacts.

Given the urgency of understanding and mitigating these impacts, this assessment aims to evaluate the effects of climate change on forest and grass vegetation ecosystems in the HKH. Through the analysis of climate change indicators such as temperature, land surface temperature (LST), and precipitation across different altitude zones, this study seeks to quantify the relationship between climate variables and FVC. The assessment utilizes statistical methods, including the calculation of the Pearson correlation coefficient, to determine the extent of correlation between climate change indicators and vegetation dynamics. By comprehensively assessing these relationships, the study aims to provide valuable insights into the vulnerability of HKH ecosystems to climate change and inform targeted adaptation and conservation strategies.

5.3.1 Methodology

Mountain forest and grassland ecosystems are extremely sensitive to global climate change. In this study, we explored the relationship between indicators representing climate change, such as the annual average LST, annual average temperature, and annual average precipitation in the HKH mountainous region, and the FVC. This work is expected to offer provide valuable insights into the impact of climate change on forest and grass ecosystems in the HKH region.

The Pearson correlation coefficient (r), a statistical measure commonly used to evaluate the linear correlation between two variables, is employed to quantify the relationship between climate change indicators and FVC. The Pearson correlation coefficient is calculated using Eq. 5.6:

$$r = \frac{\sum_{i=1}^{n}(X_i - \overline{X})(Y_i - \overline{Y})}{\sqrt{\sum_{i=1}^{n}(X_i - \overline{X})^2}\sqrt{\sum_{i=1}^{n}(Y_i - \overline{Y})^2}} \tag{5.6}$$

where X_i is the vegetation coverage of the forest (or grassland), Y_i is the climate related indicator; $\overline{X}$ and $\overline{Y}$ are the average values of the corresponding variables respectively.

By calculating the Pearson correlation coefficient, we can assess the degree of correlation between climate change indicators and vegetation dynamics, providing valuable insights into the impact of climate change on forest and grass ecosystems in the HKH region.

5.3.2 Impact of Climate Change on Forest Ecosystems

Figure 5.27 illustrates the correlation coefficient matrix between forest FVC and climate-related factors, including annual average temperature, annual average LST, and annual average precipitation in the HKH region for the years 2000, 2010, and 2020. The calculation results of the correlation coefficient reveal that precipitation, surface temperature, and air temperature exhibit varying degrees of influence on forest vegetation coverage. Additionally, climate change affects the evolution of mountainous forest vegetation coverage differently across different periods. Notably, changes in temperature and precipitation emerge as the two most crucial factors impacting forest coverage in the HKH region, with their effects on changes in vegetation coverage in forested areas being particularly significant.

Figure 5.28 illustrates the correlation coefficients among temperature, precipitation, LST, and forest FVC in mountainous areas at various altitudes. This correlation reveals that different microclimates are easily formed in different areas of the mountains due to changes in terrain and illumination, which consequently impact the distribution of forest vegetation coverage. There exists a negative correlation between FVC and LST. Below 2,500 m above sea level, this correlation is greater than 0.4, but it gradually weakens with increasing altitude. The HKH area boasts diverse forest vegetation, characterized by large canopy and leaf areas. Transpiration from this vegetation helps regulate surface soil temperature, promoting vegetation growth. As altitude increases beyond 2,500 m, there is a decrease in forest FVC, transitioning from broadleaf forest to coniferous forest. Most of the coniferous forest are evergreen types which can response to increased temperature round the year. Forests

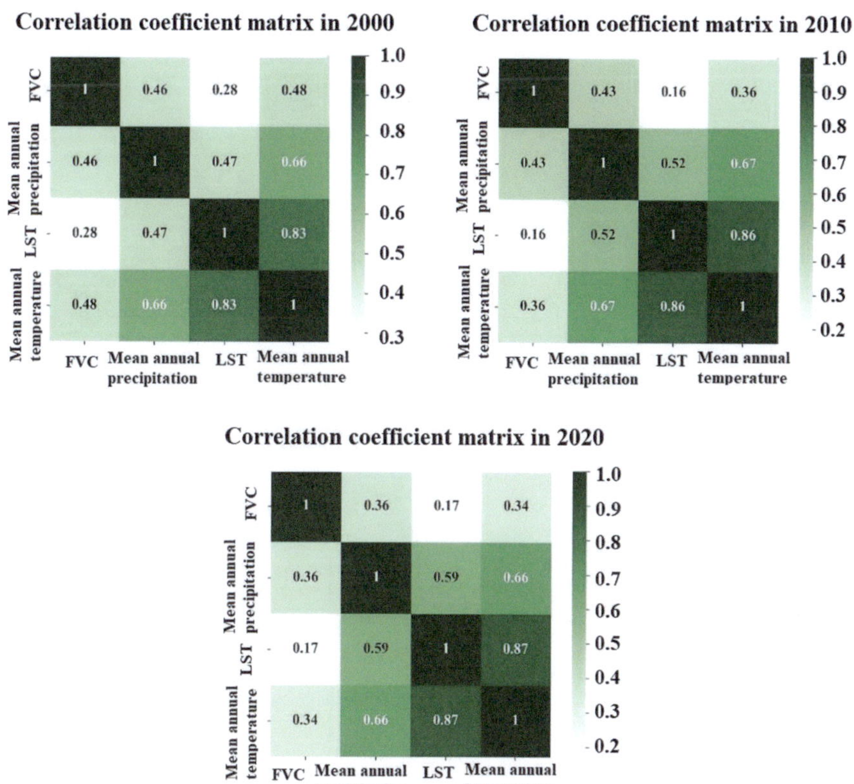

Fig. 5.27 Correlation coefficient matrix between forestland FVC and climate variables in the HKH region in 2000, 2010, and 2020

exhibit varying responses and adaptability to LST, leading to a gradual weakening of the correlation between surface temperature and FVC.

Precipitation in mountainous areas plays a significant role in the positive correlation with forestland vegetation cover. Additionally, it serves as a primary source of water, replenishing soil moisture and fostering a favorable environment for vegetation growth. At times, there is a time lag correlation between precipitation and vegetation,

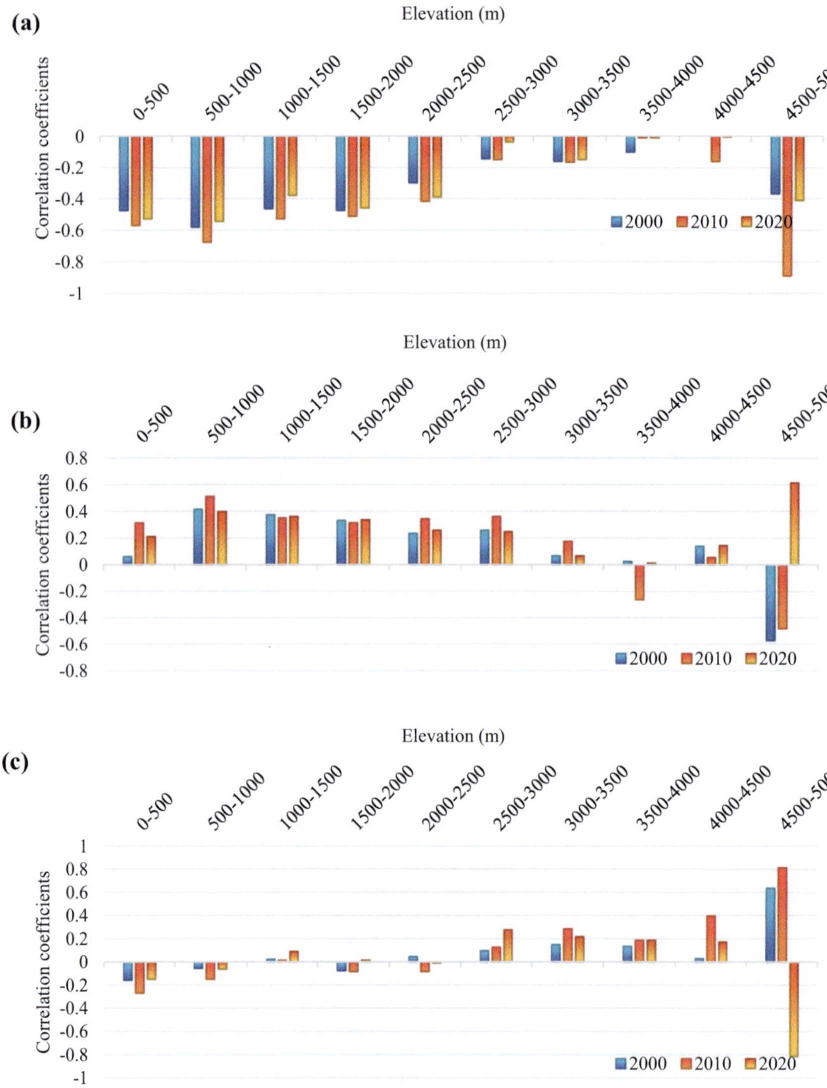

Fig. 5.28 Bar chart of correlation coefficients between: **a** LST and forest FVC, **b** annual precipitation and forest FVC, and **c** annual temperature and forest FVC at different altitude ranges

but the temperature has directly influences to vegetation cover in the HKH region. Below 3000 m above sea level, there's a stable correlation between precipitation and FVC, ranging from 0.2 to 0.4. As altitude increases, temperatures drop, facilitating precipitation formation and thereby promoting forest vegetation growth. However, in areas above 3000 m, where the climate is harsher, forest vegetation decreases, leading to a notable reduction in the interaction between FVC and precipitation.

Temperature changes and their relationship with forest FVC in mountainous regions exhibit a more intricate pattern. Below 2500 m above sea level, there's either no correlation or a weak negative correlation between temperature and FVC. However, in higher altitude areas, rising temperatures have a positive impact on forest coverage. Within the HKH area below 2500 m, predominantly tropical and subtropical evergreen vegetation thrives, mainly distributed in low-altitude mountainous areas. Here, temperature changes are minimal, and the vegetation's response to temperature is not pronounced. Yet, as altitude increases, temperature and precipitation decrease, causing shifts in forest vegetation types, making them more sensitive to temperature changes and gradually increasing the correlation between temperature and FVC.

5.3.3 Impact of Climate Change on Grassland Ecosystems

Figure 5.29 illustrates the correlation coefficient matrix between grassland FVC and climate factors such as annual average temperature, annual average LST, and annual average precipitation in the HKH region for the years 2000, 2010, and 2020. Like forest vegetation, precipitation and temperature are key factors influencing grassland FVC. However, the response of grassland vegetation to climate change is more complex and sensitive at different altitudes compared to forest vegetation.

The bar chart in Fig. 5.30 depicts the correlation coefficients between grassland FVC and various climate variables at different altitudes. It reveals a series of correlations between air temperature, precipitation, LST, and grassland FVC in mountainous areas across various altitudes. In regions below 3,000 m above sea level, there is a negative correlation between grassland FVC and LST ($-0.8 \leq r \leq -0.5$), which weakens as altitude increases. Higher LST can inhibit grassland growth by reducing water availability, particularly because grassland plants are short with small leaf canopies, resulting in weak shading on the ground surface and increased soil moisture evaporation. Between 3,000 and 5,000 m above sea level, the vegetation transitions from forest to grassland, favoring the growth of alpine grasslands with FVC remaining relatively stable.

Furthermore, there is a positive correlation between precipitation and grassland FVC. From sea level to 5,000 m above sea level, this correlation first increases and then gradually decreases with increasing altitude. In areas below 3,500 m above sea level, the correlation coefficient can reach $r \geq 0.4$. Precipitation serves as a crucial

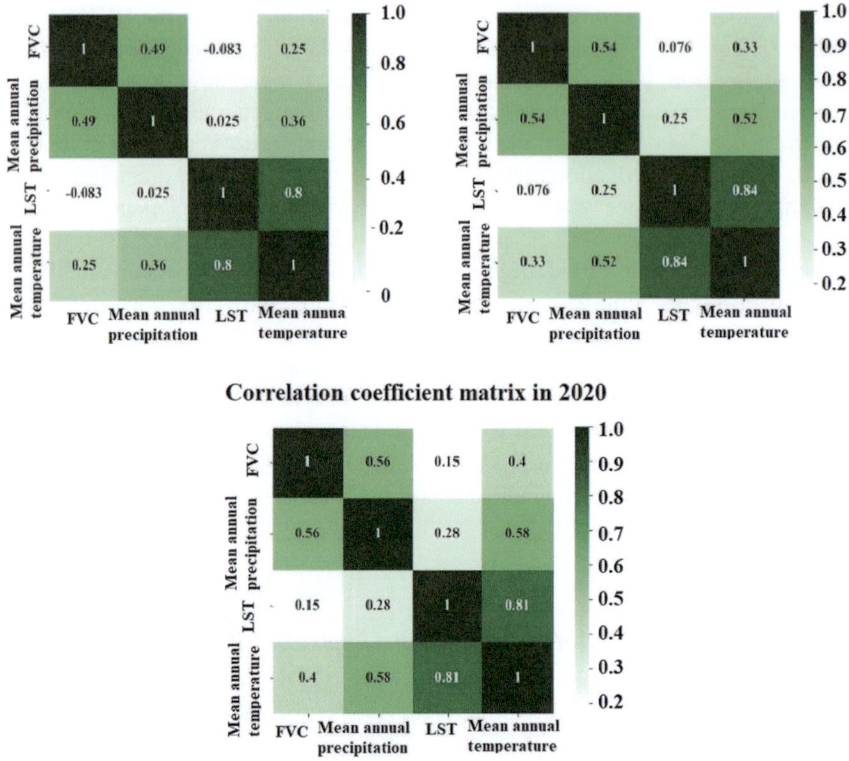

Fig. 5.29 Correlation coefficient matrix between grassland FVC and climate variables in the HKH region in 2000, 2010, and 2020

water source in mountainous regions, replenishing soil moisture and creating favorable growing conditions for grassland growth. As altitude rises, temperatures generally decrease, which reduces the evaporation process, allowing soil and vegetation to use precipitation more efficiently. Consequently, even in regions with relatively low precipitation, grassland vegetation at higher altitudes can maintain a certain degree of coverage, leading to a reduction in the correlation between precipitation and FVC.

Compared to forests, mountain grasslands are more sensitive to temperature changes. In the area between 500 and 3000 m above sea level, the impact of temperature on grassland coverage is relatively small ($r \leq 0.2$). This altitude range experiences a climate suitable for the growth of various types of forest and grass vegetation. Forest canopy reduces direct sunlight on the surface, thereby lowering the temperature of the grassland. This temperature regulation effect helps reduce stress on grassland plants, maintaining grassland coverage, and thus, the impact of temperature remains relatively low. However, as altitude increases beyond 3,000 m, grassland becomes the dominant vegetation type, which is more adapted to warmer environments and positively correlates with temperature.

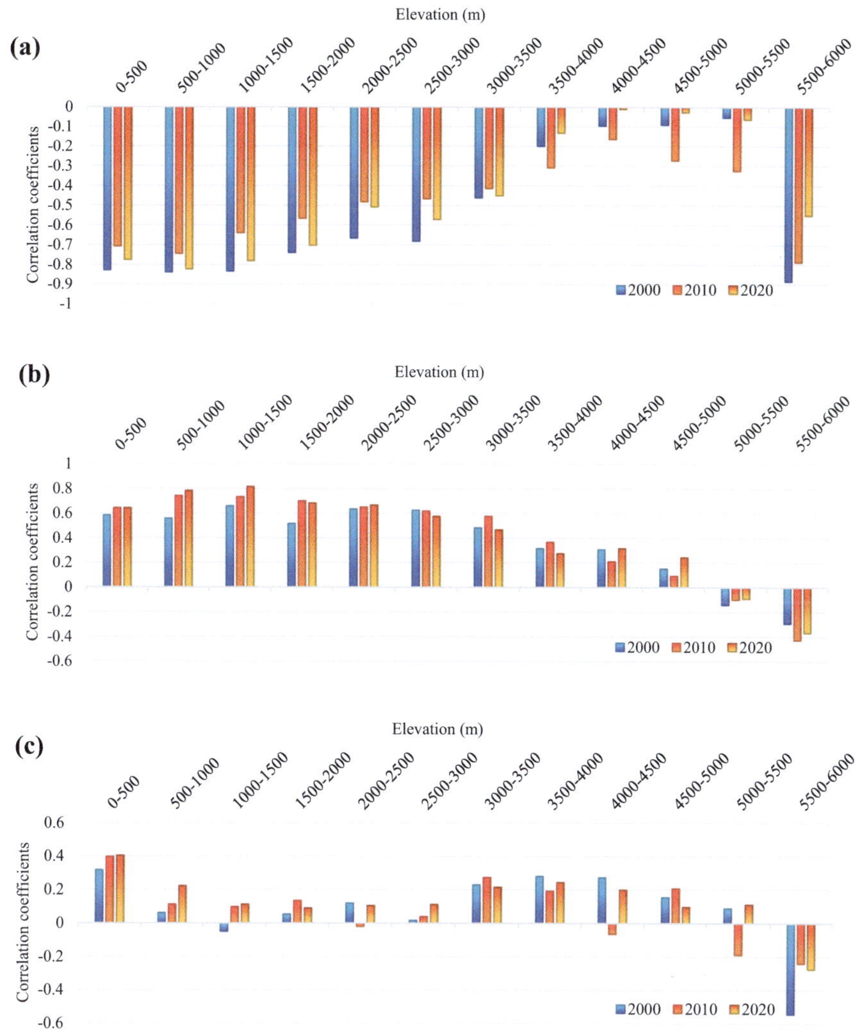

Fig. 5.30 Bar chart of correlation coefficients between: **a** LST and grassland FVC, **b** annual precipitation and grassland FVC, and **c** annual temperature and grassland FVC at different altitudes

Based on these results, in mountainous areas at different altitudes, climate-related factors such as air temperature, precipitation, and LST have varying correlations with forestland and grassland FVC in HKH mountainous areas, with grassland showing greater sensitivity to climate change compared to forests. Mountainous areas with different altitude range often form complex microclimates, leading to varied impacts on forest and grass vegetation coverage and different correlations. Therefore, to protect mountainous vegetation ecology with complex surface and environmental conditions, it is essential to consider factors such as terrain areas, vegetation types, and climatic conditions. Adopting various methods tailored to local conditions can maximize the protection of mountainous vegetation ecosystem conditions.

5.3.4 Summary

The assessment of climate change impacts on the HKH region's forest ecosystems reveals significant variability in how forest vegetation coverage responds to climate factors such as temperature, LST, and precipitation. From 2000 to 2020, precipitation and temperature emerged as critical influencers of forest coverage, with notable changes across different periods. At altitudes below 2,500 m, there is a strong negative correlation between LST and forest coverage, which weakens as altitude increases. This is due to the dense forest canopy regulating LST and promoting vegetation growth. However, at altitudes above 3,000 m, forest types shift from broadleaf to coniferous, and their correlation with LST and precipitation becomes less pronounced. The data indicate that forest ecosystems in the HKH are diverse and their response to climate change varies significantly with altitude, necessitating tailored conservation strategies.

Grassland ecosystems in the HKH region exhibit even greater sensitivity to climate change compared to forests. The correlation between grassland vegetation coverage and climate related factors such as temperature and precipitation is complex and varies with altitude. Between 500 and 3,000 m, temperature has a relatively small impact on grassland FVC, as forest canopies help mitigate temperature extremes. However, above 3,000 m, grasslands become the predominant vegetation type, showing a positive correlation with temperature. Grassland FVC is strongly negatively correlated with LST at lower altitudes, while precipitation positively influences grassland FVC up to 3,500 m. The intricate interplay of these climate factors creates varied microclimates across different altitudes, impacting FVC differently. Therefore, effective protection of mountainous vegetation in the HKH requires considering these complex interactions and adopting strategies that address the unique conditions of different terrain areas and vegetation types to ensure the sustainability of these vital ecosystems.

5.4 Summary

The spatiotemporal analysis of forest and grassland distribution in the HKH region from 2000 to 2020 reveals significant spatial and temporal variations. Low-altitude areas were predominantly covered by forests and grasslands, with grassland coverage increasing from 2000 to 2010 but declining by 2020 due to forest encroachment. Conversely, high-altitude areas saw significant increases in grassland coverage due to heightened precipitation and rising temperatures from 2000 to 2020. Spatial change analysis showed severe deforestation in regions like Southeast Asia and parts of

China, converting forests to grasslands, while the north and northeastern Tibetan Plateau experienced increases in grassland coverage from 2000 to 2020. Overall, approximately 75% of the HKH region showed an increasing trend in forest and grassland FVC from 2000 to 2020. The northeastern Tibetan Plateau and the Himalayas emerged as primary areas of vegetation restoration, while Southeast Asia, particularly Myanmar and northeastern India, faced challenges due to excessive logging and urban development. These findings highlight the dynamic nature of land cover changes in the HKH.

Overall, the HKH region demonstrated complex patterns of land cover change over the past two decades, underscoring the sensitivity of grasslands to temperature changes and the varied responses of vegetation to climatic and anthropogenic pressures. Grasslands showed significant degradation in overgrazed central regions but notable recovery in areas with improved precipitation and conservation measures. Forest areas faced extensive losses due to logging, particularly in Southeast Asia, while regions like western Sichuan and southeastern Tibet saw forest regrowth due to policy interventions. Understanding these shifts is crucial for managing natural resources and promoting sustainable land use in the HKH region. Policymakers and conservationists must consider these multifaceted influences to develop effective strategies for preserving the ecological integrity and resilience of this critical landscape.

Based on these insights, the HKH region requires a multi-faceted approach to address the complex challenges of vegetation conservation. Enhanced ecological protection efforts, sustainable land use practices, and community-based conservation initiatives are essential to mitigate the impacts of climate change and human activities. Tailored strategies that consider different terrain areas, vegetation types, and climatic conditions are necessary to maximize the protection of mountainous vegetation ecosystems. Furthermore, frameworks for local climate change adaptation planning can coordinate efforts across this transboundary region, ensuring the sustainability of the HKH's vital ecosystems and the well-being of the populations that depend on them.

References

Ahlström, A., Xia, J., Arneth, A., Luo, Y., & Smith, B. (2015). Importance of vegetation dynamics for future terrestrial carbon cycling. *Environmental Research Letters, 10*(5), 054019. https://doi.org/10.1088/1748-9326/10/5/054019

Anderson, K., Fawcett, D., Cugulliere, A., Benford, S., Jones, D., & Leng, R. (2020). Vegetation expansion in the subnival Hindu Kush Himalaya. *Global Change Biology, 26*(3), 1608–1625. https://doi.org/10.1111/gcb.14919

Bayle, A., Carlson, B., Thierion, V., Isenmann, M., & Choler, P. (2019). Improved mapping of mountain shrublands using the sentinel-2 red-edge band. *Remote Sensing, 11*(23), 2807. https://doi.org/10.3390/rs11232807

Behera, M. D., Murthy, M. S. R., Das, P., & Sharma, E. (2018). Modelling forest resilience in Hindu Kush Himalaya using geoinformation. *Journal of Earth System Science, 127*(7), 95. https://doi.org/10.1007/s12040-018-1000-x

Breiman, L. (2001). Random forests. *Machine Learning, 45*(1), 5–32. https://doi.org/10.1023/A:1010933404324

Carlson, T. N., & Ripley, D. A. (1997). On the relation between NDVI, fractional vegetation cover, and leaf area index. *Remote Sensing of Environment, 62*, 241–252.

Chen, J. (2020). Research on vegetation coverage changes and environmental influencing factors in alpine grasslands on the Qinghai-Tibet Plateau. *Journal of Surveying and Mapping, 04*(49), 533.

Chen, X., & Kuang, W. (2019). Analysis of ecosystem change characteristics of the Tibetan Plateau from 1990 to 2015. *Journal of Southwest University for Nationalities (Natural Science Edition) [CN], 45*(3), 233–242.

Franklin, J., Serra-Diaz, J. M., Syphard, A. D., & Regan, H. M. (2016). Global change and terrestrial plant community dynamics. *Proceedings of the National Academy of Sciences, 113*(14), 3725–3734. https://doi.org/10.1073/pnas.1519911113

Geng, X., Wang, X., Fang, H., Ye, J., Han, L., Gong, Y., & Cai, D. (2022). Vegetation coverage of desert ecosystems in the Qinghai-Tibet Plateau is underestimated. *Ecological Indicators, 2022*(137), 108780. https://doi.org/10.1016/j.ecolind.2022.108780

Huang, Y. (2024). *The retrieval of forest and grass fractional vegetation coverage in mountain regions based on spatio-temporal transfer learning*. Aerospace Information Research Institute Chinese Academy of Sciences.

Huang, Y., Zhou, X., Lv, T., Tao, Z., Zhang, H., Li, R., Zhai, M., & Liang, H. (2023). The retrieval of forest and grass fractional vegetation coverage in mountain regions based on spatio-temporal transfer learning. *Remote Sensing, 15*(19), 4857. https://doi.org/10.3390/rs15194857

Jacquemoud, S., Verhoef, W., Baret, F., Bacour, C., Zarco-Tejada, P. J., Asner, G. P., François, C., & Ustin, S. L. (2009). PROSPECT+SAIL models: A review of use for vegetation characterization. *Remote Sensing of Environment, 113*, S56–S66.

Jiang, H., Chen, A., Wu, Y., Zhang, C., Chi, Z., Li, M., & Wang, X. (2022). Vegetation monitoring for mountainous regions using a new Integrated Topographic Correction (ITC) of the SCS + C correction and the shadow-eliminated vegetation index. *Remote Sensing, 14*(13), 3073. https://doi.org/10.3390/rs14133073

Keenan, T. F., & Riley, W. J. (2018). Greening of the land surface in the world's cold regions consistent with recent warming. *Nature Climate Change, 8*(9), 825–828. https://doi.org/10.1038/s41558-018-0258-y

Li, J., Tang, Z., Deng, G., Sang, G., & Wang, J. (2023). Remote sensing monitoring of grassland phenology changes in the Tibetan Plateau from 2001 to 2020. *Soil and Water Conservation Research [CN], 30*(4), 265–274.

Lu, Q., Wu, S., & Zhao, D. (2017). Changes in alpine grassland coverage in the Tibetan Plateau from 1982 to 2013 and its relationship with climate. *Geographical Science [CN], 02*(37), 292–300.

Meng, X., Ge, J., Hou, M., Feng, Q., Jin, Z., Gao, J., & Liang, T. (2022). Research on spatiotemporal variation characteristics of natural grassland coverage in the Tibetan Plateau based on machine learning. *Journal of Grassland Science [CN], 10*(30), 2652–2662.

Ougahi, J. H., Cutler, M. E. J., & Cook, J. S. (2022). Assessment of climate change effects on vegetation and river hydrology in a semi-arid river basin. *PLoS ONE, 17*(8), e0271991. https://doi.org/10.1371/journal.pone.0271991

Panday, P. K., & Ghimire, B. (2012). Time-series analysis of NDVI from AVHRR data over the Hindu Kush-Himalayan region for the period 1982–2006. *International Journal of Remote Sensing, 33*(21), 6710–6721. https://doi.org/10.1080/01431161.2012.692836

Perugini, L., Caporaso, L., Marconi, S., Cescatti, A., Quesada, B., de Noblet-Ducoudré, N., House, J. I., & Arneth, A. (2017). Biophysical effects on temperature and precipitation due to land cover change. *Environmental Research Letters, 12*(5), 053002. https://doi.org/10.1088/1748-9326/aa6b3f

Piao, S., Liu, Z., Wang, T., Peng, S., Ciais, P., Huang, M., Ahlstrom, A., Burkhart, J. F., Chevallier, F., Janssens, I. A., Jeong, S.-J., Lin, X., Mao, J., Miller, J., Mohammat, A., Myneni, R. B., Peñuelas, J., Shi, X., Stohl, A., … Tans, P. P. (2017). Weakening temperature control on the interannual variations of spring carbon uptake across northern lands. *Nature Climate Change, 7*(5), 359–363. https://doi.org/10.1038/nclimate3277

Rouse, J. W., Haas, R. H., Schell, J. A., & Deering, D. W. (1974). *Monitoring Vegetation Systems in the Great Plains with ERTS* (Vol. 351, p. 309). NASA Special Publications. https://ui.adsabs.harvard.edu/abs/1974NASSP.351..309R

Sharma, M., Areendran, G., Raj, K., Sharma, A., & Joshi, P. K. (2016). Multitemporal analysis of forest fragmentation in Hindu Kush Himalaya—A case study from Khangchendzonga Biosphere Reserve, Sikkim, India. *Environmental Monitoring and Assessment, 188*(10), 596. https://doi.org/10.1007/s10661-016-5577-8

Wang, S., Xu, X., Shrestha, N., Zimmermann, N. E., Tang, Z., & Wang, Z. (2017). Response of spatial vegetation distribution in China to climate changes since the Last Glacial Maximum (LGM). *PLoS ONE, 12*(4), e0175742. https://doi.org/10.1371/journal.pone.0175742

Wester, P., Mishra, A., Mukherji, A., & Shrestha, A. B. (eds.). (2019). *The Hindu Kush Himalaya Assessment*. Springer International Publishing. https://doi.org/10.1007/978-3-319-92288-1

Xu, W., Yang, D., Li, Y., & Xiao, R. (2016). Correlation analysis of Mackenzie river discharge and NDVI relationship. *Canadian Journal of Remote Sensing, 42*(3), 292–306. https://doi.org/10.1080/07038992.2016.1171135

You, H. (2013). *Remote sensing analysis of forest disturbance in Southeast Asia in the past ten years*. Nanjing University of Information Science and Technology.

Zanaga, D., Van De Kerchove, R., De Keersmaecker, W., Souverijns, N., Brockmann, C., Quast, R., Wevers, J., Grosu, A., Paccini, A., Vergnaud, S., Cartus, O., Santoro, M., Fritz, S., Georgieva, I., Lesiv, M., Carter, S., Herold, M., Li, L., Tsendbazar, N. E., … Arino, O. (2021). ESA WorldCover 10 m 2020 v100. Zeonodo

Chapter 6
Assessment of Sustainable Development in the HKH Region

This study evaluates sustainable development in the Hindu Kush Himalaya (HKH) region by assessing agricultural production, water resources, and ecological and environmental sustainability. Aligned with the objectives of SDG 2 (Zero Hunger), SDG 6 (Clean Water and Sanitation), and SDG 15 (Life on Land), the research employs a comprehensive methodology to analyze key indicators of sustainability. The approaches detailed in Sects. 6.1.1, 6.1.2, and 6.1.3 provide a structured framework for examining trends in agricultural productivity, water resource management, and ecological state of the region. By integrating these assessments, the study aims to inform policy decisions and promote sustainable land and water management practices in the HKH region, contributing to long-term environmental resilience and socio-economic stability.

6.1 Sustainable Development Indicator System and Evaluation Methods

6.1.1 Sustainable Development Evaluation Index System and Evaluation Method for Agricultural Production

We assessed unit labor output as a measure of production efficiency and conducted an analysis of the population distribution in key grain-producing areas for the years 2000, 2010, and 2020, utilizing data sourced from the WorldPoP official website. WorldPop, a research initiative affiliated with the University of Southampton in the United Kingdom, offers comprehensive and openly accessible data resources. Leveraging the population distribution dataset available through this platform, which is instrumental in examining age demographics across various regions (Steele et al.,

S. Lu et al., *Remote Sensing Assessment of Sustainable Development in the Hindu Kush-Himalayan (HKH) Region*, https://doi.org/10.1007/978-981-96-8675-9_6

2018), we integrated regional population distribution maps with a 1 km grid resolution obtained from WorldPop with national population distribution maps from the World Bank database. This combined dataset enabled us to derive insights into the age structure composition, thereby facilitating an initial estimation of labor force population density across different regions. Internationally, individuals aged 15–64 are classified as the working-age population. Consequently, we expressed unit labor output as the ratio of yield data per unit area to the labor force population within each respective region.

$$CPPL = CP/Ls \tag{6.1}$$

Product per unit of labor (CPPL) serves as a crucial metric for gauging the efficiency and productivity of labor within the production process. This measure is derived by dividing total production (CP) by the quantity of labor (Ls) employed, providing insight into the value of production yielded per unit of labor input. Essentially, CPPL indicates the efficiency of labor in generating products within the production system. Through vigilant monitoring and analysis of unit labor output, both enterprises and governments can meticulously assess labor utilization efficiency, fine-tune production processes, and enhance overall production efficiency. Such endeavors not only pave the way for economic growth but also foster sustainable development by ensuring optimal resource utilization and minimizing waste.

Land use intensity, on the other hand, is encapsulated by the concept of land use intensity per production output (LIPP). This metric delves into the utilization of land resources concerning the output generated. By scrutinizing alterations in cultivated land across different regions and integrating yield estimation outcomes, analysts can delve into the intensity of land use. This comprehensive analysis sheds light on the optimal utilization of land resources vis-à-vis the output achieved, offering invaluable insights into land management practices and agricultural productivity.

$$A = \sum_{i=1}^{n} P_i \times PS \tag{6.2}$$

$$LIPP = A/CP \tag{6.3}$$

LIPP, or Land Use intensity per production output, serves as a crucial metric in assessing the efficiency of land utilization within a given region. It is derived from the ratio of the total Yield area of regional crops (A) to the regional crop production estimate (CP). It quantifies the relationship between the land allocated for cultivation and the resulting agricultural output. A lower LIPP calculation indicates a higher efficiency in land utilization, suggesting that the region is maximizing its agricultural productivity relative to the area of land dedicated to farming. This metric thus provides valuable insights into the optimization of agricultural resources and can guide policymakers and agricultural stakeholders in enhancing productivity and sustainability within the region's agricultural sector.

The average unit labor output, average land intensity, and cultivated land area were ranked by importance using the hierarchical analysis method. This process led to the establishment of an evaluation system aimed at achieving zero hunger in the Hindu Kush-Himalaya (HKH) region. To quantify this evaluation, we developed the Zero Hunger Index (ZHI). This index involved weighting and normalizing the indicators for each agricultural production output within the HKH region. The key components of the ZHI include CPPL, representing the average unit labor output of the region, LIPP, denoting the average land intensity, and A, indicating the proportion of cultivated land area.

$$ZHI = Normalization(0.5 \times CPPL_{avg} + 0.3/LIPP_{avg} + 0.2 \times A) \tag{6.4}$$

The Zero Hunger Index serves as a crucial measure of food security in the HKH region. A higher ZHI value signifies better food security, indicating that the region is closer to achieving zero hunger. Conversely, a lower ZHI value points to more severe food security issues, underscoring areas where significant improvements are needed. The ZHI thus provides a comprehensive and systematic approach to assess and address the food security challenges in the HKH region.

6.1.2 Evaluation Methods and Index System for Assessing Sustainable Development of Water Resources

The sustainable development of water resources in the HKH region, as aligned with SDG 6 on water resources, is evaluated through two primary dimensions: changes in the spatial extent of water bodies and water quality. These dimensions correspond to specific SDG indicators designed to track progress in water resource management and conservation. The first dimension, monitoring changes in the spatial range of water bodies, aligns with SDG 6.6.1, which focuses on "tracking changes in water-related ecosystems such as lakes, rivers, wetlands, and mangroves over time with the help of earth observation technology." This involves using satellite imagery and remote sensing technologies to observe and measure fluctuations in the size and distribution of water bodies, providing crucial data on the hydrological dynamics within the HKH region.

The second dimension involves monitoring environmental water quality, corresponding to SDG 6.3.2, which measures "the proportion of water bodies with good environmental water quality." This indicator is essential for assessing the health of aquatic ecosystems and ensuring that water bodies meet the standards required to support biodiversity and human use. To evaluate water quality, researchers focus on water transparency, a key parameter closely linked to several core water quality parameters. Studies, such as those by Shen et al. (2020), have analyzed the relationship between water transparency and comprehensive water quality, and in the process revealed that water bodies with transparency (Secchi depth, SD) below 0.5 m were generally found to have poor water quality, indicating higher levels of pollutants and reduced ecological health.

For practical application, the rate of change in surface water is calculated by comparing the area of water bodies extracted in the target year against baseline data, as defined by SDG 6.6.1 (Lu et al., 2021). This calculation helps in understanding the temporal changes in water body sizes, which can be indicative of broader environmental shifts, such as glacial melt, changes in precipitation patterns, or anthropogenic activities.

$$Ration\ of\ surface\ water\ area\ change = \frac{Water\ area_{T1} - Water\ area_{T2}}{Water\ area_{T2}} \tag{6.5}$$

For water quality monitoring, transparency serves as a proxy for assessing overall water health. According to research by Bigham Stephens et al. (2015) and Shen et al. (2020), water bodies with transparency exceeding 0.5 m generally exhibit better environmental conditions. Furthermore, the United Nations Water Mechanism stipulates that water bodies where more than 80% of the water maintains good quality are in good environmental health (UN-Water, 2017). Therefore, combining these methodologies, a sustainable development status monitoring framework targeted to correspond to SDG 6.3.2 is constructed based on water transparency to monitor water bodies with good water quality. Pixels with transparency greater than 0.5 m are classified as having good water quality (Fig. 6.1). If more than 80% of the pixels within a given water body meet this criterion, the entire water body is deemed to have good environmental water quality.

$$Ratio\ of\ water\ with\ good\ quality = \frac{Number\ of\ water\ bodies\ with\ good\ quality}{Total\ Number\ of\ water\ bodies} \tag{6.6}$$

This systematic approach allows for consistent and accurate tracking of water resource sustainability in the HKH region, facilitating informed decision-making and effective resource management to meet the targets set by SDG 6 (Fig. 6.1).

6.1.3 Evaluation Index System and Methodology for Ecological and Environmentally Sustainable Development

The eco-environmentally sustainable development evaluation index is a vital tool for assessing the health and resilience of the HKH regional ecosystem under the pressures of social and economic activities. This evaluation aligns with SDG 15, which focuses on life on land, aiming to protect, restore, and promote the sustainable use of terrestrial ecosystems, manage forests sustainably, combat desertification, and halt biodiversity loss. To gauge this sustainability, this research employs two primary indicators:

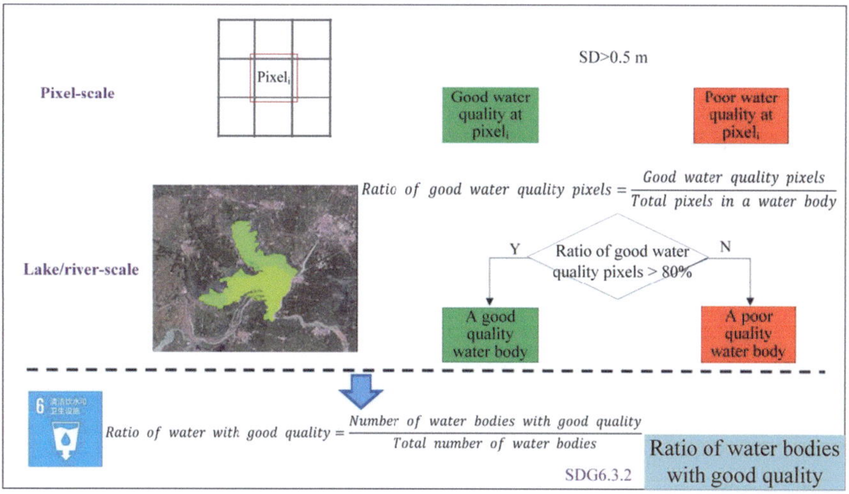

Fig. 6.1 Water resources sustainable development monitoring indicator method system based on water transparency

the annual change rate of forest and grassland area and the annual change rate of fractional vegetation cover (FVC). By monitoring these indicators, the study aims to provide a comprehensive assessment of how well the HKH region is maintaining its ecological integrity in the face of human-induced pressures. The annual change rate of forest and grassland area reflects the shifts in land use and vegetation cover, offering insights into deforestation, reforestation, and land degradation processes. Meanwhile, the annual change rate of FVC serves as a vital indicator of vegetation health and productivity, revealing changes in the density and vitality of plant life across the region. More details about the indicators include:

(1) **The annual change rate of forest and grassland areas** is a crucial metric for understanding the quantitative shifts in regional forest and grassland resources over time. By comparing data on forest and grassland coverage at different intervals, specifically for the years 2000, 2010, and 2020, this study calculates the rate of change in these ecosystems. This indicator provides an intuitive reflection of how forest and grassland areas are evolving, serving as a fundamental measure for assessing the overall status of regional ecological resources. The methodology involves using distribution grid data to analyze forest and grassland areas within each designated zone. By employing zoning statistics, the study determines the area of forest and grassland in each region for the specified years. Using the forest and grassland coverage rate in 2000 as a baseline, changes in coverage rates for 2010 and 2020 are calculated. This approach enables the identification of trends and patterns in forest and grassland dynamics, providing essential insights for ecological evaluation and resource management. The calculation formula used in this analysis is designed to offer a clear and precise measurement of these changes over the specified periods.

$$CR = \frac{F_a - F_{2000}}{F_{2000}} \tag{6.7}$$

where: CR is the change rate, F_a is the forest and grass coverage rate in the αth year, and F_{2000} is the forest and grass coverage rate in 2000.

(2) **The annual change rate of forest and grass coverage** is a critical parameter that reflects the health and stability of surface vegetation, directly influencing the ecosystem's overall function. By analyzing the coverage of forest and grass resources, we can gauge the growth and status of these vital resources, making it an essential metric for assessing their sustainable utilization. In this study, we calculated the annual change rate of forest and grass coverage using raster data from the years 2000, 2010, and 2020. Employing a zoning statistical method, we derived the average forest and grass coverage for each region. Using the year 2000 as a baseline, we then determined the proportional change in coverage for the years 2010 and 2020. This calculation provides a clear understanding of how forest and grass coverage has evolved over time, offering valuable insights into the effectiveness of conservation efforts and the resilience of these ecosystems. The calculation formula for determining these changes is as follows:

$$FCR = \frac{FVC_a - FVC_{2000}}{FVC_{2000}} \tag{6.8}$$

where FCR is the change rate of the forest and grass coverage; FVC_a is the forest and grass coverage in year a; FVC_{2000} is the forest and grass coverage in 2000.

Together, these indicators offer a detailed and nuanced understanding of the ecological sustainability of the HKH region. They inform conservation efforts by highlighting areas of concern and success, guiding sustainable development policies that balance ecological health with human needs. By integrating these measures, the project supports the achievement of SDG 15, promoting sustainable land management and biodiversity conservation in one of the world's most ecologically significant regions.

6.2 Remote Sensing Data and Preprocessing for Sustainable Development Assessment

6.2.1 Remote Sensing Data and Processing for Sustainable Development Assessment of Agricultural Production in the HKH

This study aimed to support the achievement of SDG 2.4.1, which focuses on increasing the proportion of agricultural land under sustainable management, by conducting a comprehensive assessment of agricultural trends in the Hindu Kush Himalaya (HKH) region from 2000 to 2020. Utilizing the methodological approach outlined in Sect. 3.1.1, we moved beyond merely monitoring agricultural productivity to providing insights that support sustainable land management and policy formulation. Our research integrated multiple analytical frameworks, particularly those discussed in Sects. 3.2.1 and 3.3.1, to assess changes in grain production, land use intensity, and production efficiency over time. This multidimensional approach allowed us to capture the dynamics of agricultural transformation in the region, highlighting key areas where sustainable practices have been adopted and where improvements are needed.

By leveraging these methodologies, our study facilitated predictive analyses of crop yields and their correlation with sustainable agricultural practices, thereby offering valuable indicators for long-term food security and environmental resilience. The insights generated serve as a foundation for regional and national policymakers to promote sustainable agricultural strategies, ultimately advancing SDG 2.4.1 and supporting broader sustainable development goals in the HKH region.

6.2.2 Remote Sensing Assessment and Data Processing for Sustainable Development of Water Resources

For monitoring SDG 6.6.1, which focuses on changes in surface water area, this research utilizes a surface water distribution dataset specifically developed for the HKH region using Landsat time series data (Sect. 3.2.2). This dataset boasts a 30 m high spatial resolution covering three significant years: 2000, 2010, and 2020. The study meticulously calculates the total water coverage area for each country within the HKH region for these years to determine the rate of water area change over the two decades. This comprehensive analysis is crucial for evaluating the SDG 6.6.1, which aims to track changes in surface water to ensure the sustainable management and protection of water-related ecosystems.

Similarly, for monitoring SDG 6.3.2, which pertains to the proportion of water bodies with good surface water quality, the research employs the HKH surface water transparency dataset, also derived from Landsat data with the same spatial resolution and temporal coverage. This dataset, combined with lake and river data from OpenStreetMap, enables the calculation of water quality for individual lakes and rivers across the HKH region. By assessing the transparency of these water bodies, the study determines the proportion that meets the criteria for good water quality. This evaluation is essential for understanding the health of water bodies in the region through the level of transparency of the water and ensuring that a significant proportion maintains high-quality standards, which is vital for the well-being of both ecosystems and human populations dependent on these water resources.

6.2.3 Remote Sensing and Data Processing for Sustainable Development Assessment of Forests and Grasslands

To achieve SDG 15.4.2 which focuses on Mountain Green Cover Index, the study utilized a comprehensive process for extracting forest and grassland data in the HKH region using Landsat and DEM images as illustrated in Sects. 6.1.1 and 6.2.1. Different Landsat datasets from 2000, 2010, to 2020 were employed alongside elevation data from NASA DEM to enhance the accuracy of vegetation information extraction. The data preprocessing involved using Google Earth Engine's "pixel_qa" band to identify and mark pixels containing clouds, followed by calculating the normalized difference vegetation index (NDVI) for non-cloud pixels. The study constructed a mountain forest and grass vegetation coverage sample system, a vegetation radiative transfer model (PROSAIL), and a deep spatiotemporal transfer learning inversion model to produce and analyze vegetation coverage sample data.

The forest and grass vegetation coverage inversion model developed in this study integrates convolutional neural network (CNN) and long short-term memory network (LSTM) architectures. The CNN extracts spatial features from the sample library, while the LSTM captures temporal patterns, enabling simultaneous data transfer across various spatiotemporal scales. This combined approach facilitates dynamic and seasonal analysis of vegetation index features, accounting for changes in vegetation growth, while constant terrain features are utilized for spatial analysis. By leveraging both temporal and spatial data, the model offers a detailed understanding of vegetation coverage changes in the mountainous regions of the HKH, supporting the assessment and monitoring of Mountain Green Cover in alignment with SDG 15.4.2, which measures the Mountain Green Cover Index and the proportion of degraded mountain areas.

6.3 Software System of Remote Sensing Assessment for Sustainable Development

Assessing sustainable development through remote sensing entails intricate and large-scale processes including remote sensing data acquisition, extraction of key elements, statistical analysis, and map visualization. To effectively conduct such assessments, a software system is essential, characterized by its ability to handle large-scale data processing, automation, and user-friendliness. ArcGIS, ENVI, and QGIS are commonly used desktop GIS and remote sensing software for sustainable development assessment. In contrast, cloud-based software systems leverage the resources and computing power of cloud platforms, allowing rapid and efficient processing of massive data volumes. EarthDataMiner, which is developed by the Chinese Academy of Sciences, suitable for online calculation of SDG indicators, and SDG big data platform, which focuses Zero Hunger, Clean Water and Sanitation, Sustainable Cities and Communities, Climate Action, provide 'one-stop' services for data computation, analysis, visualization, and sharing. With reference to the experience of using these platforms, this study integrates remote sensing inversion methods for key elements of sustainable development, geospatial data processing methods, thematic analyses and calculation methods for indicators of sustainable development in a cloud computing environment to realize sharing. Based on the cloud computing, it is often possible to utilize its elastic allocation of resources and high-performance computing capability to process massive data quickly and efficiently. Additionally, computing and storage resources can be dynamically adjusted for large-scale computations based on actual demand. Leveraging the flexible computing power of cloud platforms and big data processing technologies like distributed batch processing and real-time processing, along with hierarchical tile map display technology, we've developed a software system of remote sensing assessment for HKH sustainable development. This system, serving as a cloud service, is the first of its kind supporting regional sustainable development progress assessment in the HKH region. It offers a comprehensive cloud computing service platform for data product acquisition, thematic analysis, and model calculation for studying sustainable development conditions in the HKH region. By utilizing powerful cloud computing facilities for automated data acquisition, integration, processing, analysis, and model calculation, significant improvements in data production efficiency are achieved, leading to savings in labor costs and yielding economic benefits.

6.3.1 System Overview

The software system of remote sensing assessment for HKH sustainable development is comprised of four distinct modules: data sharing, online analysis, sustainable development assessment, and online AI model calculation. The data sharing module primarily facilitates online access to, query, visualization, and download of key element inversion datasets of the HKH region derived from satellite remote sensing. The online analysis module conducts further thematic analysis on these key element inversion datasets, providing insights into various applications within the HKH region. Utilizing statistical charts and maps, this module offers visual representations and analyses of the data, aiding in understanding and decision-making processes. The sustainable development assessment module computes evaluation indicators based on the inversion datasets of key elements of the HKH region, focusing on three main themes: agricultural conditions, water resources, and ecological environment. These indicators are used to assess the current state and trends of sustainable development within the region. The assessment results are then displayed online, both for the HKH region and for individual member countries within the region. The online AI model calculation module is primarily dedicated to performing online calculations using deep learning models for remote sensing inversion. This module provides fully automated remote sensing inversion services, allowing users to access a range of models for various purposes such as land cover classification, vegetation health monitoring, and environmental change detection. The functions of the remote sensing assessment cloud service software system for sustainable development in the HKH region are depicted in Fig. 6.2. This figure illustrates the interconnectedness and functionalities of each module within the system.

The software system of remote sensing assessment for HKH sustainable development merges advanced cloud-native computing technology with traditional

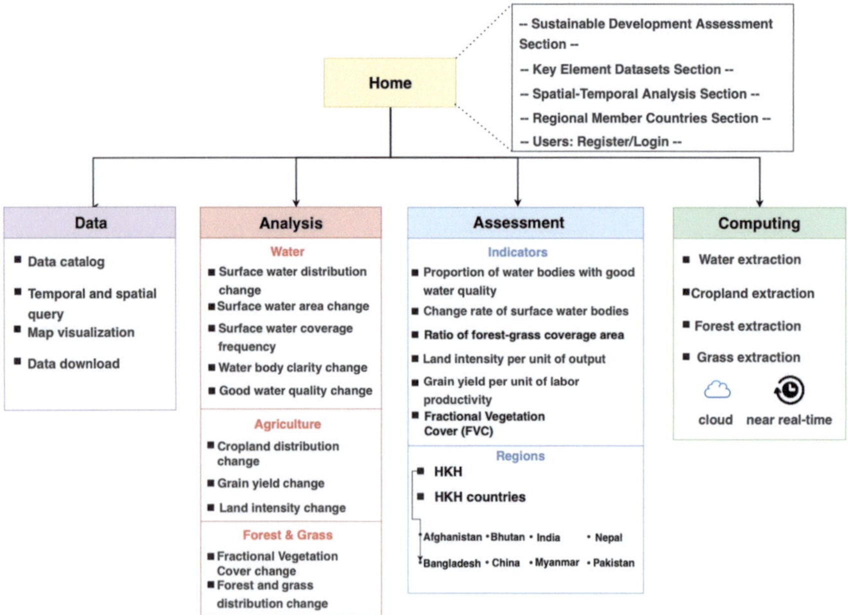

Fig. 6.2 Functional framework of the HKH regional sustainable development remote sensing assessment cloud service software functional framework

Geographic Information System (GIS) technology, representing a cutting-edge evolution in GIS technical systems. As depicted in Fig. 6.3, the technical framework of this software system includes several integral layers such as the infrastructure layer, computing framework layer, algorithm model layer, cloud service layer, and the application layer. This multi-layered framework underscores the system's capacity to harness the power of cloud-native computing technologies to enhance environmental monitoring and decision-making processes.

The infrastructure layer of the software system of remote sensing assessment for HKH sustainable development is built on a robust public cloud platform, integrating essential computing, storage, and networking resources. This comprehensive layer features computing services that offer Web interfaces and system portals for user interaction, along with container services that facilitate the deployment of inversion model algorithms and sustainable development status assessment algorithms. Additionally, it incorporates storage services to manage and store remote sensing images and data products, database services to handle metadata and software business logic functions, and network services to ensure efficient communication, traffic monitoring, and load balancing. These components work together to create a scalable

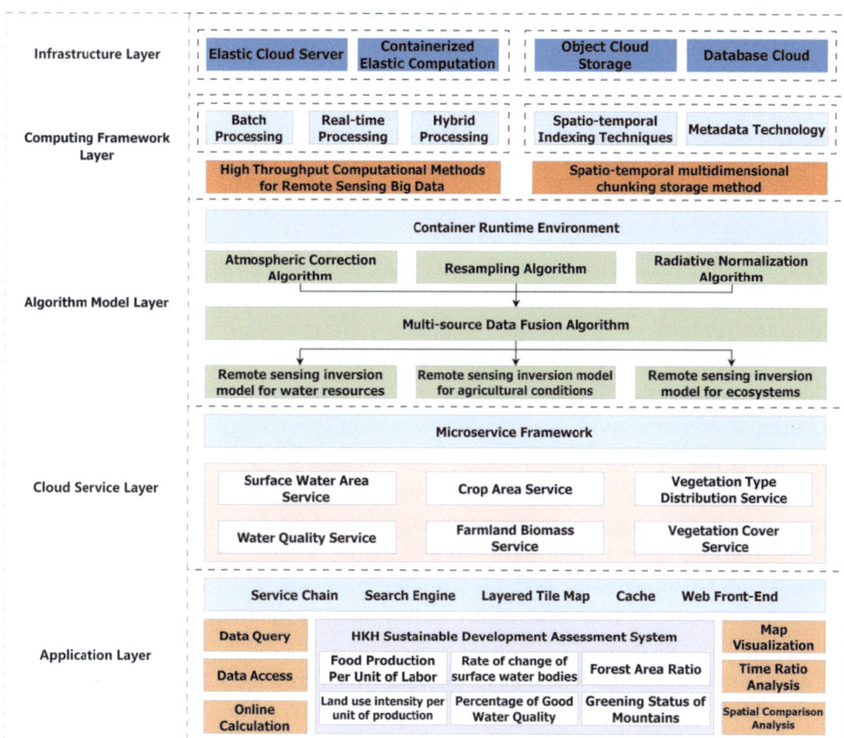

Fig. 6.3 Technical framework diagram of HKH regional sustainable development assessment cloud service software

and reliable foundation for supporting the software system's diverse functionalities and enabling effective monitoring and assessment of sustainable development in the HKH region.

The computing framework layer integrates a dual parallel computing strategy, encompassing both single-node "process level" and multi-node "cluster level" modes. Leveraging the open-source distributed computing framework Ray, this layer is designed to handle extensive remote sensing data processing and model computations efficiently. Ray's capabilities include managing million-level and millisecond-level task concurrency through features such as global control storage, distributed scheduling, memory-distributed object storage, and nested parallel batch processing. To accommodate the high-throughput, high-reliability, and scalability requirements inherent to remote sensing big data, this layer supports elastic container orchestration services. These services enable the rapid production and dynamic updating of extensive, long-term remote sensing datasets, ensuring that the computing framework can adapt to evolving data processing demands.

The storage layer system employs a global control storage model that integrates both object storage and block storage services to cater to diverse data management needs. Object storage excels in offering scalable and secure solutions ideal for cloud-native workloads, archival needs, and high-performance computing, ensuring low-latency and interactivity required by elastic container computing in cloud-native applications. It supports various storage tiers and features automated lifecycle management, which enhances efficiency and provides real-time access to spatio-temporal multi-dimensional remote sensing big data. This approach ensures that large volumes of data can be stored, managed, and retrieved swiftly and securely, facilitating advanced data analytics and processing in complex computing environments.

The algorithm model layer is designed to encapsulate remote sensing inversion models pertaining to agricultural conditions, water resources, and ecological parameters. By leveraging container technology, these algorithms are packaged into reusable and deployable units known as algorithm model containers. This containerization ensures that the algorithms can be easily reused and deployed across various computing environments. Container orchestration technology is then employed to facilitate the automatic deployment of these models on cloud platforms, allowing for scalable and elastic computing capabilities. This setup includes a comprehensive suite of algorithms such as data preprocessing algorithms for initial data preparation, multi-source data fusion algorithms for integrating diverse data sets, and inversion models for deriving valuable insights from remote sensing data. This robust framework enhances the efficiency, flexibility, and scalability of environmental monitoring and management processes.

The cloud service layer leverages a microservice framework to facilitate the independent development and deployment of model network applications, enhancing their modularity and flexibility. By using microservice-based inversion models, which communicate through standard protocols such as HTTP/HTTPS, the framework ensures seamless interaction between different services. Key components like elastic containers, API gateways, and message buses streamline the deployment and management processes, while also providing robust security, scalability, and agility for cloud-based microservice applications. This architecture supports a range of cloud services, including the analysis of surface water area, water transparency, crop area, farmland biomass, vegetation type, and vegetation coverage, thereby enabling efficient and scalable environmental monitoring and data analysis.

Lastly, the application layer integrates crucial aspects of agricultural conditions, water resources, and the ecological environment into modular microservices, facilitating a comprehensive sustainable development assessment for the HKH region. This is achieved through a Web portal application developed using the Browse/Client framework, which offers a user-friendly interface. Key modules include metrics such as the rate of change in water bodies, the ratio of good water quality, forest area ratio, and food output per unit of labor force. Additionally, the application supports spatiotemporal queries, change analysis, and map visualization, enabling detailed examination and monitoring of regional sustainability indicators. This sophisticated toolset aids stakeholders in making informed decisions to promote sustainable development in the HKH region.

6.3.2 Key Technical Methods

The fundamental technical methods utilized in the software system of remote sensing assessment for HKH sustainable development can be delineated as follows: geospatial data retrieval technology, geospatial data visualization technology, geospatial data online analysis technology, and AI-driven online inversion technology integrated with cloud computing.

(1) **Geospatial data retrieval technology**

The organization of remote sensing inversion data for sustainable development follows a hierarchical structure termed "category-dataset-data," which is exemplified by the open Spatiotemporal Asset Catalogs (STAC) specification, now an industry standard. Renowned entities such as USGS, NASA, Microsoft, AWS, ESRI, and Planet adopt the STAC standard for distributing spatiotemporal geographic data. STAC offers a standardized approach for describing and querying geospatial data, facilitating retrieval based on temporal, spatial, and other attributes. At its core, STAC comprises a network of interconnected JSON files, each conforming to strict specifications tailored to various STAC component levels. The service architecture, according to STAC specifications, is divided into three primary levels: Catalogs, Collections, and Items (illustrated in Fig. 6.4). This tripartite structure ensures a systematic and logical framework for the storage and organization of geospatial data, enhancing accessibility and usability for diverse applications in sustainable development.

The directory hierarchy serves as the primary access point (Endpoint) for data, organizing and directing users to lower-level collections and projects. Catalogs, which do not directly contain geospatial data, act as containers that link to multiple

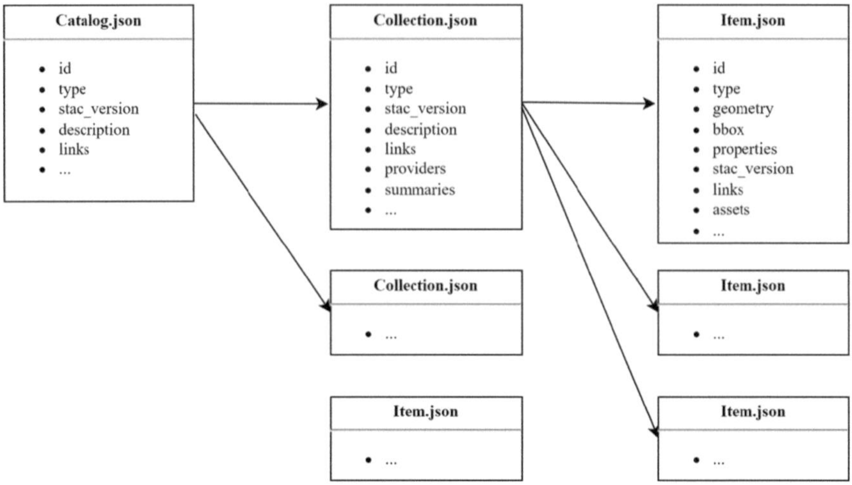

Fig. 6.4 Composition of STAC core components

collections or other directories, creating a browsable hierarchy. The collection hierarchy encompasses sets of geospatial data sharing common characteristics, each accompanied by metadata that details the dataset's spatial and temporal extent, data type, quality, and access information. These collections provide summary descriptions and essential information for interpreting and processing the geospatial data they contain. The fundamental unit of data in the STAC (SpatioTemporal Asset Catalog) architecture is the item (Item), representing a single geospatial data item. Each item (Item) includes comprehensive metadata such as geometric information, timestamps, and related links, facilitating specific data queries and retrievals. An item can describe multiple spatiotemporal assets (Asset). For example, a remote sensing image is an item, with each data band within the image considered an Asset. This structured approach enhances the organization, accessibility, and usability of geospatial data.

Furthermore, the scalability of the SpatioTemporal Asset Catalog (STAC) enhances the description and retrieval of diverse remote sensing information sources, thereby fostering significant advancements in data sharing and collaboration within the field of remote sensing big data. Upon constructing the STAC directory, the Item Search specification outlined in STAC is adhered to, enabling efficient and standardized querying. By leveraging the FastAPI framework, a robust network query interface for remote sensing data is provided, which allows users to easily obtain 'Collections' and 'Items' that match specific search criteria. This streamlined process not only simplifies data access but also enhances interoperability and collaboration among researchers and organizations, ultimately advancing the field of remote sensing through improved data utilization and sharing.

(2) **Geospatial data visualization technology**

Geospatial data visualization relies primarily on pyramid tile map technology, utilizing a hierarchical model known as the tile pyramid model. This model arranges map tiles in a multi-resolution structure, organizing them from the bottom to the top. As the pyramid ascends, the resolution of each tile decreases while maintaining consistent geographic coverage. Each map tile is uniquely identified by a level and row number identifier within this organizational structure. During map navigation, such as panning and zooming, the browser follows pyramid rules to calculate and retrieve necessary tiles from the tile server, seamlessly assembling them for display. This approach ensures smooth rendering of geospatial data across various scales and sizes, providing different levels of detail to users navigating the map interface. Traditional methods of pyramid tile map generation involve pre-generating all map tile files, resulting in static tile maps. Modifications to data require regeneration of static tile files, which limits flexibility and is time-consuming since all tiles must be generated simultaneously. However, recent advancements in cloud computing and the introduction of Cloud Optimized GeoTIFF (COG) technology have revolutionized tile map generation techniques. Dynamic tile map generation techniques based on cloud storage utilize the COG storage format and employ real-time dynamic generation of tiles, leveraging memory caching to expedite access to dynamically generated map tiles. This eliminates the need for static tile file generation, and COG image files can now be visualized online in real-time. The key components of this dynamic tile map technology include the generation of Cloud Optimized GeoTIFF (COG) files and the establishment of dynamic tile map services.

Generation of Cloud Optimized GeoTIFF (COG) files: Cloud Optimized GeoTIFF (COG) stands as a pivotal innovation in raster data formatting, strategically tailored for optimized storage in cloud environments and streamlined transmission across networks. Its architecture offers a range of technical advantages critical for modern data handling. Primarily, COG employs a tile-based structure, effectively partitioning raster data into independent tiles. This design allows for the retrieval of only pertinent data sections via HTTP requests, circumventing the need to download entire datasets. Particularly beneficial for extensive remote sensing datasets, this approach significantly enhances data access speed and efficiency. Moreover, COG integrates multiple layers of preview images, each representing a down-sampled iteration of the original image across various zoom levels. This ingenious feature facilitates swift retrieval of low-resolution previews, facilitating rapid data previewing even when detailed resolution is not immediately required. Consequently, users experience seamless transitions to full-resolution images as needed, thereby optimizing user experience and operational efficiency in data handling tasks. Moreover, COG (Cloud Optimized GeoTIFF) files boast a self-describing nature, encapsulating all essential metadata within the file structure itself. This inherent characteristic eradicates the necessity for separate metadata files, streamlining data management processes and ensuring self-sufficiency of the data. As a result, COG significantly diminishes network bandwidth requirements and minimizes superfluous data transmission, rendering it an optimal storage format particularly advantageous for dynamic tile map services where the retrieval of data on demand is paramount. This streamlined approach not only enhances efficiency but also contributes to a more sustainable and cost-effective data storage and transmission ecosystem.

Establishment of dynamic tile map services: The development of dynamic tile map services relies heavily on the open-source Titiler development package, serving as a versatile toolkit for generating such maps using Python programming language and the FastAPI framework. Titiler expedites the reading and distribution of raster data tiles by leveraging formats like COG (Cloud-Optimized GeoTIFF), STAC (SpatioTemporal Asset Catalog), and MosaicJSON, thus eliminating the preprocessing steps typically associated with static tile methods. With its lightweight API interface, Titiler enables seamless access to maps and interactive visualization through tile links, providing users with instant access to dynamic and up-to-date geospatial data without the need for complex data processing pipelines. In utilizing the Titiler development kit, the workflow initiates by obtaining the internal metadata of COGs (Cloud-Optimized GeoTIFFs). This metadata serves as the foundation for subsequent processes. By specifying the tile distribution range in the HTTP request and referencing the tile size and sequence format outlined in the metadata, the necessary raster data within the requested spatial range is retrieved. Following this retrieval, an RGB rendered image is systematically constructed in a tile format, ensuring efficient handling and display of large-scale geospatial datasets. Throughout this retrieval and construction process, users benefit from customization options tailored to their specific needs, including adjustments for tile size, resolution, and color table. Moreover, these customization options extend to support various projections and multiband combinations, ensuring versatility and adaptability across diverse geospatial applications and scenarios. Dynamic tile services that implement the MosaicJSON specification offer a powerful solution for visualizing large-scale remote sensing data by facilitating the mosaic of multiple Cloud-Optimized GeoTIFF (COG) formats. MosaicJSON leverages Web Mercator quadtree keys to index files, abstracting Web Mercator map tiles into a Z-X-Y format string, where Z represents the zoom level, and X and Y denote the row and column coordinates, respectively. This indexing mechanism enables the seamless linking of COGs to Web Mercator map tiles in XYZ format, facilitating frontend rendering. The relationship between zoom levels and COG files is established in JSON format, with each quadtree key corresponding to a list of minimum zoom level COG files. This approach ensures efficient and scalable visualization of remote sensing data, as illustrated in Fig. 6.5.

(3) **Cloud-native geospatial analysis and model computing technology**

Cloud-native methodology represents a specialized approach tailored specifically for constructing and deploying applications on cloud platforms, with the aim of fully leveraging the capabilities of cloud computing. Cloud-native means that applications are originally designed to run in the cloud in the most optimal way. Cloud-native applications exhibit distinct characteristics that enable rapid and frequent development, release, and deployment cycles. By leveraging the features offered by cloud computing, such applications achieve decoupling from underlying hardware and operating systems, ensuring scalability, availability, portability, and other

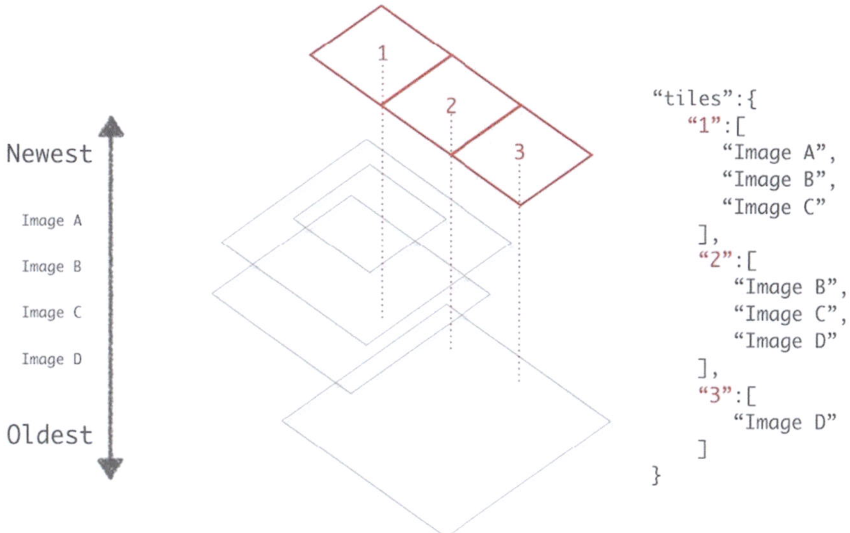

Fig. 6.5 MosaicJSON indexing mechanism

essential requirements are met. Ultimately, this approach leads to improved cost-effectiveness, as organizations can dynamically adjust resources based on demand, thereby optimizing resource utilization and minimizing operational costs.

According to the Cloud Native Computing Foundation (CNCF), cloud-native applications are defined by three key technical characteristics. Firstly, application containerization involves bundling an application and its dependencies into a container image, facilitating the dynamic separation of computing resources and environmental dependencies across different applications. Secondly, the microservices-oriented framework (Fig. 6.6), as opposed to traditional monolithic architecture, offers higher autonomy and better horizontal scalability. This architecture is advantageous for independent and scalable teams, as it allows for the development and deployment of individual services, enabling flexibility and agility in software development. Lastly, container orchestration and scheduling, supported by applications, ensure efficient management of containerized applications, including deployment, scaling, and lifecycle management, thus enhancing overall operational efficiency and reliability in cloud-native environments.

The cloud-native geospatial analysis and model calculation technology system, as depicted in Fig. 6.7, exemplifies a sophisticated approach to sustainable development assessment through remote sensing inversion. This system integrates various components, including cloud remote sensing data integration technology, multi-source remote sensing data fusion, and remote sensing inversion algorithms for agricultural

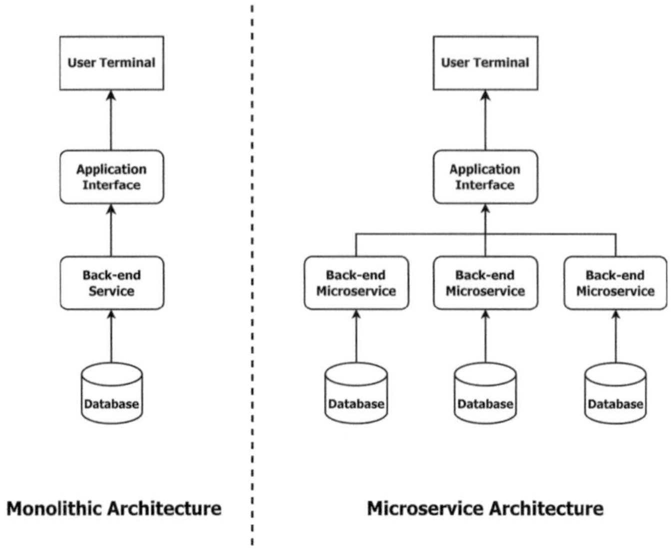

Fig. 6.6 Comparison between monolithic architecture and microservice architecture

conditions, water resources, and ecological parameters. Central to this system is the containerization of algorithm models, enabling the creation of reusable and rapidly deployable model container images. This containerization facilitates an industrial pipeline-style approach to remote sensing data acquisition, processing, calculation, and output, leveraging a workflow engine for seamless integration and automation of tasks.

The technical architecture of this system relies on cloud computing services, specifically within a public cloud environment. Elastic containers, API gateways, and message buses form the backbone of the infrastructure, providing the necessary resources for deploying and running the algorithm models. Utilizing a microservices architecture, the system enables the development of independent model online computing applications, allowing for horizontal scaling, self-healing, and load balancing of model containers. This architecture ensures the security, scalability, and agility of the cloud model microservices application system, catering to the dynamic demands of sustainable development assessment through geospatial analysis and remote sensing inversion.

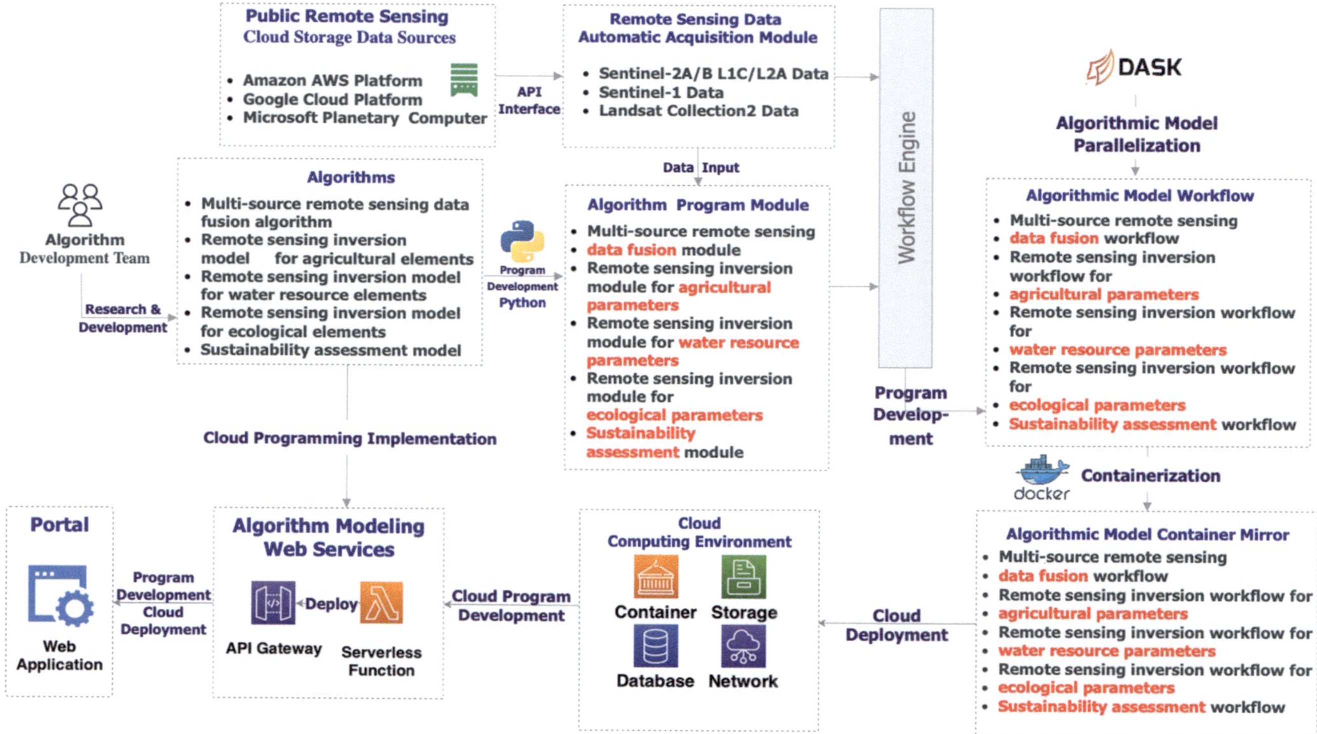

Fig. 6.7 Model calculation technology system using remote sensing inversion for sustainable development assessment as an example

6.3.3 *Online Inversion of Key Elements of Sustainable Development of the HKH: A Case Study on the Spatial and Temporal Distribution Inversion of Surface Water*

(1) **Design principle of inversion algorithm for key elements of sustainable development based on transformer**

The "Transformer" network, rooted in deep learning principles and inspired by the "Attention" mechanism, has emerged as a powerful framework, particularly in natural language processing. This mechanism mirrors human cognitive attention, enabling the network to prioritize relevant information within input data. By assigning higher weights to important parts and lower weights to less relevant ones, the Attention mechanism facilitates focused processing, thereby enhancing the network's efficiency. As a result, Transformer networks have gained prominence for their ability to handle complex tasks in various domains, including language processing and, more recently, computer vision.

In response to the expanding landscape of deep learning applications in computer vision, researchers have adapted Transformer networks for visual tasks, giving rise to visual Transformer networks. These networks integrate the Attention mechanism into image analysis, enabling them to concentrate on critical regions within images. Unlike traditional convolutional neural networks (CNNs), visual Transformer networks offer a more flexible and adaptive approach to visual processing, as they can dynamically attend to different parts of an image based on their relevance to the task at hand. This adaptability makes them well-suited for a wide range of visual tasks, from object detection to image classification.

One notable application of visual Transformer networks is in the design of networks for complex tasks such as the spatiotemporal distribution inversion of surface water. In such tasks, where capturing intricate details and nonlinear features is crucial, the Swin Transformer network stands out as a foundational architecture. Leveraging a pyramid structure, the Swin Transformer can extract and process features at multiple levels, ranging from overall scene context to fine-grained textures within images. Furthermore, its incorporation of window-based and moving window attention mechanisms contributes to improved efficiency and linear complexity, making it a versatile choice for handling diverse visual tasks. Thus, the Swin Transformer serves as an effective foundation for designing networks tailored to challenging visual analysis tasks like surface water distribution inversion.

(2) **Remote Sensing Inversion Model of Key Elements of Sustainable Development Based on Transformer**

The remote sensing inversion model of key elements of sustainable development based on Transformer is an encoder-decoder structure, as shown in Fig. 6.8.

The encoder section of our model is structured around the Swin Transformer network, organized into four main stages, each anchored by the Swin Transformer module. This module incorporates two distinct self-attention mechanisms:

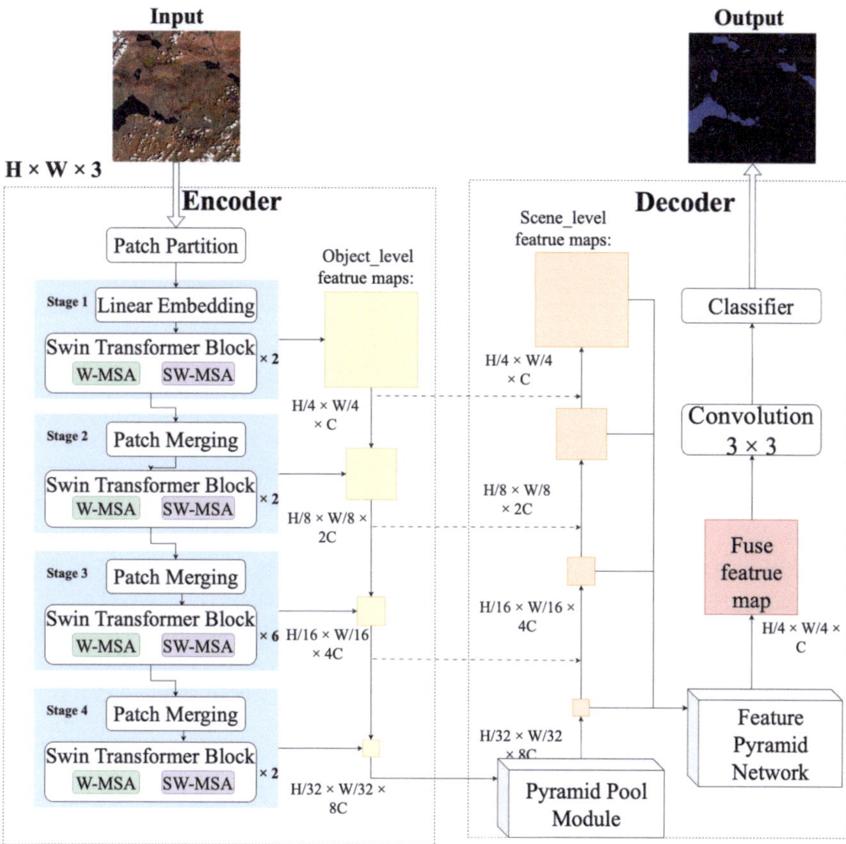

Fig. 6.8 Remote sensing inversion model structure of key elements of sustainable development based on transformer

The Window Multihead Self-Attention (W-MSA) module and the Shifted-Window Multihead Self-Attention (SW-MSA) module. The W-MSA facilitates self-attention computation within non-overlapping windows, while the SW-MSA extends this capability by introducing cross-window connections, thereby enhancing efficiency, particularly for large remote sensing images. This approach enables swift identification of surface water features in such images. Prior to entering the Swin Transformer module, the original image undergoes processing through a Patch Partition module, which reshapes it into 2-dimensional patches. These patches are first converted into patch tokens utilizing the Linear Embedding module before being inputted into the Swin Transformer module. Additionally, a Patch Merging module is interposed between Swin Transformer modules to generate hierarchical feature maps by amalgamating the patches generated by the Swin Transformer module.

In the process of generating object-level surface water feature maps, the encoder undergoes four distinct stages, each contributing to the refinement of these maps at varying scales. These maps are produced at scales of 1/16, 1/32, 1/64, and 1/128 relative to the original input image size. Subsequently, in the decoder section, the surface water spatiotemporal distribution inversion network integrates the output of the encoder through a combination of the pyramid pooling module (PPM) and the feature pyramid network (FPN). The PPM facilitates the generation of scene-level feature maps, aligning with the encoder's outputs, while the FPN amalgamates the four-scale feature maps from the encoder with those generated by the PPM, thereby forming a multi-scale integrated feature map. Towards the conclusion of the decoder process, a segmentation head is introduced, comprising a 3×3 convolutional layer and a pixel-level classifier. This component serves to up-sample the multi-scale integrated feature map to the original image size and classifies pixels into surface water and non-surface water categories, thereby producing the final output image of the model.

Regarding the specific parameter configurations, critical settings are established to optimize the functionality of the model. The image patch size is standardized at 2, with an embedding dimension of 96 designated for the three-band channel input (SWIR, NIR, and Red). The shift window size is set to 9, enhancing the model's ability to capture spatial relationships effectively. Across the four encoder stages, the number of Swin Transformer modules varies, specified as (Farrukh et al., 2020; Shen et al., 2020), while the number of Multihead Self-Attention heads ranges from [3, 6, 12, 24]. In the decoder section, the PPM is configured with four distinct sizes, corresponding to scales of 1, 2, 3, and 6. These scales denote average patch sizes calculated by the PPM, facilitating nuanced analysis of surface water features at various spatial resolutions. Through meticulous parameter tuning and the integration of sophisticated modules, the model exhibits enhanced proficiency in delineating surface water distribution across diverse landscapes with improved accuracy and granularity.

(3) **Architecture and implementation of the online remote sensing inversion system based on AI model and cloud computing**

Online remote sensing inversion, powered by AI models and cloud computing, revolutionizes the processing of remote sensing data by providing a streamlined and automated workflow. This sophisticated system encompasses a range of tasks, including data querying and acquisition, image preprocessing, model prediction, time-series data synthesis, spatial integration, and result visualization on tile maps. By harnessing the capabilities of cloud computing and big data technologies, this system efficiently performs large-scale remote sensing inversion and monitoring tasks. Illustrated in Fig. 6.9, the architecture of this system comprises six backend modules and a frontend web interface, facilitating seamless operation from image query to tile mapping. This innovative approach not only enhances the speed and accuracy of remote sensing data processing but also enables comprehensive analysis and visualization, contributing to more effective monitoring and management of environmental changes.

The image processing system encompasses several modules designed to efficiently retrieve, preprocess, and analyze remote sensing data. The image query module allows users to specify parameters such as image type, spatial range, time range,

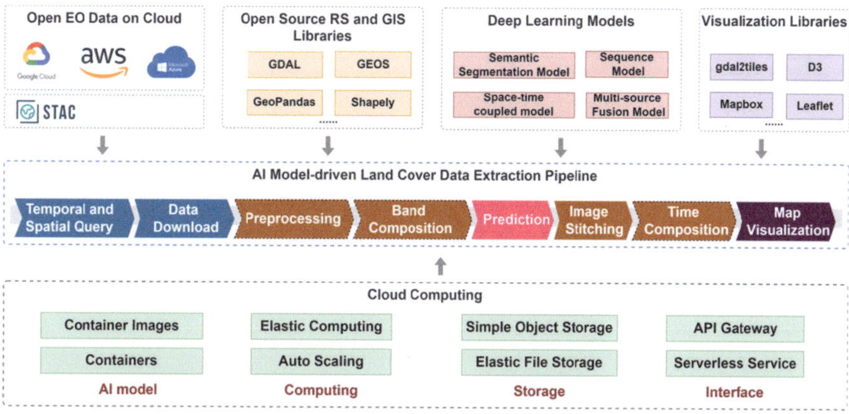

Fig. 6.9 Online remote sensing inversion system architecture based on AI model and cloud computing

and cloud cover to retrieve relevant remote sensing images. These images are then obtained through cloud storage channels like Microsoft Azure, Google Cloud, and Amazon AWS Data. The automatic image preprocessing module focuses on preprocessing Sentinel-1 GRD images, while the key element extraction module employs trained deep learning models to predict target areas within the images. Subsequently, the extracted results are synthesized and spliced temporally and spatially according to user preferences. The tile mapping module then generates tile maps from these synthesized results, enabling fast spatial visualization using a frontend map rendering engine. An illustrative example of online automatic extraction, demonstrating surface water spatiotemporal distribution inversion, is provided in Fig. 6.10, along with an interface diagram for browsing and downloading the extracted results, showcasing the system's capabilities in efficiently processing and visualizing remote sensing data.

6.3.4 *Data Integration, Analysis and Evaluation of Key Elements of the HKH's Sustainable Development*

(1) **HKH data integration and sharing of key elements of sustainable development**

The integration of remote sensing inversion data for key elements of the Hindu Kush Himalaya (HKH) sustainable development comprises three hierarchical levels: data classification, metadata lists, and data entities. The data categories encompass topics related to water resources, agricultural conditions, and the ecological environment. Each category is associated with a specific metadata list, and each metadata item corresponds to a distinct data entity, as illustrated in Fig. 6.11. The metadata items for

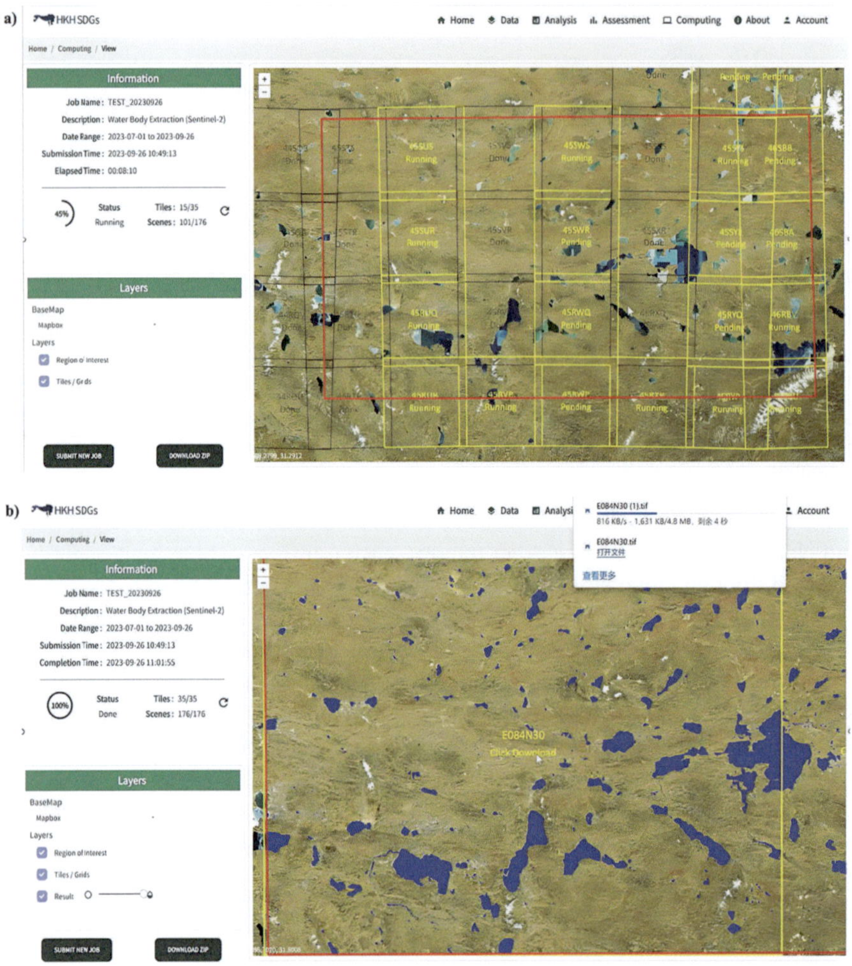

Fig. 6.10 Online remote sensing inversion system architecture based on AI model and cloud computing. **a** Online automatic extraction interface for surface water spatiotemporal distribution; **b** real-time visualization and downloading of surface water spatiotemporal distribution extraction results

the HKH sustainable development key elements dataset include the title, description, category, year, region, geographic extent, spatial resolution, author, and organization. The comprehensive list of data sets for key elements of HKH sustainable development is presented in Table 6.1.

The HKH's sustainable development data entities leverage object storage services within cloud computing environments. Object storage technology, integral to cloud-native applications, is designed for the storage and management of unstructured data in Key-Value formats. The core functionality of object storage involves storing data as Key-Value pairs, where the Key serves as a unique identifier to access the

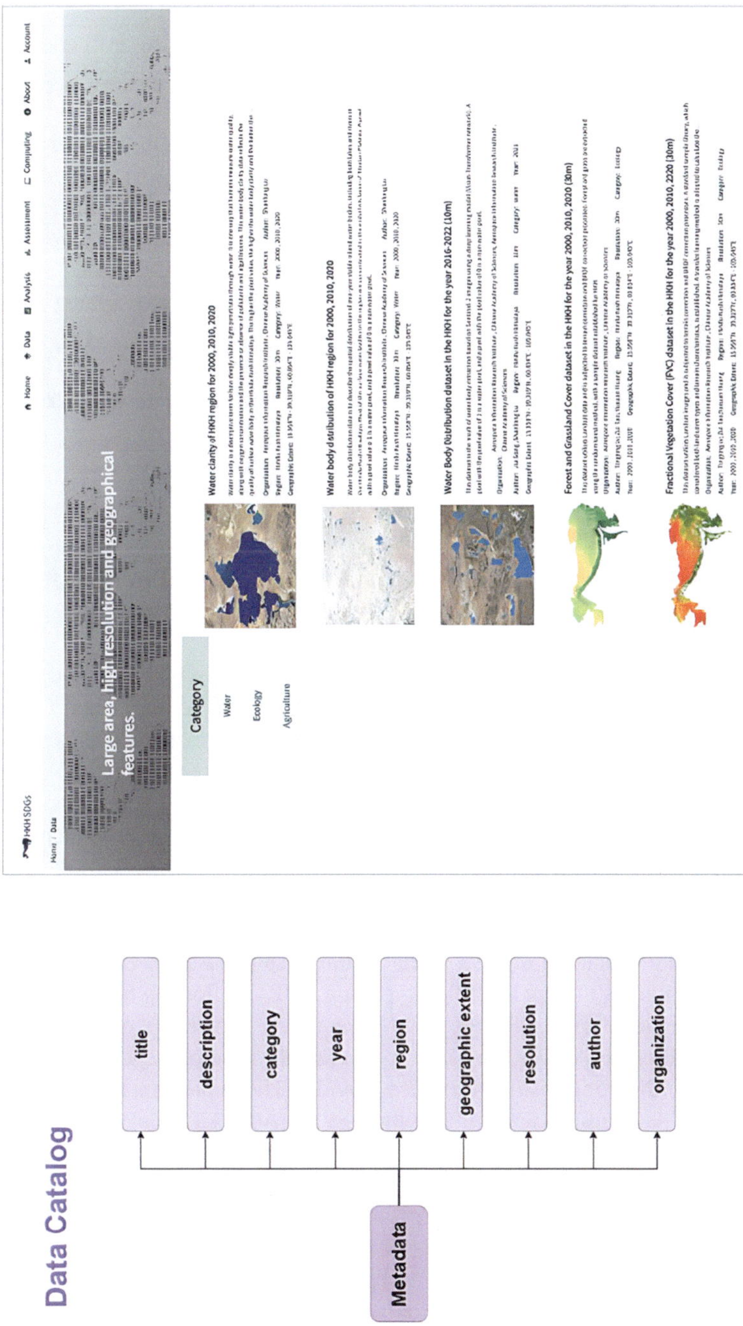

Fig. 6.11 HKH key elements of sustainable development data integration and sharing page

Table 6.1 List of HKH key elements of sustainable development data sets

Category	Dataset name
Water	Water body distribution dataset in the HKH for the year 2000, 2010, 2020 (30 m)
	Water clarity dataset in the HKH for the year 2000, 2010, 2020 (30 m)
	Water body distribution dataset in the HKH for the year 2016–2022 (10 m)
Agriculture	Main crop distribution dataset in the HKH for the year 2000, 2010, 2020 (30 m)
	Labor output in the HKH main agricultural area for the year 2000, 2010, 2020 (30 m)
	Land use intensity in the HKH main agricultural area for the year 2000, 2010, 2020 (30 m)
	Agricultural biomass dataset in the HKH for the year 2000, 2010, 2020 (30 m)
Forest and grass	Fractional Vegetation Cover (FVC) dataset in the HKH for the year 2000, 2010, 2020 (30 m)
	Forest and grassland cover dataset in the HKH for the year 2000, 2010, 2020 (30 m)

associated Value. This creates a flat storage architecture, distinct from traditional hierarchical systems, facilitating data storage in buckets. Users can then retrieve and analyze objects of any type.

Key benefits of object storage include its near-infinite scalability and cost-effectiveness, essential for handling vast amounts of remote sensing data. Object storage operates on a distributed storage system, dispersing large data volumes across multiple physical devices while providing efficient access through a unified virtual repository. This distribution across various geographic locations enhances storage capacity and disaster recovery capabilities, ensuring global accessibility, increased data resilience, and availability.

Due to its flexibility, scalability, and efficiency, object storage is the preferred solution for large-scale geospatial data storage. It eschews traditional directory tree structures, managing data as objects within a flat system instead. Each object is accessible via a globally unique identifier (typically a URL), allowing for storage and retrieval from any location worldwide without requiring knowledge of its physical storage site or directory path. The HKH's sustainable development data entities utilize AWS S3 as their object storage service provider. The implementation of object storage for these data entities is facilitated through the AWS CLI tool, as illustrated in Fig. 6.12.

The data sharing system for key elements of the HKH's sustainable development provides extensive capabilities, including spatial and temporal data queries, map browsing, and data downloading. Through a user-friendly filtering interface based on the SpatioTemporal Asset Catalog (STAC) protocol, users can precisely select the required spatiotemporal range for specific datasets (as illustrated in Fig. 6.13). Spatial filtering methods include either drawing a selection on the online map or choosing

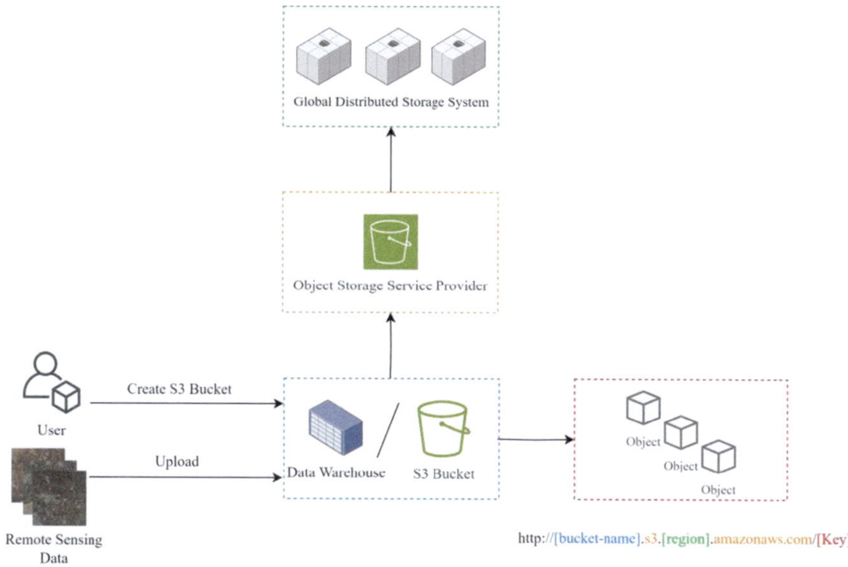

Fig. 6.12 Schematic diagram of object storage for data entities of key elements in the HKH's sustainable development

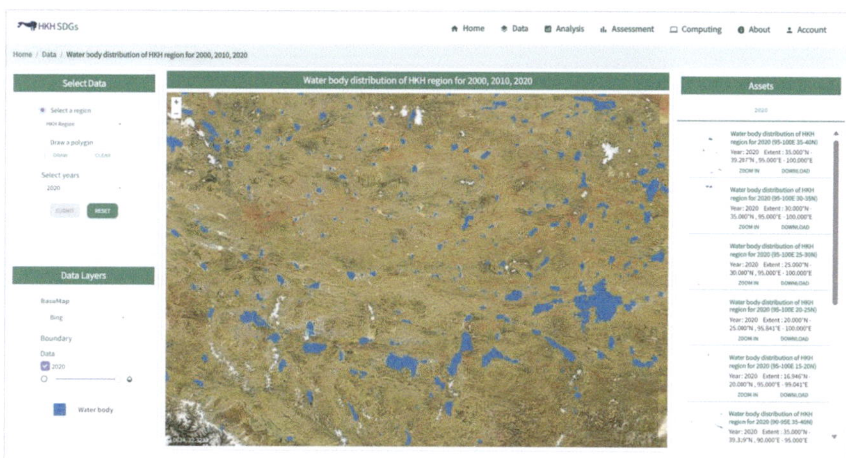

Fig. 6.13 Spatio-temporal screening interface for key elements of sustainable development in the HKH

from a predefined list of typical areas. Once the spatial and temporal parameters are specified, users can visualize the filtered data in real time using the dynamic tile service, enhancing their ability to analyze key elements dynamically (as shown in Fig. 6.14). Additionally, all key element data, or data filtered by specific criteria,

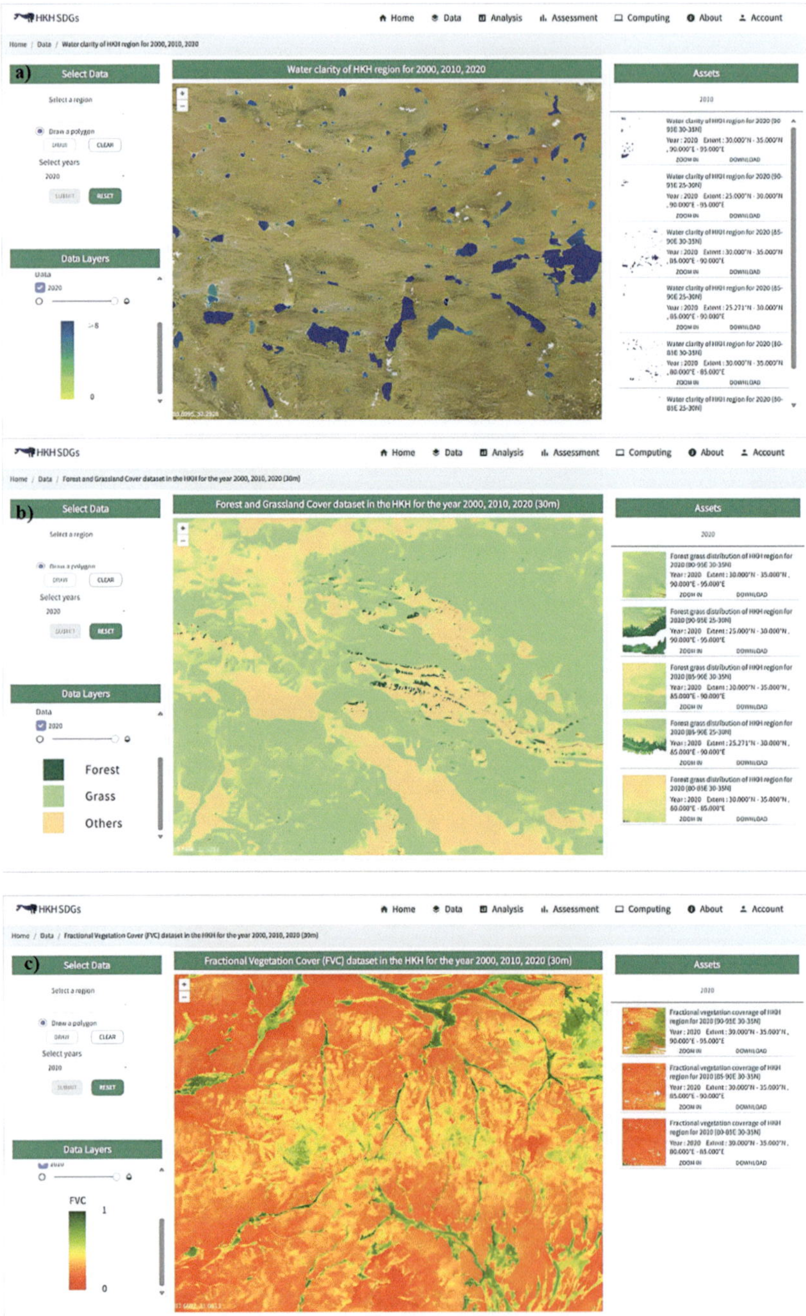

Fig. 6.14 Visual browsing and online download of HKH key elements of sustainable development: water transparency, forest and grass coverage, and vegetation coverage data map. **a** Water clarity data in the HKH; **b** forest and grassland cover data in the HKH; **c** fractional vegetation cover (FVC) data in the HKH

are accessible for download via HTTP from a Nginx server. The download service supports segmenting data based on a 5-degree longitude and latitude grid, ensuring efficient and manageable data retrieval. This robust system facilitates detailed and accurate analysis of the HKH region's sustainable development elements, supporting better-informed decision-making and research.

(2) **Online analysis of HKH key elements of sustainable development data**

All analyses are based on key element inversion products and structured according to specific evaluation indicators, which ensures a systematic and thorough approach to understanding the region's dynamics. These analyses are divided into three primary themes: water resources, agricultural conditions, and the ecological environment. Each theme encompasses a variety of specific analysis tasks tailored to address the unique aspects of the theme. The following sections will delve into the specific algorithms used for each thematic area, illustrating how they contribute to a comprehensive understanding of the region's environmental and agricultural status.

(1) Spatiotemporal analysis of surface water changes

The key element of the inversion product based on the spatiotemporal analysis of surface water changes is the water body distribution area data. This analysis is further divided into two parts: surface water spatiotemporal change map analysis and surface water area change statistical map analysis.

The surface water spatiotemporal change map analysis focuses on pixel-scale changes in surface water, encompassing three types of changes: from non-existence to existence, from existence to non-existence, and no change. The results are displayed as a tile map. The calculation method, illustrated in Fig. 6.15, determines changes in surface water between different years based on pixel values (0 represents non-surface water, and 1 represents surface water). If the pixel value changes from 0 in the previous year to 1 in the following year, it indicates an increase; if it changes from 1 in the previous year to 0 in the following year, it indicates a decrease; if the pixel values remain the same (either both 0 or both 1), it indicates no change.

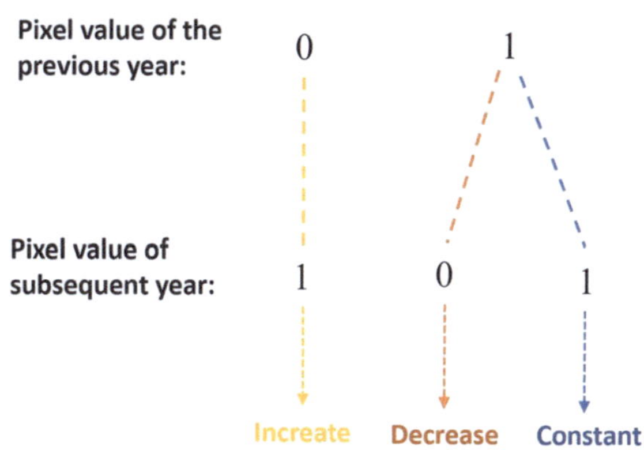

Fig. 6.15 Schematic diagram of surface water spatiotemporal change map analysis algorithm

User input parameters include the spatial area and time range. By default, the spatial area is set to the Hindu Kush-Himalaya (HKH) region. Users can select other sub-regions, upload, or manually draw the Area of Interest (AOI). The time range options are 2000, 2010, and 2020. Users can select any two or all three years. If all three years are selected, three independent map layers will display the changes between each pair of years.

The statistical map analysis focuses on changes in the surface water area and the rate of these changes. The analysis results are presented as a line chart. The input parameters for this analysis are the same as those for the surface water spatiotemporal change map analysis. Quantitative results from the statistical map analysis are output together with the results from the spatiotemporal change map analysis. Figure 6.16 illustrates the results of the surface water spatiotemporal change analysis.

(2) Spatio-temporal analysis of surface water coverage frequency

The key inversion product derived from the spatio-temporal analysis of surface water coverage frequency focuses on mapping and statistically analyzing the distribution area of water bodies. This analysis consists of two main components: the surface water coverage frequency map analysis and the surface water area statistical map analysis. The surface water coverage frequency map analysis details how often surface water appears at each pixel over a given period. As illustrated in Fig. 6.17, this calculation uses the cumulative count of pixels identified as water pixels to determine the pixel value. The resulting data is presented as a tile map, where for a one-year period, each pixel can have a value of either 0 or 1, indicating whether water was present. For a two-year period, pixel values can be 0, 1, or 2, and for a three-year period, they can range from 0 to 3, reflecting the frequency of water coverage at each location. This methodology provides a detailed and temporal understanding of water body distribution, crucial for water resource management and environmental monitoring.

The analysis allows users to define their input parameters, specifically the spatial area and the time range. By default, the spatial area is set to the Hindu Kush-Himalaya region, but users can select other sub-regions, upload their own area of interest (AOI), or manually draw the AOI on a map interface. The time range options include the years 2000, 2010, and 2020, enabling comparisons across one, two, or all three years. The primary focus of the analysis is on the area of stable surface water for the selected year and its change ratio over the different years, with results displayed in a bar chart format within a statistical chart. These input parameters align with those used in the spatio-temporal analysis of surface water coverage frequency, ensuring consistency in data processing. Both the statistical and spatio-temporal analysis results are provided as quantitative data, and the spatio-temporal analysis outcomes of surface water

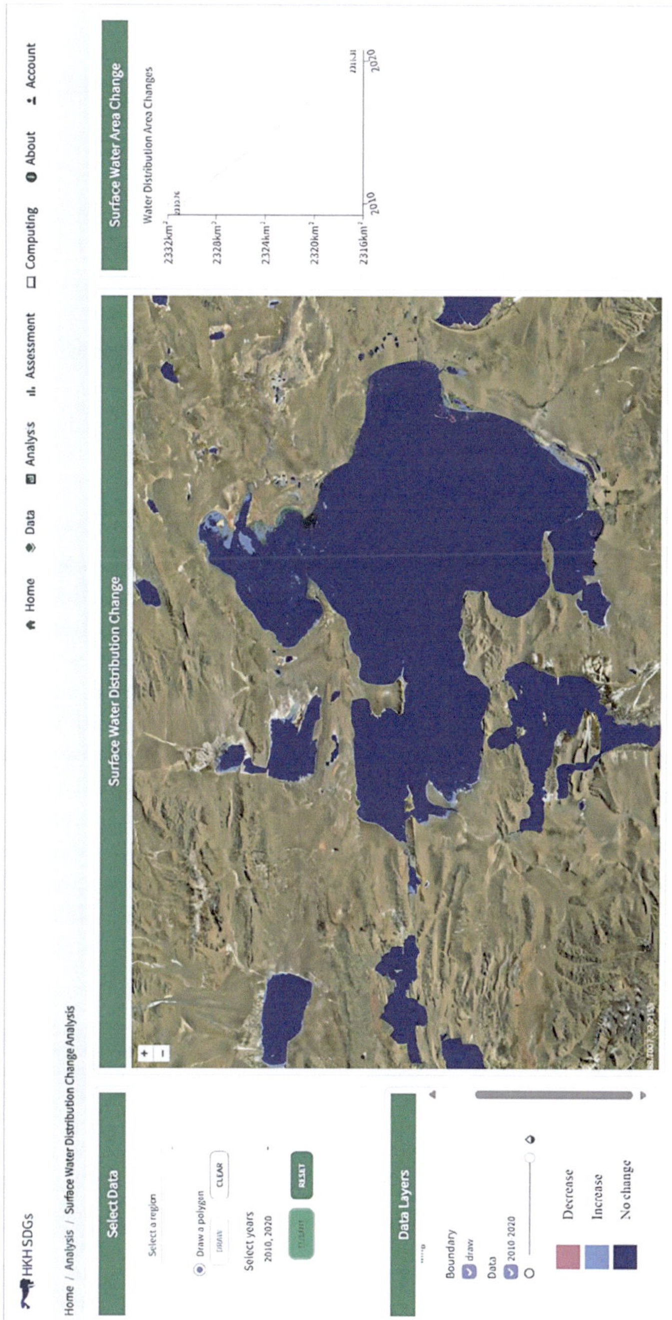

Fig. 6.16 Schematic diagram of analysis results of surface water spatiotemporal changes

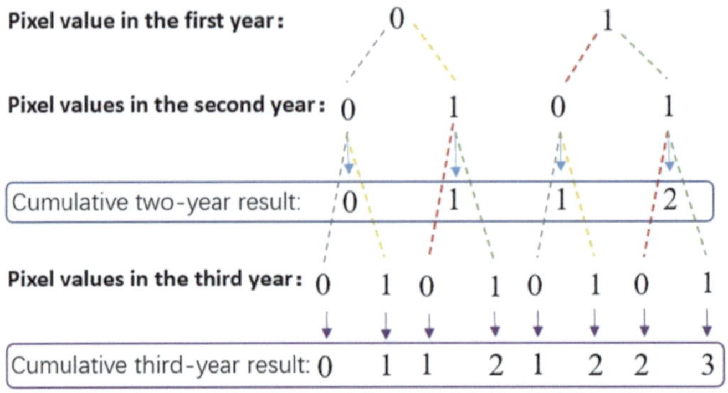

Fig. 6.17 Schematic diagram of surface water coverage frequency map analysis algorithm

coverage frequency are illustrated in Fig. 6.18, offering a comprehensive view of water coverage dynamics over time.

(3) Spatiotemporal analysis of water transparency changes

The key inversion product derived from the spatiotemporal analysis of water body transparency changes is the water body transparency data. This detailed analysis is divided into two main components: a map analysis of water body transparency changes and a statistical map analysis of the areas impacted by these changes. The map analysis operates at the pixel scale, identifying alterations in water transparency as either increased, decreased, or unchanged, thereby indicating shifts in water quality. The results are visualized using a tile map. The methodology involves filtering out pixel groups that have non-zero values in both compared years. The changes in water transparency are then determined for these pixels by comparing their values over the years. Pixels are classified based on whether their transparency has increased, decreased, or remained unchanged. The extent of change is quantified by the difference in pixel values, following the formula outlined in Eq. 6.9. This approach provides a granular and comprehensive view of how water body transparency evolves over time, aiding in the assessment and management of water quality.

$$\begin{aligned}\text{Water Transparency}_{\text{change}} = & \ \text{Water Transparency}_{\text{subsequent year}} \\ & - \text{Water Transparency}_{\text{previous year}}\end{aligned} \tag{6.9}$$

The input parameters for the analysis include spatial area and time range. By default, the spatial area is set to the Hindu Kush-Himalaya (HKH) region. However, users have the flexibility to select other sub-regions within the HKH, upload a specific area of interest (AOI), or manually draw the desired area on the map. The time range parameter allows users to choose any two years among 2000, 2010, and 2020, enabling a comparative analysis of changes over these years. Additionally, users can select all three years simultaneously, using three independent map layers to visualize the changes across the selected time periods.

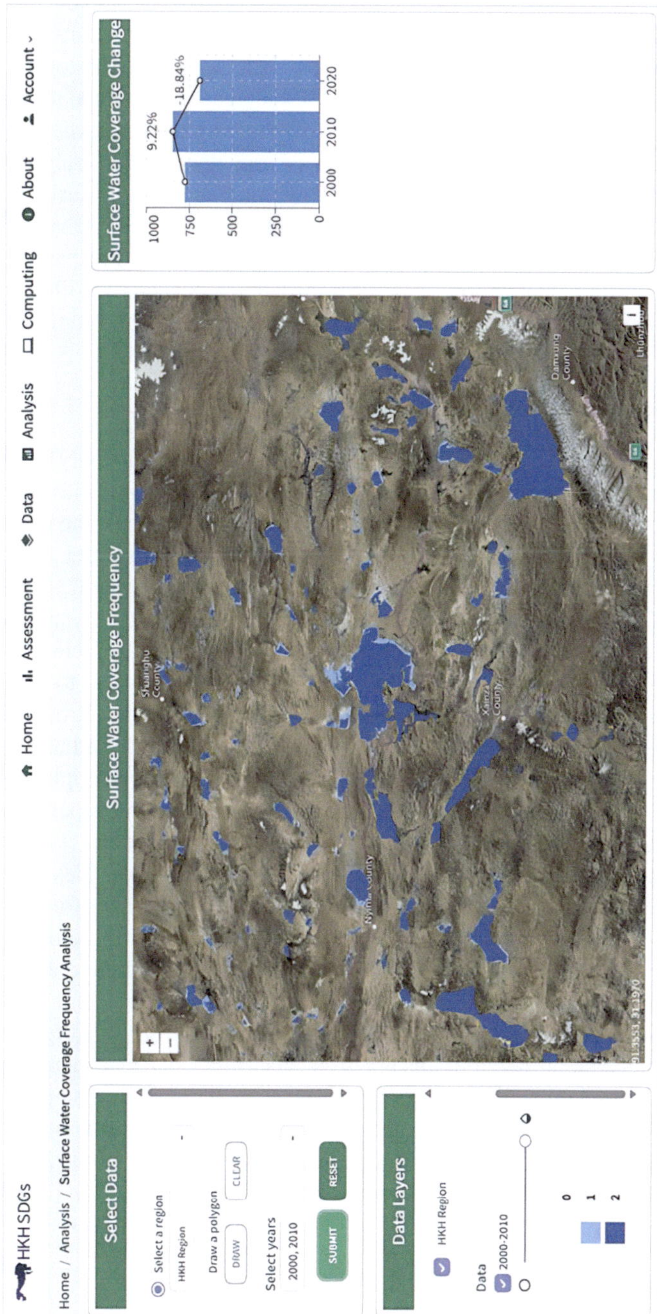

Fig. 6.18 Schematic diagram of spatiotemporal analysis results of surface water coverage frequency

The analysis of water body transparency changes includes detailed statistical charts that illustrate the area and proportion of water body transparency variations for the selected years. These results are presented in the form of histograms and pie charts, providing a clear and quantitative depiction of changes. In the histogram, each column can represent areas where transparency has increased, remained unchanged, or decreased. The pie chart complements this by showing the proportion of areas experiencing different types of changes. The input parameters for this analysis mirror those used for the water body transparency change map analysis, ensuring consistency in data handling. The results, displayed alongside transparency change maps, offer a comprehensive view of spatiotemporal changes in water transparency. The detailed analysis and visualization of water transparency changes are illustrated in Fig. 6.19, showcasing the utility of this approach in environmental monitoring and assessment.

(4) Spatial and temporal analysis of good water quality

The inversion product's key element, which is the spatio-temporal analysis of good water quality, hinges on water body transparency data. This analysis is bifurcated into two primary components: a distribution map and a statistical map of the proportion of good water quality. The distribution map showcases surface water areas with transparency values greater than 0.5, indicating good water quality, and is displayed as a tile map. The method involves setting a threshold of 0.5 for good water quality, retaining pixels with values above or equal to this threshold, while excluding those with lower values. Users can customize the spatial area, typically set to the Hindu Kush-Himalaya region by default, or select other sub-regions, upload, or manually draw their Area of Interest (AOI). Additionally, the time range can be adjusted to any of the years 2000, 2010, or 2020, or a combination thereof. When multiple years are chosen, each year's good water quality distribution is presented in separate layers, allowing a comparative temporal analysis of water quality in different regions.

The statistical map of the proportion of good water quality provides a detailed visual representation of surface water quality within a selected area for a specific year. This map highlights areas where surface water maintains a transparency value greater than 0.5, indicating good water quality. The data is presented as a pie chart, showing the proportion of surface water pixels that meet this quality standard relative to all surface water pixels in the area. The same input parameters used for creating the distribution map are utilized in this analysis, ensuring consistency and accuracy in the results. Quantitative results accompany the map to provide a comprehensive understanding of water quality distribution. Additionally, the spatio-temporal analysis results, depicted in Fig. 6.20, illustrate the changes and trends in water quality over time, offering valuable insights into environmental conditions and informing water management practices.

(5) Analysis of changes in cultivated land distribution

The inversion product's key element, based on analyzing farmland distribution changes, hinges on detailed crop area data. This analysis is bifurcated into two primary components: the map analysis of cultivated land distribution changes and the examination of cultivated land area and its rate of change. The map analysis

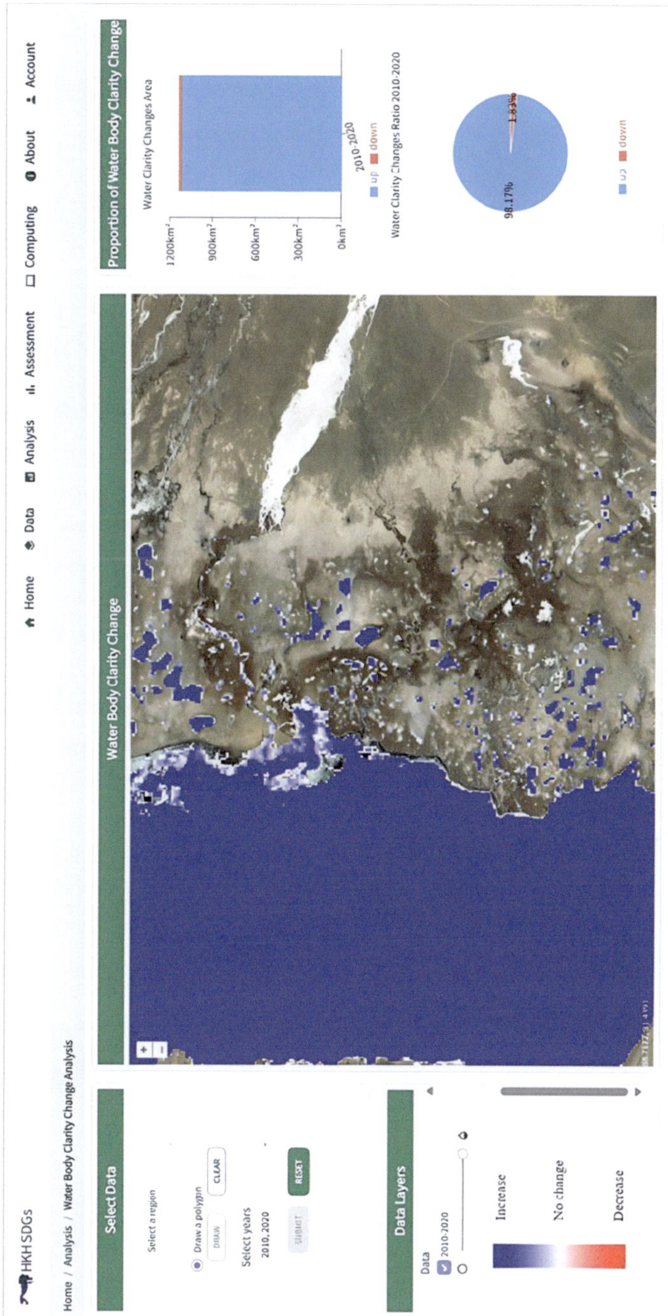

Fig. 6.19 Schematic diagram of spatiotemporal analysis results of water transparency changes

Fig. 6.20 Schematic diagram of spatio-temporal analysis results of good water quality

component provides a visual representation of how cultivated land distribution shifts between two selected years. It categorizes these changes into four scenarios: cultivated land transforming into non-cultivated land, non-cultivated land becoming cultivated land, land consistently remaining as cultivated land, and land consistently remaining as non-cultivated land. The results are depicted through a tile map, where the methodology, as demonstrated in Fig. 6.21, calculates the alterations in pixel values across the two years. A transition from a pixel value of 1–0 signifies that cultivated land has reverted to non-cultivated land, while a shift from 0 to 1 indicates the conversion of non-cultivated land into cultivated land. Consistent pixel values of 1 or 0 represent persistent cultivated land or non-cultivated land, respectively. Users can input parameters such as the spatial area and time range, defaulting to the Hindu Kush-Himalaya region but allowing for the selection of other sub-regions or manually drawn Areas of Interest (AOI). The temporal range can span any two years from 2000, 2010, to 2020, or include all three years, with changes displayed in three independent map layers.

The analysis of cultivated land area and its change rate complements the map analysis by providing statistical evaluations of cultivated land area within the selected region and its rate of change over the specified period. The results are illustrated using bar charts and line charts; the bar charts display the total area of cultivated land, while the line charts depict the rate of change. The input parameters for this analysis are consistent with those used for the cultivated land distribution change map analysis. The output includes both quantitative data and visual representations, facilitating a comprehensive understanding of how agricultural land use is evolving. This dual-faceted approach allows for an in-depth examination of farmland dynamics, providing valuable insights into the patterns and rates of change in cultivated land areas, which is crucial for agricultural planning and policymaking in the Hindu Kush-Himalaya region.

(6) Analysis of changes in grain output per unit labor force

The inversion product's core component, which focuses on analyzing changes in grain yield per unit of labor, relies heavily on crop area data. This comprehensive analysis is conducted in two distinct parts: a map-based examination of changes in grain output per unit of labor, and an evaluation of the rate at which this output changes over time.

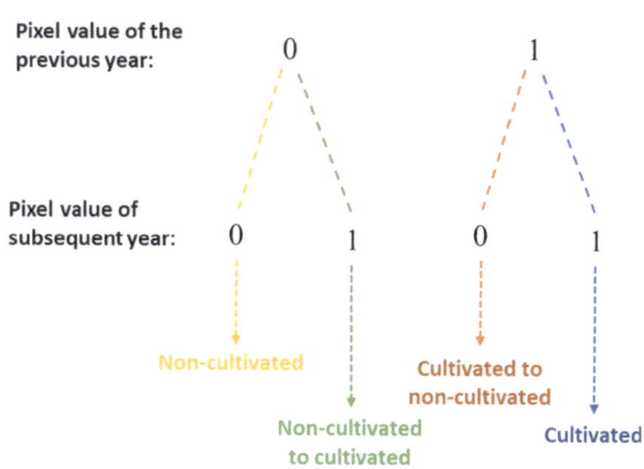

Fig. 6.21 Schematic diagram of map analysis algorithm for cultivated land distribution changes

The map analysis operates at the pixel level, capturing granular details about whether grain output per unit of labor has increased, decreased, or remained constant, along with the specific degree of change. These results are visually represented in a tile map format, facilitating easy interpretation. The methodology involves calculating the difference in pixel values, which represent grain output per unit of labor, between selected years (2000, 2010, and 2020). A positive difference indicates an increase in grain output, a negative difference signifies a decrease, and a zero value shows no change. The absolute value of this difference quantifies the extent of the change, providing a precise measure as illustrated by Eq. 6.10.

$$\begin{aligned}\text{Labor Output}_{\text{change}} &= \text{Labor Output}_{\text{subsequent year}} \\ &\quad - \text{Labor Output}_{\text{previous year}}\end{aligned} \tag{6.10}$$

The input parameters for this analysis include the spatial area and time range. By default, the spatial area is set to the Hindu Kush-Himalaya (HKH) region, but users can opt to select other sub-regions by either uploading a specific file or manually drawing their area of interest (AOI). For the time range, users have the flexibility to choose any two years from 2000, 2010, to 2020, or they can select all three years to create three distinct map layers that depict the changes between each pair of years. The primary focus of the analysis is on the change rate of average unit labor force grain output, which measures the variation in grain output per labor unit within the selected region over the specified years. The results are presented in a statistical line chart format, showing the change rates. The calculation involves determining the average grain output per labor unit for the chosen area and then computing the change rate between the selected years based on these averages. These input parameters are the same as those used for the map analysis, providing a comprehensive quantitative assessment alongside the visual map analysis of changes in grain output per labor unit.

(7) Analysis of changes in land intensity per unit output

The inversion product primarily comprises land intensity data per unit output, obtained through an analysis of changes in land intensity per unit output. This analysis consists of two main components: mapping changes in land intensity per unit output and assessing the rate of change in average land intensity per unit output. The mapping process on land intensity per unit output at the pixel level focuses on identifying three types of changes—increase, decrease, and no change along with quantifying the degree of change. Results are presented in a tile map format, with calculations involving the difference in pixel values (land intensity per unit output) between two years as illustrated in Eq. 6.11. A positive value indicates an increase, a negative value indicates a decrease, and zero signifies no change, with the absolute value representing the magnitude of the change.

$$\text{Land Use Intensity}_{\text{change}} = \text{Land Use Intensity}_{\text{subsequent year}} - \text{Land Use Intensity}_{\text{previous year}} \quad (6.11)$$

The user input parameters for this analysis tool encompass spatial area and time range specifications. By default, the spatial area focuses on the Hindu Kush-Himalaya region, although users have the flexibility to select alternative sub-regions or define their Area of Interest (AOI) by uploading or drawing it. Regarding the time range, users can opt for any two years from 2000, 2010, to 2020, or encompass all three years. Selecting three years triggers the generation of three distinct map layers illustrating the change between two consecutive years.

For the average change rate analysis of land use intensity per unit output, the tool computes the change rate of land use intensity per unit output within the chosen region across different years. The resulting analysis is depicted as a statistical line chart. The calculation methodology involves ascertaining the average value of land intensity per unit production in the specified area and subsequently computing the change rate based on these average values between different years. The input parameters mirror those utilized for the map analysis of land intensity change per unit output, with the results presented as quantitative data alongside the map analysis outcomes.

(8) Analysis of farmland biomass changes

The inversion product's central element, which revolves around analyzing shifts in farmland biomass, comprises farmland biomass data. This analysis entails two primary components: examining changes in farmland biomass maps and scrutinizing farmland biomass values and change rates. The assessment of alterations in farmland biomass maps zooms in on pixel-level variations, delineating increases, decreases, or stability, coupled with specific degrees of change. These findings are depicted in a tiled map format. The calculation process involves determining the disparity in pixel values (representing field biomass) between two time points, as indicated by Eq. 6.12. A positive value denotes augmentation, a negative value signifies reduction, and a value of 0 indicates stagnation. The absolute value conveys the magnitude of change, offering insights into the dynamics of farmland biomass over time.

$$\text{Above-ground biomass}_{\text{change}} = \text{Above-ground biomass}_{\text{subsequent year}} - \text{Above-ground biomass}_{\text{previous year}} \quad (6.12)$$

The user input parameters for the analysis platform encompass spatial area and time range specifications. By default, the spatial area is set to the expansive Hindu Kush-Himalaya region, offering a comprehensive overview of this vital landscape. However, users have the flexibility to opt for other sub-regions within the HKH or even upload or manually delineate their Area of Interest (AOI), tailoring the analysis to specific geographic contexts. Furthermore, the time range feature allows users to select any two years from 2000, 2010, to 2020, or all three years for a more comprehensive assessment. Opting for three years generates three independent map layers, each displaying the change map between two consecutive years, facilitating a nuanced understanding of temporal trends and transitions within the selected area.

Central to the analysis is the examination of farmland biomass values and their rate of change over time, serving as crucial indicators of agricultural dynamics and land use patterns. This analysis offers valuable insights into the quantitative values and change rates of farmland biomass across different years within the chosen area. The presentation of results is facilitated through histograms and line charts embedded within statistical charts, providing a visual representation of the data. The histogram depicts the average value of farmland biomass in the selected area, offering a snapshot of overall biomass trends, while the line chart illustrates the rate of change of the average farmland biomass between different years, enabling users to discern patterns of growth or decline. With input parameters mirroring those of the farmland biomass change map analysis, users can seamlessly compare quantitative results alongside spatial representations, fostering a comprehensive understanding of agricultural dynamics and informing land management strategies.

(9) Analysis of changes in vegetation coverage

The primary focus of the inversion product lies in the detailed analysis of vegetation coverage change, which constitutes a crucial component of the assessment. This analysis is bifurcated into two main aspects: the examination of vegetation coverage changes maps and the assessment of vegetation coverage change rates. The evaluation of vegetation coverage changes maps, depicted as tile maps, emphasizes alterations in vegetation coverage at the pixel level over a specific time frame, delineating between areas of increase, decrease, and stability in vegetation coverage, while also quantifying the extent of change. This computation as shown in Eq. 6.13 involves comparing pixel values representing vegetation coverage between two designated years, with positive values denoting augmentation, negative values indicating reduction, and zero values indicating no change. The magnitude of change is represented by the absolute value obtained through this calculation, thus facilitating a comprehensive understanding of vegetation dynamics over time.

$$\begin{aligned} Frendonal\ Vegetation\ Cover_{change} &= Fractional\ Vegetation\ Cover_{subsequent\ year} \\ &\quad - Fractional\ Vegetation\ Cover_{previous\ year} \end{aligned} \tag{6.13}$$

The user input parameters for this analysis tool encompass spatial area and time range specifications. By default, the spatial area encompasses the Hindu Kush-Himalaya region, though users retain the flexibility to opt for alternative sub-regions or manually delineate their area of interest (AOI) through uploading or drawing features. The time range option enables users to select two years from 2000, 2010, to 2020, or all three years for a comparative analysis. In the latter scenario, three independent map layers depict changes between consecutive years. The analysis yields vegetation coverage values and change rates for the chosen area, presented via a line chart format. This calculation method involves averaging vegetation coverage across the selected area and determining change rates based on these average values across different years. The input parameters mirror those of the vegetation coverage change map analysis, with quantitative results displayed alongside the mapped outputs in Fig. 6.22.

(10) Analysis of forest and grass distribution changes

The inversion product primarily focuses on analyzing changes in forest and grass distribution, which entails examining forest and grassland distribution data. This analysis is segmented into three key components: forest and grass distribution change map analysis, forest and grass area change rate analysis, and forest and grass conversion statistical analysis. In the forest and grass distribution change map analysis, alterations in landform types are scrutinized, encompassing six potential scenarios: forest transitioning to grassland, grassland transforming into forest, forest transitioning to other types, other types converting into forest, grassland converting into other types, and other types evolving into grassland. Additionally, three scenarios involve no change, resulting in a total of nine possible cases. Within the raster image, forest pixels are denoted by a value of 1, grassland pixels by a value of 2, and other types of pixels by a value of 3. Determining the change in surface object type involves conducting algebraic operations on the pixel values from previous and subsequent years. Equation 6.14 illustrates that different combinations of values can produce nine distinct outcomes ranging from 0 to 8, corresponding to the nine situations identified.

$$\begin{aligned} \textit{Type Pixel Value}_{change} = {} & 3 \times \textit{Type Pixel Value}_{subsequent\ year} \\ & - \textit{Type Pixel Value}_{Previous\ year} \end{aligned} \tag{6.14}$$

Users can input parameters such as spatial area and time range to customize their analysis. By default, the spatial area encompasses the Hindu Kush-Himalaya region, but users have the flexibility to select other sub-regions or manually define the area of interest (AOI) by uploading or drawing it themselves. Regarding the time range, users can choose any two years from 2000, 2010, to 2020, or select all three years for comparison. If all three years are chosen, the analysis generates three independent map layers illustrating changes between consecutive years.

The analysis primarily focuses on assessing changes in forest and grassland areas and their corresponding change rates within the selected area. The results of this analysis are presented through line charts, visually representing the extent of change in woodland or grassland area across different years. Separate charts display data for woodland and grassland, allowing users to discern trends and patterns over time. The input parameters mirror those utilized for the forest and grass distribution change map analysis, and the quantitative results are presented alongside the distribution change map outputs. These findings are visually depicted in Fig. 6.23, offering users

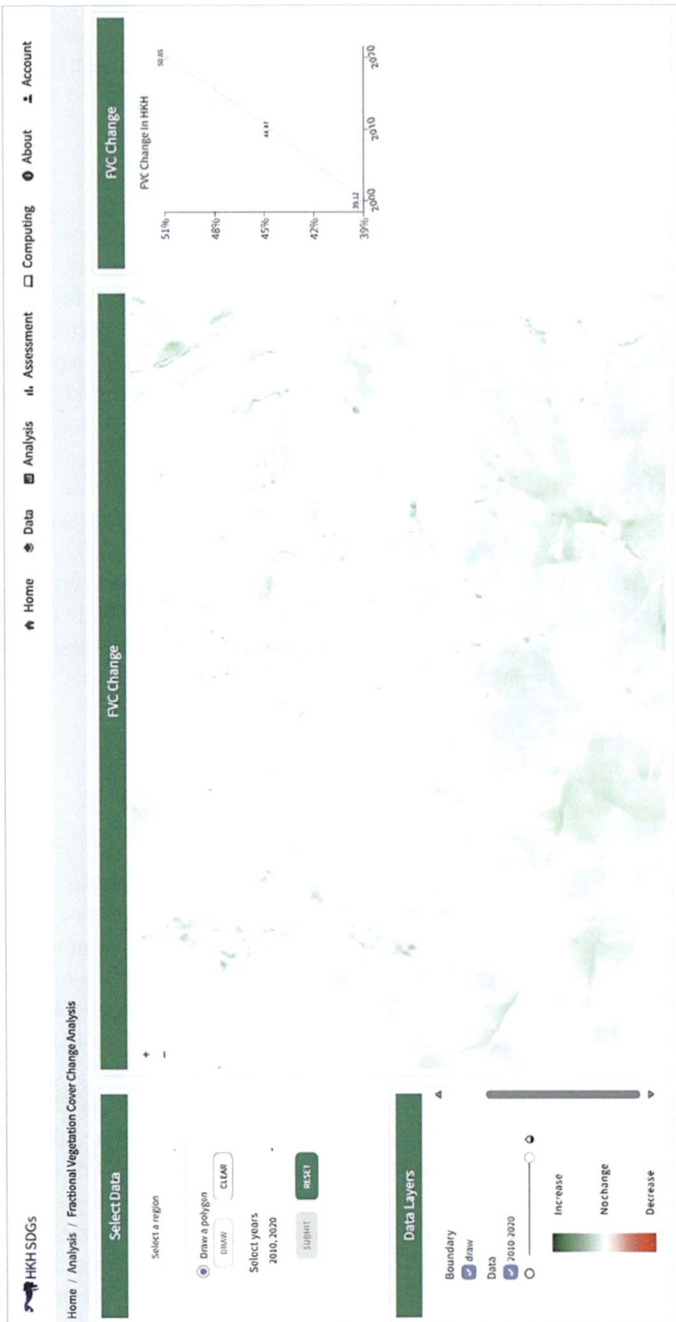

Fig. 6.22 Schematic diagram of vegetation coverage change analysis results

a comprehensive understanding of the evolving landscape within their specified area and time frame.

(3) **HKH sustainable development online assessment application**

The HKH Sustainable Development Online Assessment Application offers an extensive overview and indicator values for assessing agricultural conditions, water resources, and the ecological environment's sustainability across different years and regions from 2000 to 2020. Through the provision of statistical charts in time series, it enables users to track changes and trends over time comprehensively. The assessment of sustainable development in water resources is approached from two key perspectives: the rate of change of surface water bodies and the proportion of good water quality, both of which are essential indicators corresponding to SDG 6.6.1. The rate of change of surface water bodies illustrated in Eq. 6.15 reflects alterations around surface water bodies within a specific region, offering valuable insights into environmental dynamics and potential stressors. This assessment metric is quantified by the ratio of detected changes in water body area between the target year and the base year, providing a quantitative measure of environmental shifts and their implications for water resource management.

$$Change\ rate\ of\ surface\ water\ bodies = (Area_{target\ year} - Area_{base\ year})/Area_{base\ year} \tag{6.15}$$

where Area $_{\text{target year}}$ is the surface water area in the target year, and Area $_{\text{base year}}$ is the surface water area in the base year before the change.

The calculation equation for area is:

$$Area = N * 30m * 30m \tag{6.16}$$

where N in the Equation is the number of pixels of a certain type of feature.

The proportion of good water quality, as defined by water bodies with a transparency of over 0.5 m, serves as a crucial metric in assessing the health of ecosystems and human well-being, aligning with SDG 6.3.2. A water body boasting excellent water quality, where more than 80% of the area meets this criterion, is considered exemplary. This indicator's calculation involves dividing the area of excellent water bodies by the total water body area, as illustrated in Eq. 6.17.

$$Proportion\ of\ water\ bodies\ with\ good\ water\ quality = Area_{water\ clarity \geq 0.5} \tag{6.17}$$

The above equation delineates the measurement of area water clarity, representing the portion of water with transparency exceeding 0.5 m, within the total water body area. Evaluation of ecological resource sustainability focuses on two key aspects: the ratio of forest to grass area and the greening status of mountainous regions, indicating vegetation coverage. The forest-grass area ratio offers insights into the distribution balance between forests and grasslands, with its calculation detailed in Eq. 6.18.

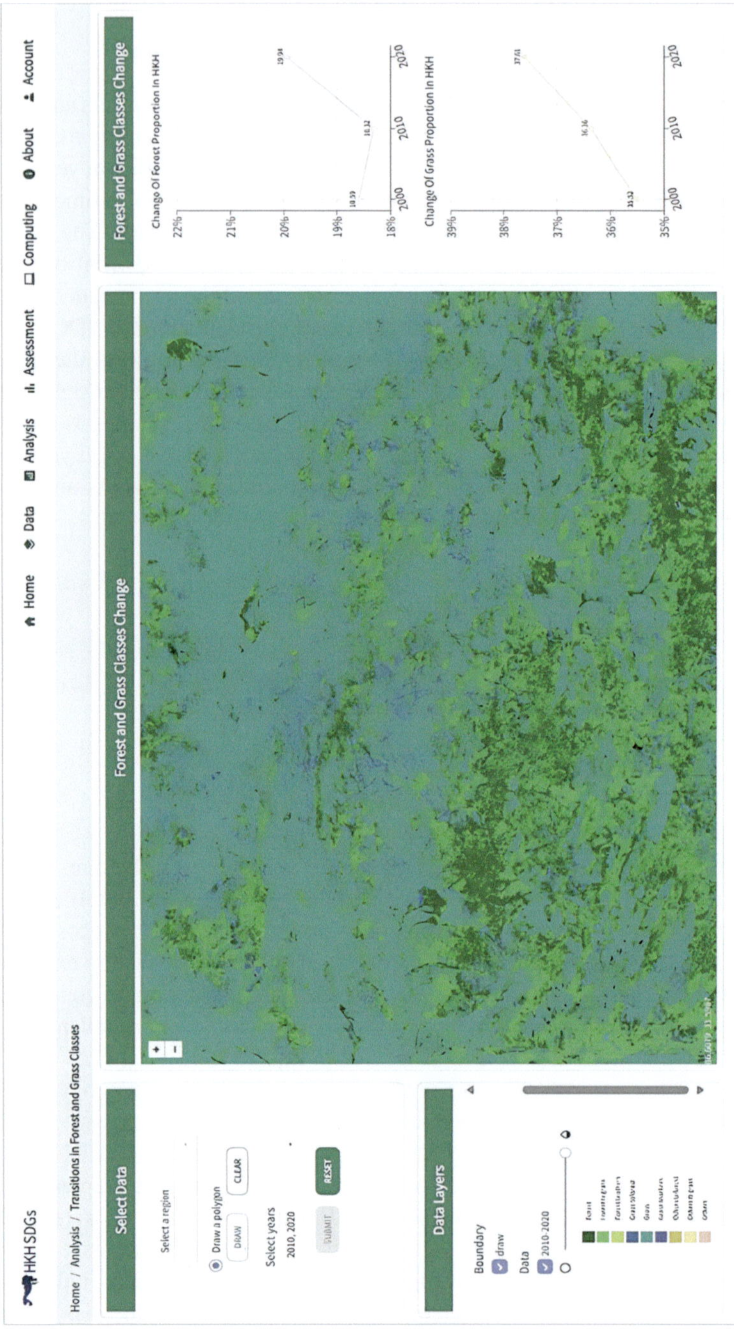

Fig. 6.23 Schematic diagram of analysis results of forest and grass distribution changes

$$Ratio\ of\ forest\text{-}grass\ coverage\ area = Area_{forest\text{-}grass}/Area_{region} \tag{6.18}$$

where Area $_{forest\text{-}grass}$ is the area of forest and grassland in the region, and Area $_{region}$ is the area of the region.

Equation 6.19 illustrates the calculation method for determining the greening status, or vegetation coverage, of mountainous areas in the Hindu Kush-Himalayan region. This metric offers valuable insights into the overall greening status of forests and grasslands within the region, serving as a crucial indicator of ecosystem health and environmental sustainability.

$$Fractional\ Vegetation\ Cover(FVC) = \sum_{k=0}^{n}(FVC_k)/n \tag{6.19}$$

where n in the Equation is the number of pixels in the area.

The evaluation of agricultural sustainability considers two key aspects: grain output per unit of labor and land intensity per unit of output. Grain production per labor unit signifies the average amount of grain produced by a single labor force member, typically representing the population aged 15–64 within the region. The calculation method for this metric is detailed in Eq. 6.20.

$$\begin{aligned} &Grain\ yield\ per\ unit\ of\ labor\ productivity \\ &\quad = Grain\ yield/Number\ of\ labor\ force\ population \end{aligned} \tag{6.20}$$

The land intensity per unit output reflects the land area required to produce one ton of grain, in hectares/ton, and its calculation method is shown in Eq. 6.21.

$$\begin{aligned} &Land\ intensity\ per\ unit\ of\ output \\ &\quad = Area_{l}and/Grain\ yield \end{aligned} \tag{6.21}$$

where Area $_{land}$ in the Equation is the farmland planting area.

In the online evaluation application interface, users have the flexibility to choose both the target area and target time. If a single year is selected, the evaluation indicators' results are presented numerically, as depicted in Fig. 6.24. Conversely, when multiple years are chosen, the evaluation indicators are displayed through a line chart, showcasing their evolving trends, as illustrated in Fig. 6.25.

6.3.5 Summary

The software system of remote sensing assessment for HKH sustainable development aims to support the sustainable development of the HKH region through technologies such as remote sensing, cloud computing, and artificial intelligence. The system includes four major modules: data sharing, online analysis, sustainable

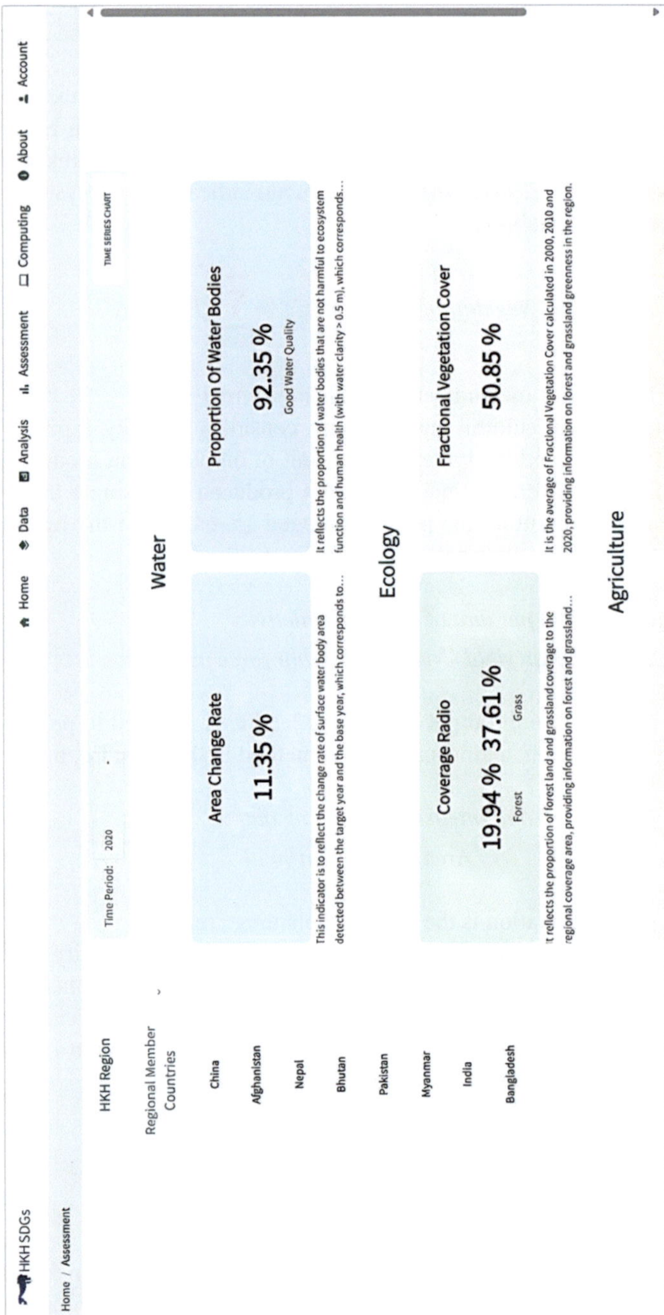

Fig. 6.24 Single-year indicator display interface for sustainable development online assessment application

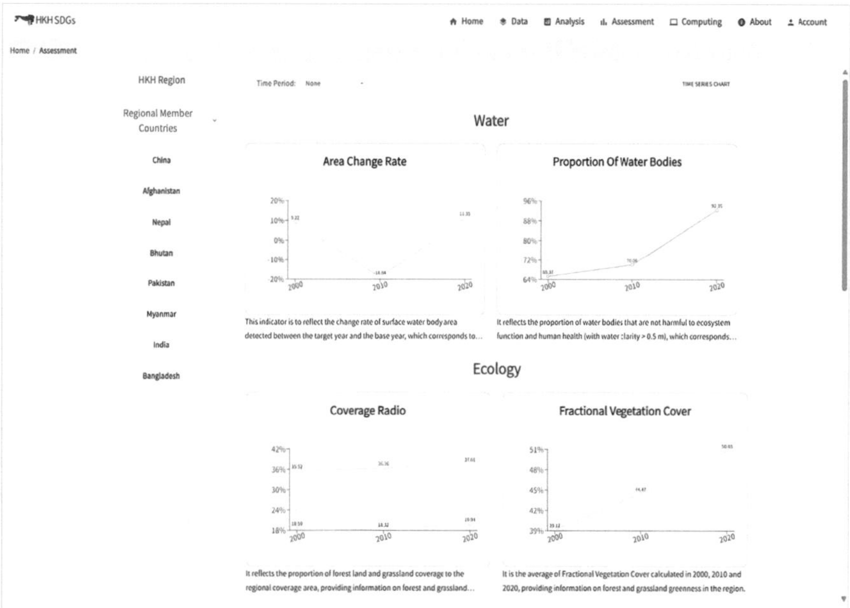

Fig. 6.25 Indicator change display interface for sustainable development online assessment application across different years

development assessment, and online AI model calculation, providing comprehensive remote sensing data processing and analysis functions. It enables online data query, download, analysis, and evaluation, and supports automated remote sensing data processing services, characterized by strong comprehensiveness, high efficiency, and intelligence. The data sharing module provides online query, map visualization, and download functions for key element inversion datasets in the HKH region; the online analysis module supports thematic application analysis of key element data and visualizes the results through statistical charts and maps; the sustainable development assessment module conducts regional and national-level assessments based on indicators defined for agriculture, water resources, and ecological environment; the online AI model calculation module offers online computation services for remote sensing inversion models driven by deep learning, achieving fully automated remote sensing inversion. This system has application scenarios and potential in supporting SDG remote sensing monitoring methods, geoscience model visual analysis techniques, and earth big data sharing services. Additionally, the system can achieve data, model, and analysis interoperability, as well as function calls with other system tools through API interfaces, providing a certain degree of flexibility and scalability.

6.4 Assessment Results of Regional Water Resources, Agriculture, and Ecological Sustainability

6.4.1 Regional Agricultural Sustainable Development Assessment Results

(1) **Land use intensity and production efficiency assessment results**

This study leverages advanced remote sensing technologies and machine learning algorithms to evaluate the resilience and sustainability of the agricultural system in the HKH region, specifically focusing on SDG 2.4.1, which aims to promote sustainable food production. A key metric in this evaluation is productivity per unit of labor (CPPL), calculated by dividing total production (CP) by the number of labor units (Ls). CPPL reflects production efficiency, providing governments with insights into labor utilization that can optimize production processes, enhance overall efficiency, and support economic growth and sustainable development. Figure 6.26 shows the distribution of land use intensity in the HKH region.

In Afghanistan, CPPL has significantly increased over the past two decades, driven by slow population growth and a substantial rise in Wheat production, leading to enhanced labor productivity. Conversely, China's wheat labor productivity has seen minimal growth, but its rice labor productivity has improved notably. On the other hand, Myanmar's rice labor productivity has stagnated, while Pakistan and Nepal have experienced significant declines in unit labor productivity, highlighting the diverse agricultural development paths across the region and the need for tailored policy interventions (Figs. 6.26 and 6.27).

From 2000 to 2020, labor productivity trends in the HKH region revealed a decline in rice productivity and an increase in Wheat productivity. Specifically, rice labor productivity decreased by 3% in Myanmar, 7% in southern China, 27% in Nepal, and 34% in Pakistan. Meanwhile, Wheat labor productivity rose by 104% in Afghanistan and 17% in Qinghai, China. These trends underscore the complex dynamics of labor productivity influenced by factors such as crop type, local agricultural practices, and external economic conditions.

Land use intensity analysis further illustrates these trends, as land use intensity typically decreases with yield improvements. From 2000 to 2020, land use intensity has generally declined across all regions. The most notable decrease in wheat land use intensity occurred in Afghanistan, with a significant reduction in rice land use intensity in southern China. Despite fluctuations, other regions showed a slow but steady decline in land use intensity, indicating more efficient land utilization alongside rising yields (Figs. 6.28 and 6.29).

Between 2000 and 2020, the land use intensity of all crop types has consistently declined. Overall, between 2000 and 2020, the land use intensity for all crops consistently declined (Fig. 6.30). For rice, land use intensity dropped by 15% in Myanmar, 6% in Pakistan, 11% in southern China, and 24% in Nepal. For wheat, Afghanistan saw a dramatic 63% reduction, while Qinghai, China, experienced a 34% decline.

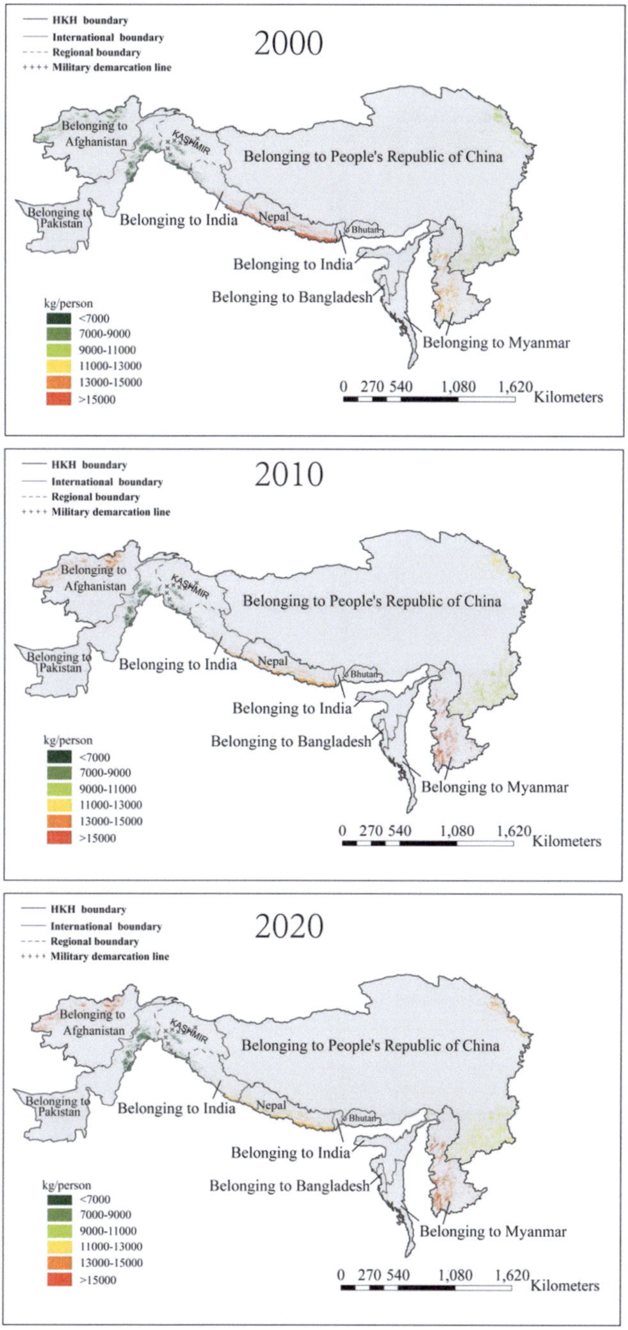

Fig. 6.26 Grain output per unit labor force in major grain-producing areas in 2000–2020

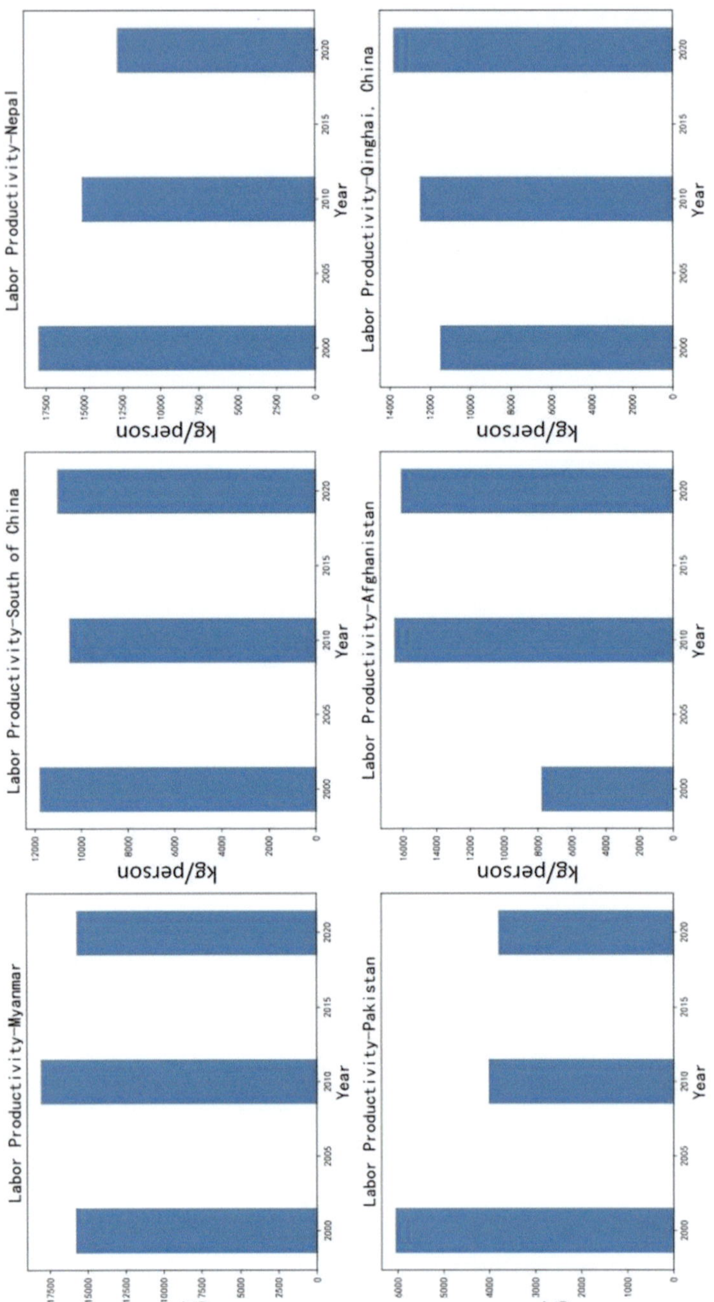

Fig. 6.27 Statistics of grain output per unit of labor force in major grain producing areas

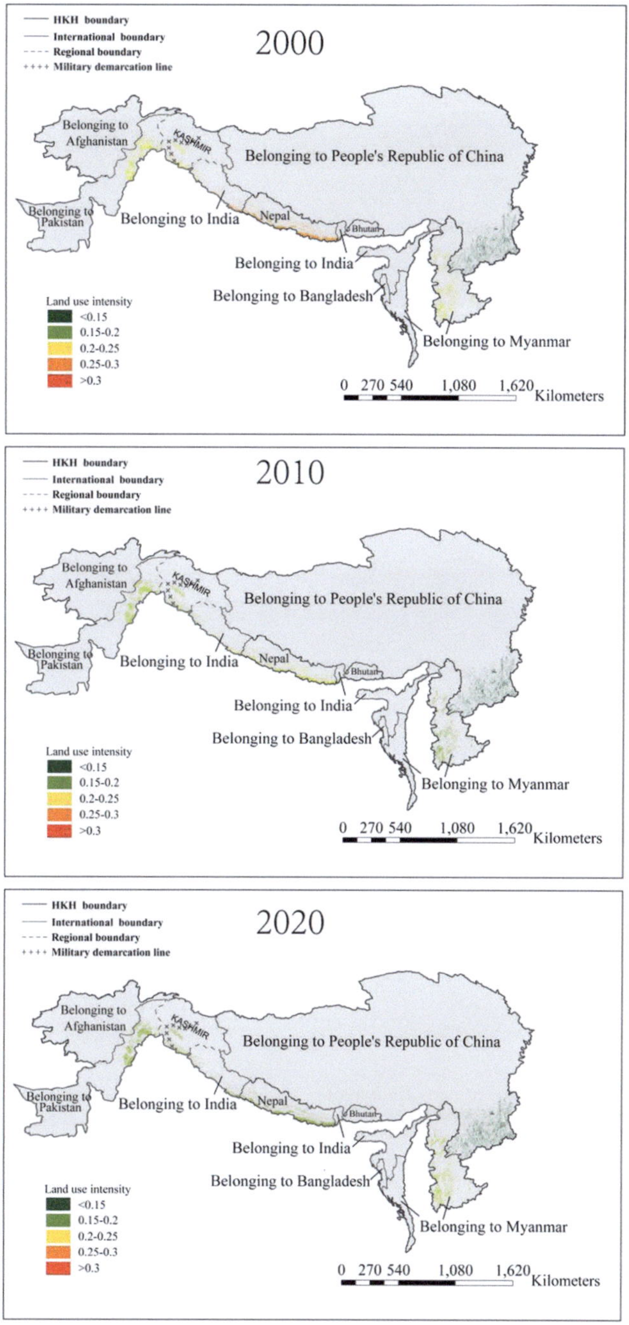

Fig. 6.28 Land use intensity in major rice producing areas in 2000–2020

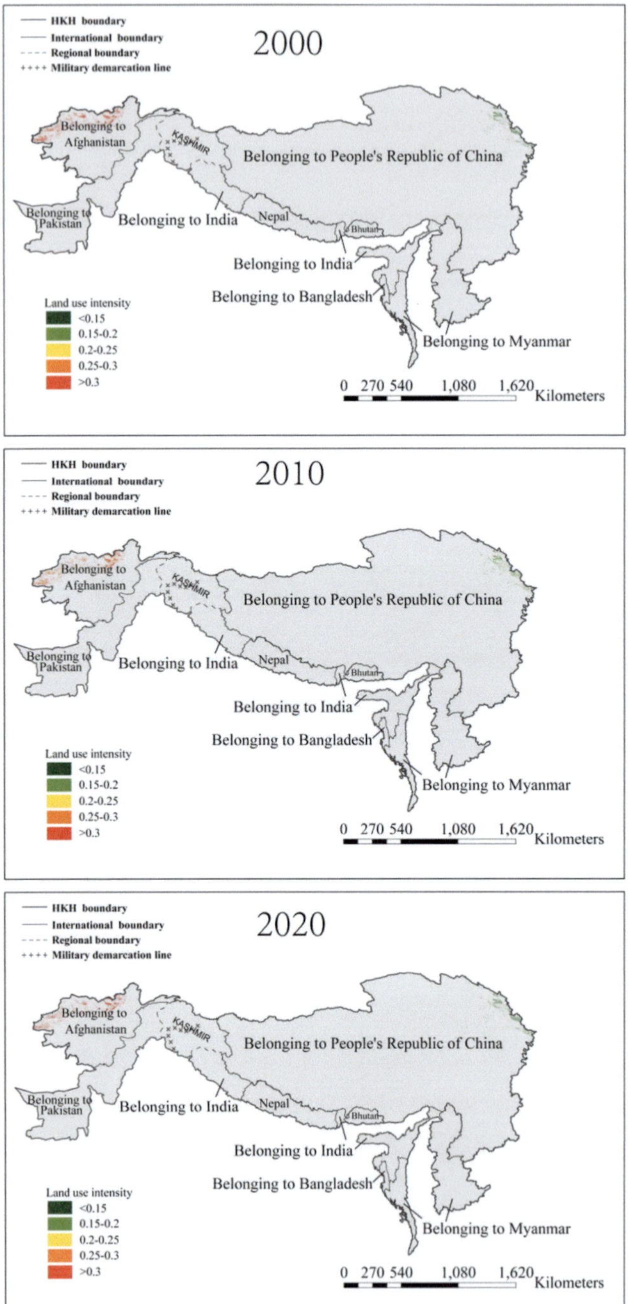

Fig. 6.29 Land use intensity in major wheat producing areas in 2000–2020

These figures reflect a broad trend towards more efficient land use in agricultural production, supporting increased productivity and the sustainable development goals of the region. The combined decline in land use intensity and improved food production indicates significant progress in achieving SDG 2.4.1 in the HKH region over the two-decade study period.

(2) **Evaluation of the impact of changes in grain production on the realization of regional zero hunger goals**

We utilized the hierarchical analysis method to prioritize key factors including average unit labor output, land use intensity, and cultivated land area, establishing a comprehensive evaluation system aimed at achieving zero hunger in the HKH region. Each indicator was weighted and normalized across major agricultural centers within the HKH region to construct the Zero Hunger Index (ZHI). Specifically, CPPL represents the average unit labor output, LIPP denotes the average land use intensity, and a signifies the proportion of cultivated land area. A higher ZHI value indicates better food security within the region, while a lower value indicates more acute food security challenges (see Fig. 6.31).

This approach enables a nuanced assessment of food security dynamics across the HKH region, facilitating targeted interventions to improve agricultural productivity and mitigate hunger. By integrating these critical metrics into the ZHI, we could provide a robust tool for policymakers and stakeholders to monitor and enhance food security outcomes in this geographically diverse and economically significant area. Based on our analysis, grain production in the HKH region has shown significant growth over the past two decades. However, achieving the goal of zero hunger necessitates substantial ongoing investment. Key areas for investment include enhancing agricultural productivity and sustainability, improving food distribution networks, ensuring food safety through education, safeguarding farmers' rights, fostering regional and global food cooperation, advancing agricultural research, and developing resilient crop varieties suited to local conditions. These efforts are crucial not only for boosting production but also for promoting the long-term sustainability of agriculture in the region.

According to statistical data and historical research on grain production in Pakistan, the country's primary agricultural areas have struggled to keep pace with population growth, leading to concerns about meeting domestic food demand (Farrukh et al., 2020). Increasing grain production hinges on optimizing resource allocation and adjusting production factors. Investing in indirect factors such as education, healthcare, and infrastructure proves more effective than solely expanding direct factors like land and fertilizers. Such investments not only enhance grain yields but also contribute to the overall sustainable development of agriculture. Moreover, Pakistan faces significant challenges from climate change, including increased flooding that affects farmland. Addressing these challenges requires strategic investments in water management infrastructure and public education on environmental resilience (Shehzad, 2023).

Over the past two decades, Nepal's primary agricultural regions have experienced varying levels of food security. While improvements in yield capacity and

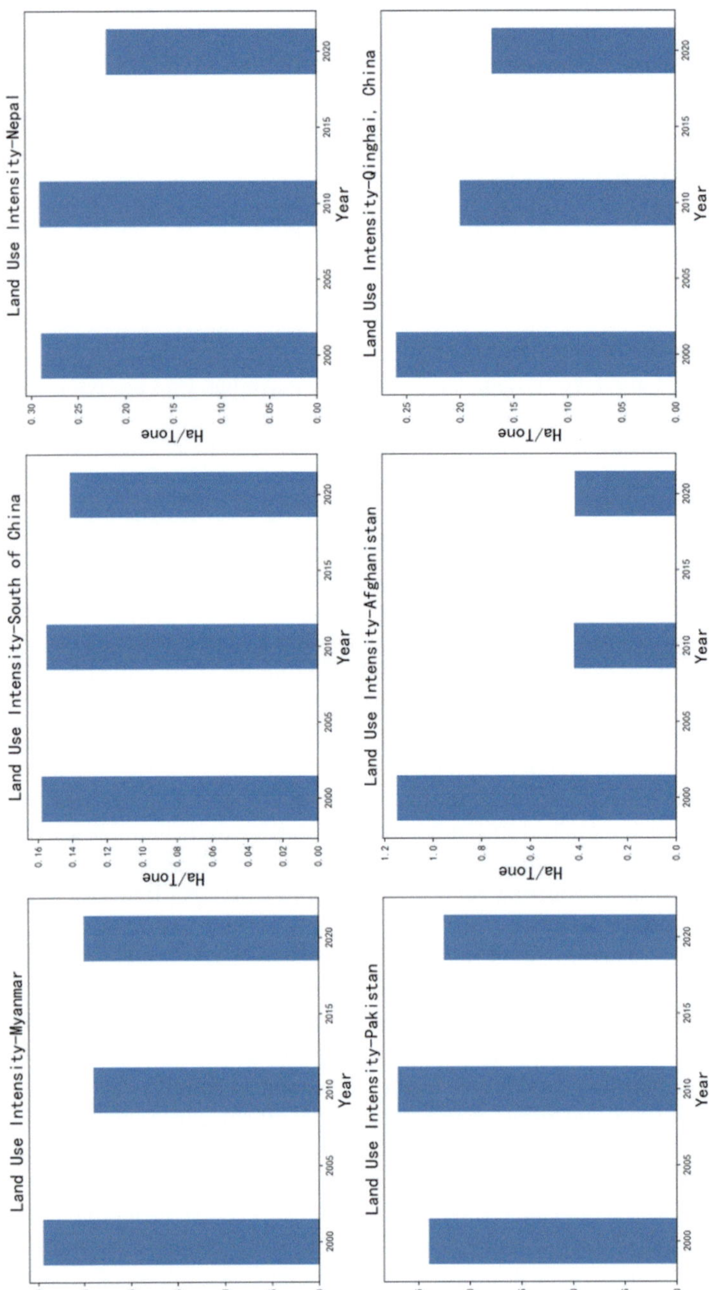

Fig. 6.30 Statistical chart of land use intensity changes in major grain production areas

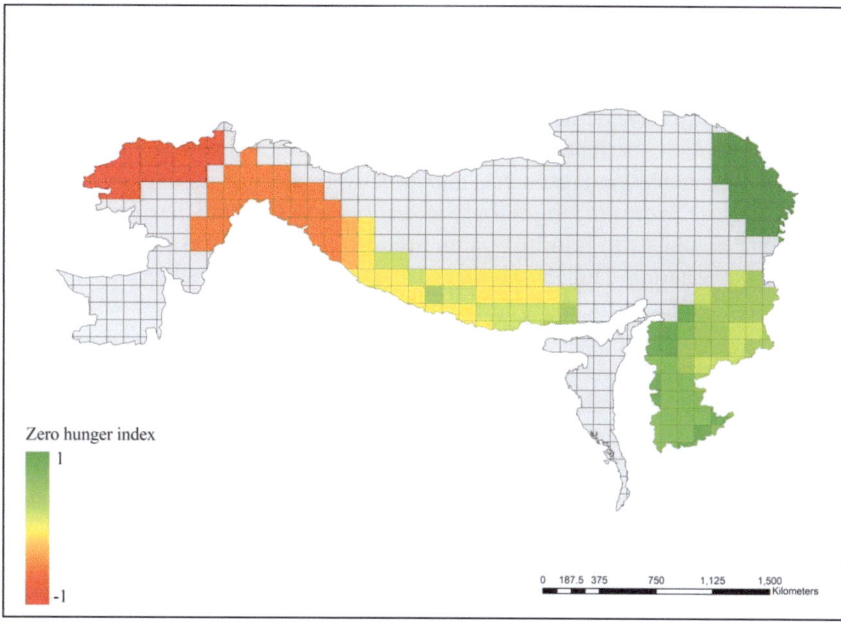

Fig. 6.31 HKH region 2020 Zero Hunger Index distribution map

land use intensity have been achieved, there has been a concerning decline in labor productivity. Climate change poses a significant threat by potentially worsening crop-growing conditions, thereby jeopardizing food production. Concurrently, population growth adds pressure to food security challenges in the country. Despite efforts, Nepal continues to grapple with persistent issues of food security and hunger. A recent 2022 study highlighted that approximately 48.2% of households have attained food security, with urban areas generally faring better than rural counterparts (Sapkota, 2022).

To address these challenges effectively, it is imperative for the government to bolster public investment in health and education. Furthermore, there is a critical need to ensure the rational allocation and efficient utilization of rural lands. Protecting and rehabilitating natural resources crucial for food security—such as land, water sources, and forests—is essential to fostering sustainable agricultural development. Collaborative efforts with international organizations like the Food and Agriculture Organization of the United Nations (FAO), the World Food Program (WFP), and the International Fund for Agricultural Development (IFAD) are crucial. Together, these partnerships can work towards achieving the ambitious goal of zero hunger, thereby ensuring a more resilient and food-secure future for Nepal. Over the past two decades, China has experienced steady growth in grain production across its key agricultural regions within the HKH region. However, significant regional disparities in output persist, posing challenges to equitable food distribution. While China's population peaked in 2021 and has since entered a period of gradual decline due to low birth rates and an aging demographic, the demand for food, particularly grains, remains strong. Changing dietary patterns, rapid urbanization, and increased consumption of meat and dairy products—which require substantial grain for animal feed—continue to drive demand. Addressing these challenges necessitates strengthening regional food distribution systems, refining agricultural policies, safeguarding and enhancing arable land resources, and effectively responding to global environmental changes. To ensure long-term food security and self-sufficiency, the Chinese government continues to prioritize grain production and reserves. These strategies are vital in bolstering China's grain production capacity and advancing regional food security goals, particularly in the pursuit of eradicating hunger. Meanwhile, Myanmar has maintained relatively stable agricultural development over the past two decades, marked by incremental increases in both per-unit output and arable land area. Despite its considerable agricultural potential, Myanmar faces challenges stemming from inadequate agricultural infrastructure. To unlock its agricultural productivity, Myanmar must prioritize international collaborations aimed at enhancing agricultural technology, refining cultivation and breeding techniques, and ultimately boosting crop yields and agricultural product quality. These efforts are pivotal in harnessing Myanmar's agricultural strengths to meet domestic food needs effectively and sustainably.

Conversely, Afghanistan has witnessed notable growth in agricultural output and labor productivity, yet achieving zero hunger remains a distant goal. The country's agricultural sector is significantly influenced by regional policies, impacting its food security landscape. To address these challenges, Afghanistan must strengthen economic and political ties with neighboring nations, fostering greater food security and stability. Comprehensive strategies encompassing improved security measures, enhanced regional cooperation, agricultural productivity enhancements, nutritional improvements, establishment of sustainable food supply systems, and intensified international partnerships are imperative to ensure widespread access to safe, nutritious, and ample food supplies for all Afghan citizens.

6.4.2 Assessment of Regional Water Resources Sustainability

In addressing SDG 6 within the HKH region, this study focuses on two specific indicators: SDG 6.6.1, which tracks the "Change in the extent of water-related ecosystems over time," and SDG 6.3.2, concerning the "Proportion of bodies of water with good ambient water quality." The analysis of these indicators provides valuable insights into both the physical changes in water bodies and the quality of water across the HKH region over two decades (2000–2020).

(1) **Changes in surface water area**

The study highlights notable changes in the surface water area across the HKH region over the two-decade period. From 2000 to 2020, the region experienced an overall 18.7% increase in water bodies. Most countries saw a net expansion in water surface area between 2000 and 2010, and this trend largely continued between 2010 and 2020, except for Pakistan and Myanmar, where reductions were more pronounced during the latter decade. Geographically, the most significant changes occurred in the central and northeastern parts of the Qinghai-Tibet Plateau in China (Fig. 6.32).

Interestingly, the rate of surface water expansion slowed from 2010 to 2020 compared to the previous decade. Bangladesh was an exception, where the rate

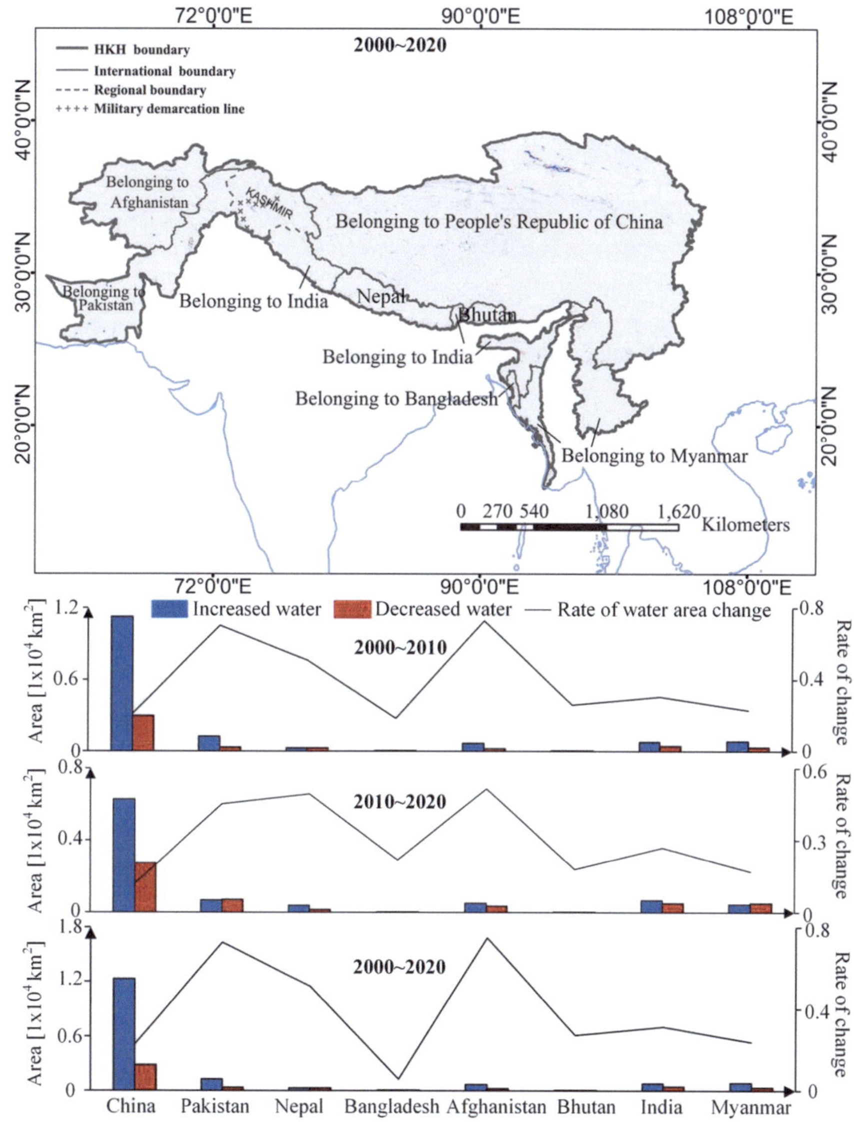

Fig. 6.32 Changes in surface water area in the HKH region from 2000 to 2020

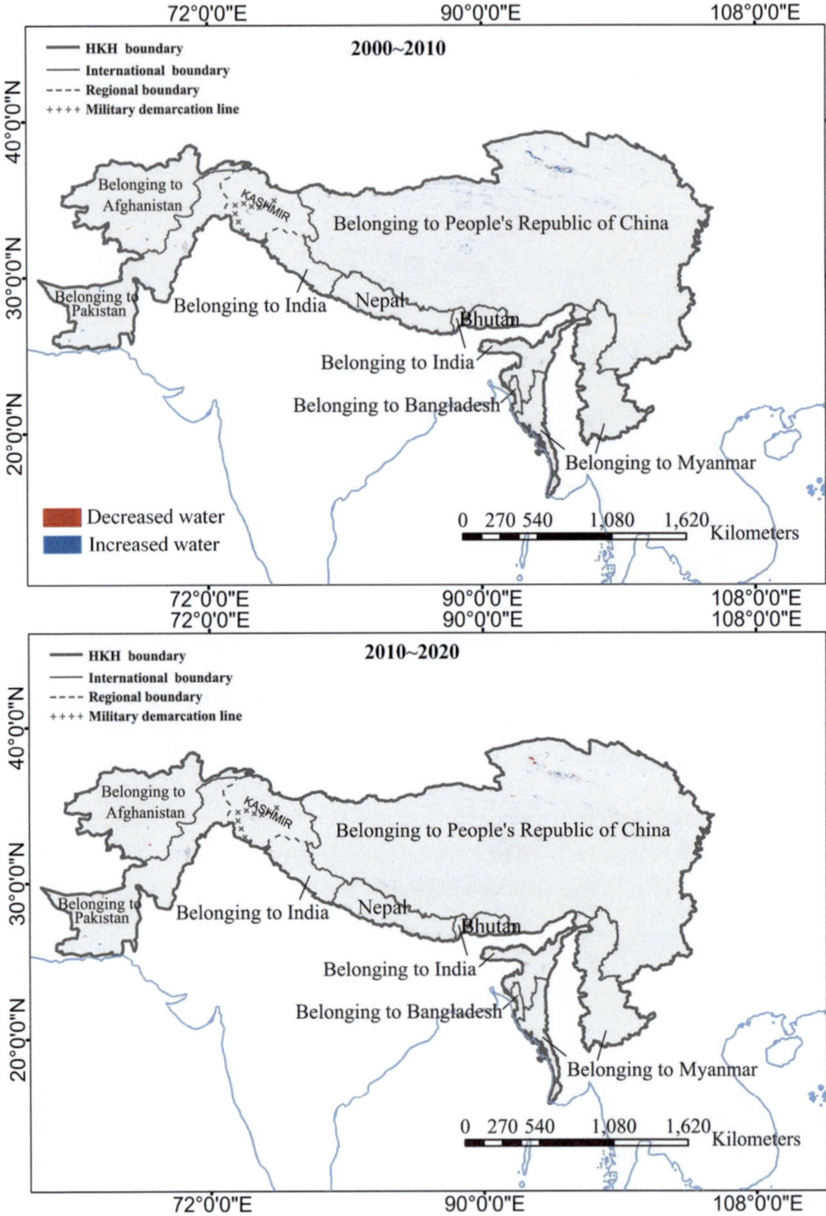

Fig. 6.32 (continued)

of increase accelerated in the latter decade. Specific rates of change reveal diverse trends across the region.

China: Saw a total 19.7% increase in water bodies between 2000 and 2020, with a higher rate of 21.25% from 2000 to 2010, which slowed to 11.98% from 2010 to 2020.

Pakistan: Exhibited a 70.23% increase in water bodies between 2000 and 2010, later slowing to 45.24%.

Nepal: Experienced steady changes with rates of 50.24 and 49.29% for the respective decades.

Bangladesh: Saw an increase in water body expansion, with rates rising from 18.28 to 21.93%.

Afghanistan: While having high rates of 72.65% in the first decade, saw a reduction to 51.40% in the second decade.

Bhutan, India, and Myanmar showed relatively moderate but consistent growth patterns, with Bhutan experiencing a drop from 25.91 to 18.04%, India from 30.52 to 26.90%, and Myanmar from 23.06 to 17.06%. It is worth noting that **Pakistan** had the largest rate of increase between 2000 and 2010, while **Nepal** led the second decade from 2010 to 2020. This reveals both regional diversity and the dynamic nature of water body changes within the HKH.

(2) **Changes in ambient water quality**

In terms of water quality, the study analyzed water transparency as a key indicator for SDG 6.3.2. Using data from 2000, 2010, to 2020, it was observed that the proportion of water bodies with good ambient water quality (transparency greater than 0.5 m) improved significantly over time, with an overall 38% improvement in water quality from 2000 to 2020 across the HKH region.

China: Saw a significant increase in the proportion of water bodies with good quality, rising from 48.13% in 2000 to 92.01% in 2020.

Pakistan: Experienced an improvement from 64.38 to 83.56% during the same period.

Nepal: Saw its proportion increase from 70.77 to 80.00%.

Bangladesh: Also saw improvement, though more moderate, from 65.79 to 71.05%.

Afghanistan and **India** experienced similar upward trends, with Afghanistan's water quality improving from 55.45 to 79.09%, and India from 72.30 to 78.38%.

Myanmar: Saw a smaller improvement, from 61.29 to 67.74%.

Bhutan: Maintained a consistently high proportion of 96.19% of water bodies with good quality throughout the period.

Despite the overall improvement, there were fluctuations in water quality, particularly in 2010, where most countries experienced a decline in the proportion of water bodies with good quality. The exception was Pakistan, which saw continued improvement even during this period. These temporary declines highlight the variability and challenges in maintaining consistent water quality improvements. However, by 2020, most countries showed a recovery and improvement, suggesting that effective management and conservation efforts had been implemented successfully across the region (Fig. 6.33).

The findings from this study provide critical insights into the state of water-related ecosystems and water quality in the HKH region. SDG 6.6.1 reveals an overall increase in water bodies, with significant spatial and temporal variations, while SDG 6.3.2 indicates marked improvements in water quality, particularly post-2010. The continued expansion of surface water and the enhancement of water quality underscore the importance of sustainable water management and international cooperation in addressing the challenges posed by climate change and population growth in the HKH region.

6.4.3 HKH Regional Ecological Environment Sustainable Development Assessment

This study extends its analysis to SDG 15 (Life on Land), focusing on mountainous ecosystems in the HKH region. It specifically examines changes in vegetation coverage, aligning with SDG 15.4.2, which monitors the Mountain Green Cover Index and the proportion of degraded mountain areas. Forests and grasslands are critical to maintaining ecological balance and ensuring the region's ecological security. The analysis spans from 2000 to 2020 and highlights significant changes in forest and grassland coverage across the HKH countries.

(1) **Changes in forest coverage (2000–2020)**

The study's findings on forest coverage reveal significant declines across the HKH region over the two-decade period. Figure 6.34 presents a statistical chart depicting the annual change rate of forest coverage in the HKH region from 2000 to 2020. The data highlights the regional variations in forest coverage trends over two distinct periods: 2000–2010 and 2010–2020.

During the 2000–2010 period, forest coverage declined in Bhutan, Afghanistan, and Pakistan, with reduction rates of 2.24, 11.47, and 18.06%, respectively. Although there was an improvement in forest coverage in these countries between 2010 and 2020, the overall change still indicate a decline: −2.04% in Bhutan, −4.86% in Afghanistan, and −12.59% in Pakistan.

In contrast, forest coverage in China and India showed declines during 2000–2010, at 2.06 and 1.72%, respectively. However, these losses were reversed in the following decade, resulting in a modest net increase of 2.58% in China and 0.40% in India from 2000 to 2020.

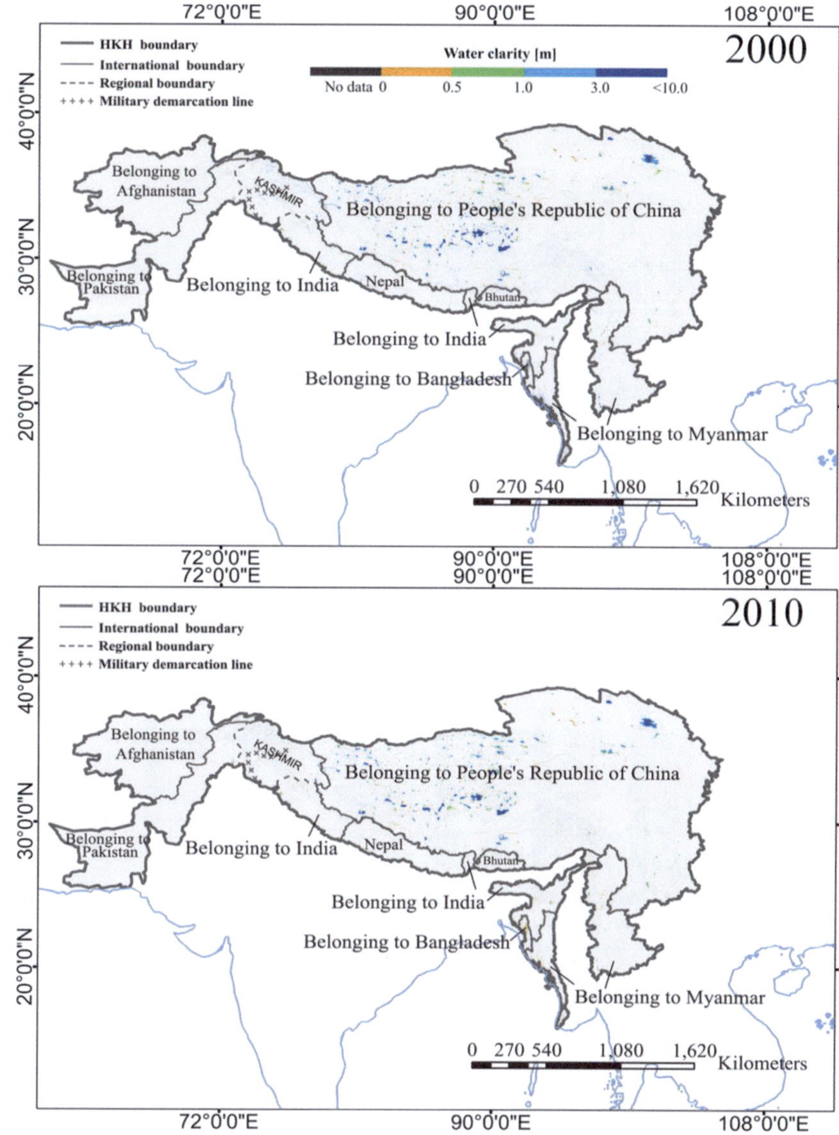

Fig. 6.33 Proportion of water bodies with good water quality in the HKH region from 2000 to 2020

Myanmar and Bangladesh experienced consistent declines in forest coverage across both decades, with significant cumulative losses of 7.74 and 16.89%, respectively. These sustained reductions underscore the challenges faced in these regions in maintaining forest ecosystems.

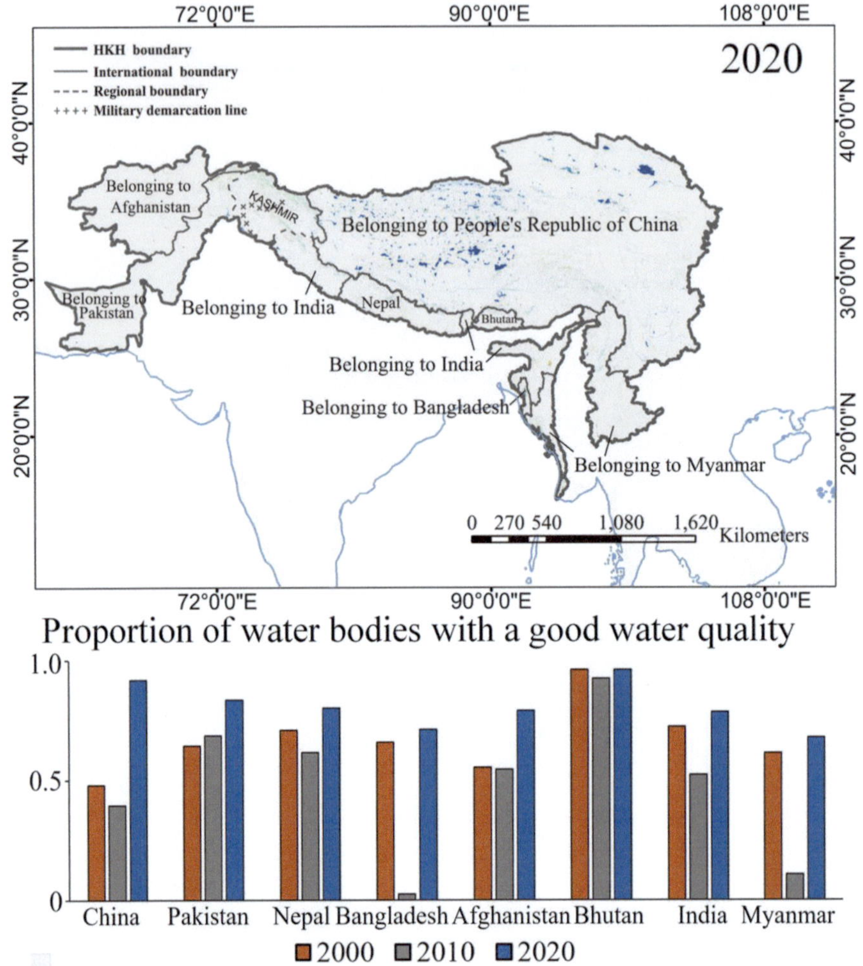

Fig. 6.33 (continued)

Nepal stands out as the only HKH country with a continuous increase in forest coverage throughout both periods, achieving a net gain of 5.89% over the two decades. This positive trajectory highlights Nepal's successful efforts in forest conservation and restoration.

Overall, the HKH region experienced a net decline of 1.53% in forest coverage from 2000 to 2020, largely driven by substantial losses (2.98%) during the 2000–2010 period. Despite some recovery in forest coverage across the region during 2010–2020, it was insufficient to offset the earlier reductions, reflecting an ongoing challenge in achieving sustainable forest management across the HKH countries. This

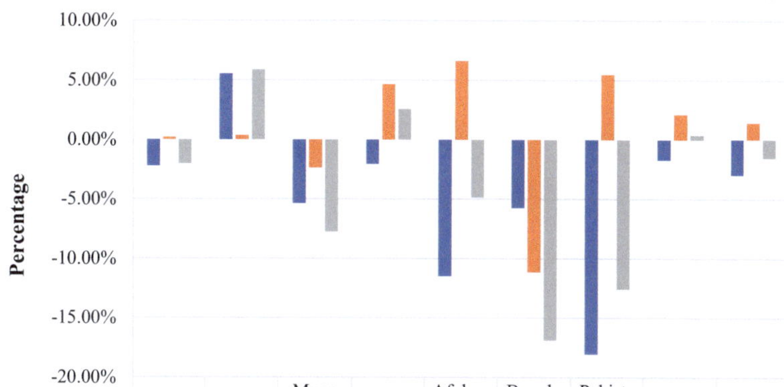

	Bhutan	Nepal	Myanmmar	China	Afghanistan	Bangladesh	Pakistan	India	Total
forest (2000-2010)	-2.24%	5.53%	-5.38%	-2.06%	-11.47%	-5.74%	-18.06%	-1.72%	-2.98%
forest (2010-2020)	0.20%	0.36%	-2.37%	4.64%	6.61%	-11.15%	5.47%	2.12%	1.44%
forest (2000-2020)	-2.04%	5.89%	-7.74%	2.58%	-4.86%	-16.89%	-12.59%	0.40%	-1.53%

Fig. 6.34 Statistical chart of annual change rate of forest coverage in the HKH region forest

deceleration may be attributed to improved forest management practices and conservation policies, but the overall decline emphasizes the need for stronger protection measures.

The consistent downward trend in forest coverage across the HKH region is concerning, as it threatens biodiversity and ecosystem resilience. Targeted efforts to halt deforestation and restore degraded forests are essential for achieving the goals set by SDG 15.

(2) **Changes in grassland coverage (2000–2020)**

In contrast to forest dynamics, grassland ecosystems exhibited a more positive trend. Figure 6.35, which illustrates the annual change rate of grassland coverage, shows an 11.09% increase in grassland areas across the HKH region between 2000 and 2020. The trends, however, were mixed.

Between 2000 and 2010, grassland coverage increased across the HKH region, except in Nepal, where it decreased by 4.19%. This period was marked by overall expansion, reflecting shifts in land use and vegetation dynamics in most countries.

During the 2010–2020 period, however, grassland coverage exhibited mixed trends. Bhutan, Afghanistan, Pakistan, and India experienced declines of 3.69, 22.71, 17.83, and 3.63%, respectively. Conversely, the other HKH countries, including China, Myanmar, Bangladesh, and Nepal, showed an increasing trend in grassland coverage, demonstrating regional variability in land cover changes. Over the entire 2000–2020 period, the HKH region experienced a significant overall increase in grassland coverage, with a net rise of 11.09%. This growth was driven primarily by substantial increases in Myanmar (59.14%) and Bangladesh (107.56%). In contrast,

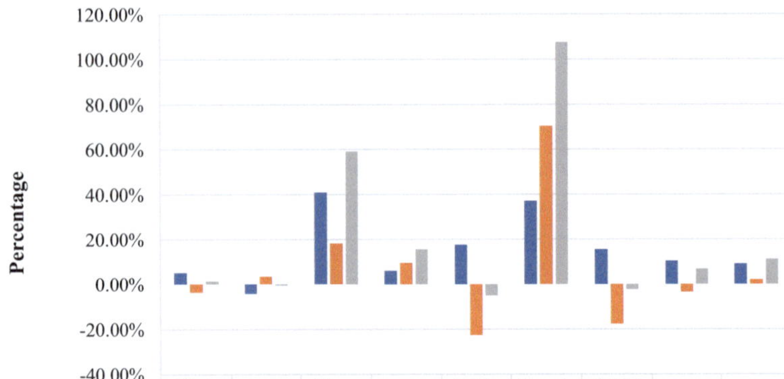

	Bhutan	Nepal	Myanmar	China	Afghanistan	Bangladesh	Pakistan	India	Total
grassland (2000-2010)	5.22%	-4.19%	40.96%	6.08%	17.50%	37.14%	15.50%	10.45%	9.06%
grassland (2010-2020)	-3.69%	3.56%	18.18%	9.51%	-22.71%	70.43%	-17.83%	-3.63%	2.03%
grassland (2000-2020)	1.53%	-0.63%	59.14%	15.59%	-5.21%	107.56%	-2.33%	6.81%	11.09%

Fig. 6.35 Statistical chart of annual change rate of grassland coverage in HKH region grassland

Afghanistan and Pakistan exhibited declining grassland coverage, with decreases of 5.21 and 2.33%, respectively. Nepal remained relatively stable, with a slight reduction of 0.63%.

The overall increase in grassland coverage indicates a significant shift in vegetation patterns, although localized declines in certain countries emphasize the importance of implementing targeted grassland management and conservation strategies. While the second decade showed mixed outcomes, the substantial gains made during the first decade were sufficient to offset the losses observed in some regions. Grasslands play a vital role in supporting biodiversity, sequestering carbon, and sustaining the livelihoods of local communities. Understanding these trends is essential for designing effective interventions to preserve and enhance grassland ecosystems, ensuring their resilience and contributions to environmental and socio-economic well-being.

(3) **Forest and grassland dynamics**

Figure 6.36 illustrates the annual change rate of forest and grassland coverage in the HKH region across three time periods. Nepal, China, and India exhibited a consistent upward trend during both 2000–2010 and 2010–2020, with increases of 0.81 and 1.92%, 4.26 and 8.42%, and 2.71 and 0.03%, respectively. Over the entire period from 2000 to 2020, these three countries showed overall growth in forest and grassland coverage of 2.73, 12.68, and 2.74%, respectively. This positive trend can largely be attributed to government-led ecological restoration initiatives such as afforestation programs and efforts to convert farmland back to forests. For instance, historical trend records in Nepal show fluctuating forest cover with 43.5% in 1960 and declining to 38.1% in 1978 and further to 29% in 1994 (Paudel et. al., 2016); After the initiation of

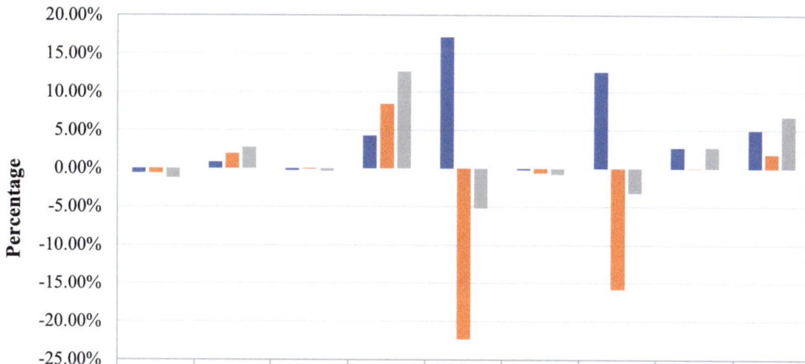

	Bhutan	Nepal	Myanmmar	China	Afghanistan	Bangladesh	Pakistan	India	Total
2000-2010	-0.61%	0.81%	-0.27%	4.26%	17.12%	-0.20%	12.61%	2.71%	4.94%
2010-2020	-0.65%	1.92%	-0.10%	8.42%	-22.32%	-0.60%	-15.82%	0.03%	1.83%
2000-2020	-1.26%	2.73%	-0.37%	12.68%	-5.20%	-0.80%	-3.21%	2.73%	6.78%

Fig. 6.36 Statistics of annual change rate of forest and grass coverage in the HKH region

community forest, forest cover has increased to 39.1% in 2010 (Uddin et al., 2015), 40.36% in 2015 (DFRS, 2015), and 44.47% in 2018 (MoFE, 2019).

Afghanistan and Pakistan experienced significant growth in forest and grassland coverage during 2000–2010, with an increase of 17.12 and 12.61% respectively. However, a notable decline during 2010–2020 led to an overall forest and grassland coverage reduction of 5.20 and 3.21%, respectively during the 2-decade study period. In contrast, forest and grassland coverage in Bhutan, Myanmar, and Bangladesh remained relatively stable across all three time periods. Overall, the HKH region experienced a 6.78% increase in forest and grassland coverage from 2000 to 2020. This positive trend was primarily driven by substantial growth during the first decade, with a remarkable increase of 4.94% between 2000 and 2010.

(4) **Fractional vegetation cover (FVC) analysis in the HKH Region**

The study analyzed Fractional Vegetation Cover (FVC) to evaluate changes in the overall greenness and vegetation dynamics of the Hindu Kush Himalayan (HKH) region over a 20-year period (2000–2020). The results reveal a complex and evolving pattern of vegetation cover across the region, highlighting significant differences among countries, as well as notable periods of loss and recovery.

Figure 6.37 presents the change rate of Fractional Vegetation Coverage (FVC) in the HKH region from 2000 to 2020. Using 2000 as the baseline, the data reveals contrasting trends during the first and second decades. Between 2000 and 2010, the FVC of six regions in the HKH region, excluding Afghanistan and Pakistan, displayed a decreasing trend. Afghanistan and Pakistan experienced increases of 5.28 and 31.41%, respectively. Myanmar exhibited the highest decline in this period, with a decrease of 22.36%. From 2010 to 2020, the overall vegetation coverage of

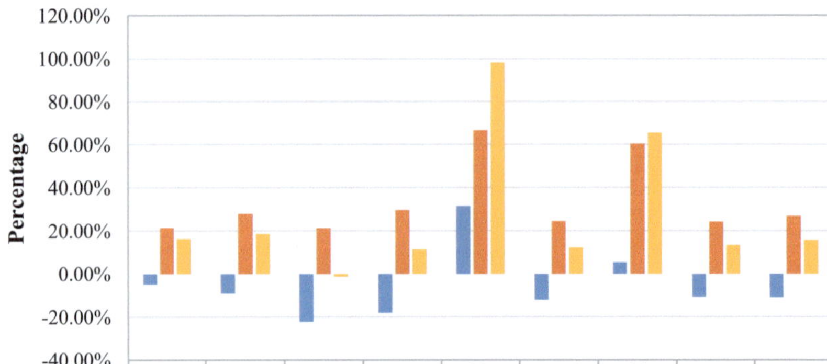

	Bhutan	Nepal	Myanmmar	China	Afghanistan	Bangladesh	Pakistan	India	Total
2000-2010	-5.06%	-9.21%	-22.36%	-18.14%	31.41%	-12.09%	5.28%	-10.82%	-11.04%
2010-2020	21.25%	27.83%	21.16%	29.53%	66.66%	24.41%	60.21%	24.16%	26.75%
2000-2020	16.19%	18.62%	-1.20%	11.40%	98.08%	12.32%	65.49%	13.34%	15.71%

Fig. 6.37 Statistical chart of FVC change rate in the HKH region

the HKH region reversed to an increasing trend. Afghanistan and Pakistan showed the most significant growth, with increases of 66.66 and 60.21%, respectively.

Over the two-decade period from 2000 to 2020, the FVC of the HKH region demonstrated a net upward trend, with an increase of 15.71%. Afghanistan and Pakistan showed the most pronounced growth during this time, increasing by 98.08 and 65.49%, respectively. However, Myanmar's sharp decline in FVC during the first decade (22.36%) contributed to an overall decrease of 1.20% in its FVC from 2000 to 2020.

While the HKH region experienced an overall decline in FVC of 11.04% from 2000 to 2010, the subsequent recovery during 2010–2020, with an increase of 26.75%, significantly offset earlier losses. This resulted in a net increase of 15.71% in FVC across the HKH region from 2000 to 2020.

The extent of the recovery highlights the success of national policies aimed at combating deforestation, promoting sustainable land management, and restoring degraded landscapes. It is also indicative of improved ecological resilience in these areas, showing that targeted interventions can reverse large-scale vegetation loss when appropriately implemented and supported by favorable environmental conditions.

This analysis provides valuable insights into the achievement of SDG 15.4.2, which focuses on the protection and restoration of forest and grassland ecosystems in mountainous areas like the HKH. The fluctuations in FVC across different countries reveal the complexity of balancing land use, economic development, and ecosystem conservation in this ecologically sensitive region.

Despite the encouraging recovery trends observed between 2010 and 2020, the study underscores the significant challenges that remain. The decline in forest coverage during the first decade and the uneven patterns of grassland recovery point to the need for more integrated approaches to sustainable land management. Countries like Myanmar, which experienced a reduction in vegetation cover, along with others that faced significant deforestation before the recovery period, will need continued support and intervention to ensure long-term ecological sustainability.

In summary, understanding the vegetation dynamics in the HKH region through the lens of FVC analysis offers critical insights for developing targeted strategies that promote ecological recovery and sustainable development. By leveraging these findings, governments, policymakers, and stakeholders can better address the environmental challenges facing the region and work toward achieving SDG 15.4.2.

References

Bigham Stephens, D. L., Carlson, R. E., Horsburgh, C. A., Hoyer, M. V., Bachmann, R. W., & Canfield, D. E. (2015). Regional distribution of Secchi disk transparency in waters of the United States. *Lake and Reservoir Management, 31*(1), 55–63. https://doi.org/10.1080/10402381.2014.1001539

DFRS. (2015). State of Nepal's Forest Resource Assessment (FRA), Nepal, Department of Forest Research and Survey (DFRS), Kathmandu, Nepal.

Farrukh, M. U., Bashir, M. K., & Rola-Rubzen, F. (2020). Exploring the sustainable food security approach in relation to agricultural and multi-sectoral interventions: A review of cross-disciplinary perspectives. *Geoforum, 108*, 23–27. https://doi.org/10.1016/j.geoforum.2019.11.012

Lu, S., Jia, L., Jiang, Y., Wang, Z., Duan, H., Shen, M., Tian, Y., & Lu, J. (2021). Progress and prospect on monitoring and evaluation of United Nations SDG 6 (clean water and sanitation) target. *Bulletin of Chinese Academy of Sciences (Chinese Version), 36*(8).

MoFE. (2019). National land forests and land cover analysis of Nepal using google earth images. Ministry of Forest and Environment, Forest Resource and Training.

Paudel, B., Zhang, Y., Wu, X., Li, S., & Khanal, N. (2016). Review of studies on land use and land cover changes in Nepal. *Journal of Mountain Science, 13*(4), 643–660.

Sapkota, B. D. (2022). State of food security and hunger in Nepal. *European Journal of Nutrition & Food Safety, 14*, 14–23. https://doi.org/10.9734/ejnfs/2022/v14i121278

Shehzad, K. (2023). Extreme flood in Pakistan: Is Pakistan paying the cost of climate change? A short communication. *Science of the Total Environment, 880*, 162973. https://doi.org/10.1016/j.scitotenv.2023.162973

Shen, M., Duan, H., Cao, Z., Xue, K., Qi, T., Ma, J., Liu, D., Song, K., Huang, C., & Song, X. (2020). Sentinel-3 OLCI observations of water clarity in large lakes in eastern China: Implications for SDG 6.3.2 evaluation. *Remote Sensing of Environment, 247*, 111950. https://doi.org/10.1016/j.rse.2020.111950

Steele, J. E., Nieves, J., Tatem, A. J., Forget, Y., Shimoni, M., & Linard, C. (2018). Worldpop—Fusion of earth and big data for intraurban population mapping. In *IGARSS 2018—2018 IEEE International Geoscience and Remote Sensing Symposium, 2070–2071*. https://doi.org/10.1109/IGARSS.2018.8518181

Uddin, K., Shrestha, H. L., Murthy, M. S. R., Bajracharya, B., Shrestha, B., & Gillani, H. (2015). Development of 2010 national land cover database for the Nepal. *Journal of Environmental Management, 148*, 82–90.

UN-Water. (2017). Integrated monitoring guide for SDG 6—Targets and global indicators.

Chapter 7
Key Findings, Limitations and Challenges, and Way Forward

7.1 Key Findings

The Hindu Kush Himalayan (HKH) region, a critical ecological zone, is increasingly impacted by climate change and anthropogenic alterations in land cover. These changes significantly disrupt the structure and functioning of terrestrial ecosystems, which are vital for the health of soil, water, and atmospheric systems, and consequently, for regional sustainable development. The complexity of these mountainous regions, exacerbated by challenges like mountain shadows, data insufficiencies, and technical limitations, has traditionally hindered research efforts aimed at promoting sustainability. This study addresses these challenges by integrating advanced methodologies and innovative tools, with a focus on key Sustainable Development Goals (SDGs) relevant to the HKH region, including SDG 2 (Zero Hunger), SDG 6 (Clean Water and Sanitation), and SDG 15 (Life on Land).

To advance SDG 2 in the HKH, this study leverages two decades of time-series remote sensing data (2000–2020), combined with population distribution information, FAO food datasets, and machine learning algorithms such as Random Forest and Gated Recurrent Unit (GRU) models (Table 7.1). Implemented on the Google Earth Engine (GEE) cloud computing platform, these tools provide robust estimates of the impacts of land use changes on food production. This approach yields critical insights into the resilience and sustainability of agricultural systems in the region, directly supporting SDG 2.4.1, which aims for sustainable food production systems.

Building on this, the study also addresses SDG 6 by employing time-series SDGSAT-1 nighttime thermal infrared and Sentinel-2 multispectral satellite data to develop a high-precision water monitoring method tailored for the challenging terrain of mountainous areas. This method not only overcomes the technical difficulties posed by mountain shadows but also contributes significantly to the creation of a 10-m resolution long-term water dynamic distribution product (DWH10) for the HKH region, covering the years 2000 to 2020. The integration of the DWH10 product with Joint Research Centre Global Surface Water (JRC-GSW) time-series

S. Lu et al., *Remote Sensing Assessment of Sustainable Development in the Hindu Kush-Himalayan (HKH) Region*, https://doi.org/10.1007/978-981-96-8675-9_7

Table 7.1 Major research achievements and the main contributions

Types	Achievements	Contributions
Model and method development	(1) Machine learning algorithms, including Random Forest, Leaf Area Index (LAI) inversion techniques, and Gated Recurrent Unit (GRU) models	(1) The integration of Machine learning, LAI and GRU model enables the prediction of wheat and rice crop yield, as well as the changes in cultivated land area in the HKH, supporting agricultural planning and resource management
	(2) Recursive Feature Elimination (RFE) model	(2) RFE effectively identifies key attributes for accurate farmland extraction
	(3) Local adaptive XGBoost water extraction model	(3) The XGBoost water extraction model facilitated the accurate identification of temporal and stable water bodies
	(4) Water transparency inversion models utilizing the Forel-Ule index (FUI) and watercolor characteristics	(4) Enabled the inversion and mapping of water transparency products in the HKH
	(5) Advanced machine learning techniques, such as the Random Forest method and a vegetation coverage transfer learning inversion model that combines convolutional neural networks (CNN) with long short-term memory (LSTM) networks	(5) A vegetation coverage transfer learning inversion model enhanced accuracy and reduced uncertainty in vegetation coverage extraction across the HKH's complex mountainous terrain ① The Random Forest method was employed to precisely extract forest and grassland areas. ② Integrating CNN with LSTM networks allows for comprehensive vegetation coverage analysis in mountainous regions by utilizing both temporal and spatial data

(continued)

Table 7.1 (continued)

Types	Achievements	Contributions
Datasets (remote sensing, in-situ and other global)	(1) Two decades of 30m resolution time-series remote sensing data (2000–2020), including SDGSAT-1 nighttime infrared, population distribution data, and FAO food datasets	(1) The remote sensing datasets were crucial for analyzing and mapping the distribution of cultivated land, water bodies, forests, and grasslands across various regions. SDGSAT-1 nighttime TIR bands were used to generate images of maximum water body coverage, effectively reducing the impact of mountain shadows
	(2) 10-m resolution long-term water dynamic distribution product (DWH10) in 2022	(2) The DWH10 product aided in extracting both stable and temporary water body areas in the HKH region
	(3) Fine-resolution, long time-series water transparency products for the HKH region (2000, 2010, 2020)	(3) The water transparency product provides a high-resolution depiction of water quality across the HKH countries, illustrating the spatial transparency of water bodies throughout the region
	(4) Cultivated land area data for major grain-producing regions from 2000 to 2020	(4) Cultivated land area data provides valuable insights into agricultural trends and land use patterns
	(5) Crop yield distribution data for rice and wheat	(5) Crop yield distribution data highlights yield trends for key crops production (rice and Wheat) in the Himalayan region from 2000 to 2020
	(6) Forest and grassland distribution data across various altitudes (7) Spatial distribution data of Fractional Vegetation Coverage (FVC) for forest and grass	(6) Forest and grassland distribution data offers crucial insights into the changing trends of forests and grasslands in the HKH, supporting natural resource management and sustainable land use (7) The FVC underscores the significant spatial variations in forest and grassland vegetation coverage across different HKH regions

(continued)

Table 7.1 (continued)

Types	Achievements	Contributions
Sustainable development remote sensing assessment cloud service platform	(1) Data sharing module	(1) The data-sharing module enables online access, querying, visualization, and downloading of key element inversion datasets for the HKH region, derived from satellite remote sensing
	(2) Online analysis module	(2) The online analysis module performs thematic analysis on these datasets, offering insights into various applications within the HKH region. It uses statistical charts and maps to visually provide representations, enhancing comprehension and supporting decision-making processes
	(3) Sustainable development assessment module	(3) The sustainable development assessment module computes evaluation indicators based on the inversion datasets of key elements of the HKH region, focusing on three main themes: agricultural conditions, water resources, and ecological environment
	(4) Online AI model calculation module	(4) The online AI model calculation module is dedicated to performing online calculations using deep learning models for remote sensing inversion

water distribution data and Digital Elevation Models (DEM) allows for effective monitoring of long-term changes in water distribution, a crucial factor in achieving SDG 6.6.1, which focuses on the "Change in the extent of water-related ecosystems over time." Additionally, by leveraging inversion models based on the Forel-Ule index (FUI) and synthetic weather-consistent cloud-free Landsat images, the study develops the first fine-resolution (30 m) time-series water transparency products for the HKH region, offering valuable insights into the change processes affecting regional and national sustainable indicators, particularly SDG 6.3.2, which concerns the "Proportion of bodies of water with good ambient water quality."

Further extending its focus to SDG 15 (Life on Land), the study employs a comprehensive approach to analyze vegetation coverage in mountainous regions. By integrating advanced machine learning techniques, such as the Random Forest method and a deep spatiotemporal transfer learning inversion model—combining convolutional neural networks (CNN) and long short-term memory (LSTM) networks—the study conducts a detailed examination of forest and grassland cover at a 30-m resolution using Landsat images from 2000, 2010, and 2020 (Table 7.1). The GEE platform plays a crucial role in facilitating this analysis, offering a clear understanding of the changes in forest and grassland coverage over the past 20 years. These findings provide essential data for sustainable development planning and assessment in the HKH region, underscoring the importance of preserving and enhancing vegetation cover to support SDG 15.4.

The progress of the HKH region toward achieving the SDGs reveals a multifaceted landscape of both accomplishments and ongoing challenges. Over a two-decade analysis (2000–2020), significant contrasts emerged in the region's vegetation coverage. Forests in the HKH countries experienced a deforestation rate of 2.98%, largely driven by unsustainable land use practices and heavy dependency on forest resource. In contrast, grassland areas grew significantly by 9.06%, highlighting a divergence in the region's ecological development.

However, when the FVC which measures vegetation relative to the total land area was considered, a more nuanced picture emerged. Most HKH countries saw overall improvements in vegetation coverage, although Myanmar slightly lagged due to a modest 1.20% reduction in its overall vegetation coverage. This slight setback underscores the uneven progress across the region in maintaining ecological balance.

The FVC analysis from 2000 to 2020 indicates that the HKH region made some strides toward achieving SDG 15.4.2 (Mountain Green Cover Index), which aims to protect and restore sustainable ecosystems in mountain areas. While the region initially fell short of meeting this target between 2000 and 2010, improvements were seen in the subsequent decade, with most countries showing positive trends except for Myanmar, which faced persistent challenges related to vegetation loss.

These findings reflect the complex nature of environmental conservation efforts in the HKH, where successes in grassland restoration are offset by ongoing issues of deforestation, land degradation, insufficient reforestation, and the impacts of a warming climate. The outcomes highlight both the achievements and the obstacles the region faces in meeting SDG 15, pointing to the need for more focused interventions in forest management and climate resilience strategies.

The warming climate also impacted the HKH hydrological system, leading to a steady decrease in permafrost. This melting permafrost contributed to an overall improvement in water resources, particularly in achieving SDG 6.6.1 and SDG 6.3.2. The HKH saw an overall improvement in the SDG 6.6.1 with an 18.7% increase in water body area. The most significant contributions came from Pakistan, China, and Afghanistan, which had change rates of 24.05, 19.70, and 15.02%, respectively. Conversely, Bangladesh and Nepal showed limited progress, with reductions in water body size of 7.70 and 0.99%, respectively. Although this study shows improvement in the HKH water body area prompted by the melting permafrost in the region, on the long run, this wouldn't be the case as the persistent deterioration in the permafrost coverage could lead to a devastating future of the HKH hydrological system.

Regarding SDG 6.3.2, the HKH region demonstrated significant improvement in water quality, with all countries making progress except Bhutan, which remained stable. For instance, the proportion of water bodies with good water quality in China rose from 48.13% in 2000 to 92.01% in 2020. Similar improvements were observed in Pakistan, Nepal, Bangladesh, Afghanistan, India, and Myanmar (Fig. 7.1). However, while the current data indicates positive trends in water resources, continued permafrost melting due to climate change could undermine these gains in the long term, potentially disrupting the region's hydrological systems.

The increased water body size during the study period also correlated with improved agricultural productivity in the HKH region, despite fluctuations in labor output. Overall grain productivity grew by 10%, supported by an 11% expansion in cultivated land, particularly in Myanmar, Nepal, and China. Although Pakistan and Afghanistan initially saw losses in cultivated land, both countries stabilized by 2020. The study revealed a complex dynamic of labor productivity for rice and wheat, influenced by factors such as crop type, local agricultural practices, and external economic conditions. Despite these fluctuations, rice and wheat yields increased significantly across the region.

This growth in grain yield aligns with the SDG 2.4.1, reflecting a sustainable trend characterized by a decrease in land-use intensity. Notable reductions in wheat land-use intensity were observed in Afghanistan (63%) and Qinghai, China (34%), while rice land-use intensity declined in southern China (11%), Nepal (24%), and Pakistan (6%). Other HKH regions showed a slow but steady downward trend in land-use intensity, indicating improved land-use efficiency alongside rising yields. This study underscores the complexity of sustainable development in the HKH region, highlighting both the progress made and the ongoing challenges that must be addressed to achieve long-term sustainability.

Based on the findings of this study, several key decision-making support suggestions can be proposed to enhance the sustainable development efforts in the HKH region.

(1) **Strengthen Vegetation Restoration Initiatives**

The analysis highlights a concerning 35.25% decline in forest coverage across the HKH region from 2000 to 2020. While Afghanistan and Pakistan exhibited positive trends in forest coverage during the first decade, both countries faced significant declines in the subsequent decade. This trend underscores the urgent need for targeted reforestation and forest conservation programs to halt and reverse the ongoing loss of forested areas.

The HKH region experienced a 166.87% increase in grassland area between 2000 and 2020, largely due to growth in the first decade. Treeline and vegetation recruitments have also found increased in the high altitudinal region of the Mountain region. This indicates the potential success of conservation efforts and the importance of implementing sustainable management practices to preserve and enhance grassland ecosystems, especially in regions where challenges persist.

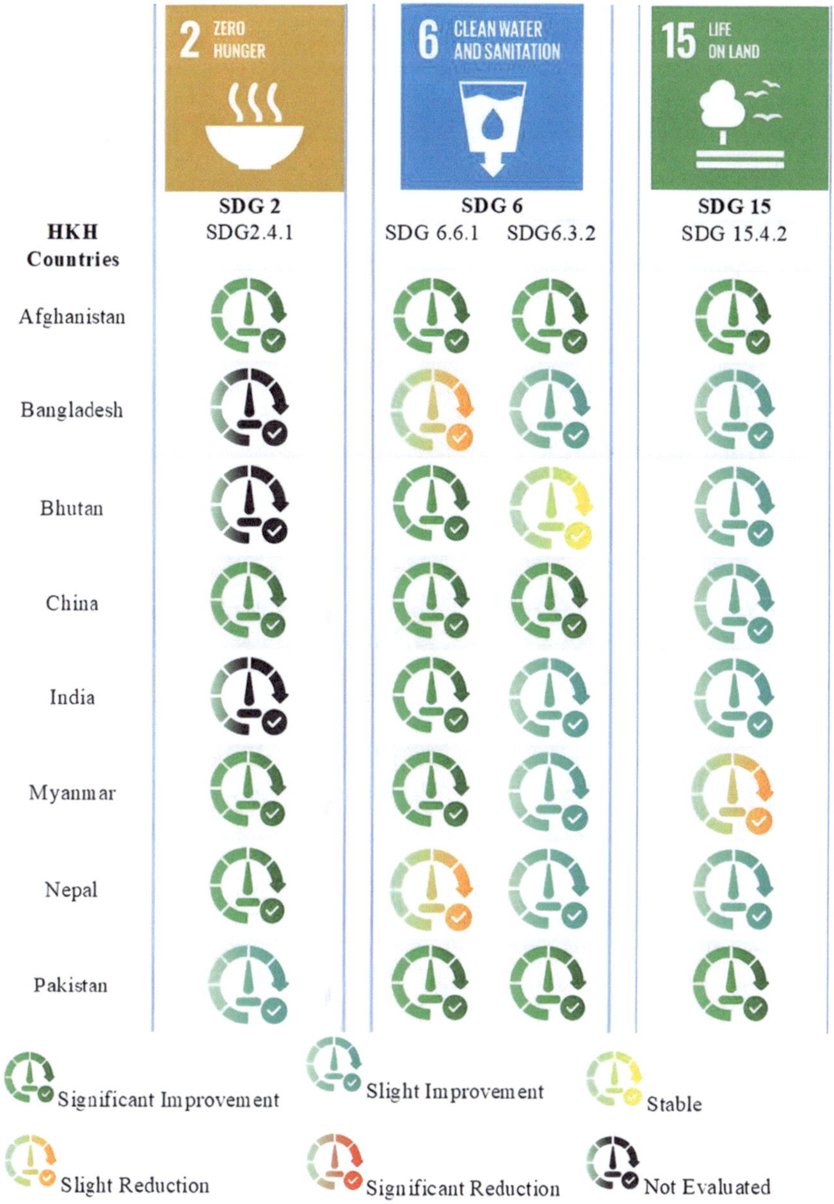

Fig. 7.1 Current SDG status of the HKH regional countries

(2) Strengthen Water Resource Management

The observed increase in water body areas in several HKH countries indicates potential for further improving water-related ecosystems. However, continuous monitoring is essential to mitigate the long-term impacts of permafrost melting, and snow and ice cover changes, which could destabilize water resources. And a future scenario analysis on the permafrost changes and the potential impact on the HKH hydrological system could be conducted.

Advanced remote sensing technologies and the development of high-resolution water monitoring tools, like the 10-m resolution long-term water dynamic distribution product (DWH10), should be expanded to improve the precision and accuracy of water resource assessments in mountainous regions.

Countries showing limited progress in water quality (SDG 6.3.2), such as Bhutan possibly prompted by eutrophication, unimproved sanitation facilities in some of its urban and rural areas and so on, should adopt more aggressive water quality management policies and strengthen their water monitoring systems.

(3) Promote Sustainable Agricultural Practices

The study's findings highlight the importance of sustainable land-use practices that balance agricultural productivity with environmental conservation. Policies should be developed to encourage the adoption of precision agriculture and sustainable farming techniques, particularly in regions where land-use intensity is still high.

Governments should incentivize the transition to lower land-use intensity farming methods while ensuring that agricultural productivity continues to grow, as demonstrated by the increased grain yields in Myanmar, Nepal, and China.

(4) Facilitate Regional Collaboration and Data Sharing

The complexities of sustainable development in the HKH region necessitate collaborative efforts across borders. Regional platforms for data sharing and joint research initiatives should be strengthened to allow for more comprehensive and coordinated responses to environmental challenges.

The use of shared platforms like the Sustainable Development Remote Sensing Assessment Cloud Service Platform developed in this study for data analysis and modeling can enhance the capacity of HKH countries to monitor progress towards the SDGs, fostering a collective approach to sustainability.

(5) Implement Adaptive Strategies for Climate Resilience

Given the ongoing impacts of climate change, adaptive strategies should be developed to bolster the resilience of the HKH region's ecosystems. This includes measures to mitigate the effects of permafrost melting and to manage the changing water and vegetation dynamics.

Policymakers should consider integrating climate adaptation into national and regional planning, ensuring that development strategies are flexible and responsive to the evolving environmental conditions in the HKH.

These above-mentioned suggestions emphasize the importance of targeted, evidence-based decision-making to address the diverse challenges facing the HKH region. By adopting these strategies, stakeholders can make significant strides toward achieving the SDGs and ensuring long-term sustainability in this critical ecological zone.

7.2 Limitations and Challenges

This detailed examination of vegetation coverage, water resources, and agricultural production in the HKH region provides valuable insights but comes with certain limitations. Despite methodological improvements enhancing data accuracy and reliability, several persistent challenges could impact the effectiveness of monitoring and management efforts. The following sections outline specific drawbacks related to vegetation sample library limitations, model optimization needs, water body distribution monitoring, water quality assessment, and data collection frequency. Addressing these issues in future research will be essential for refining sustainable development strategies in the region.

Data gaps: Remote sensing and field observations posed significant challenges to the study. Optical remote sensing data suffered from cloud cover interference, leading to gaps in areas with persistent cloudy and rainy conditions, while synthetic aperture radar (SAR) data encountered speckle noise and geometric distortions due to terrain undulations. Additionally, the interannual observation scale of Landsat data, coupled with cloud interference, restricted the feasibility of high-resolution, frequent monitoring. Field data limitations further affected the accuracy of the yield estimation model, as the scarcity of field surveys reduced its precision and reliability. Moreover, the vegetation sample library was constrained to forest and grassland types, limiting classification accuracy, particularly during seasons with sparse vegetation cover.

Model optimization needs: This was evident in several aspects of the study. The spatiotemporal transfer learning model required continuous refinement in terms of structure, parameter settings, and training sample quality to improve its adaptability across different environmental conditions. Remote sensing classification of forests and grasslands in the HKH region faced technical difficulties due to low data accuracy, incomplete vegetation representation, and inconsistencies arising from complex terrain and variable climatic conditions. Additionally, the use of uniform land classification features across the diverse HKH region failed to fully account for regional variations in landscape patterns, topography, and climate, leading to potential misclassification.

Water resource assessment limitations: While the methodology integrating SDGSAT-1 thermal infrared data with Sentinel-2 multispectral data provided insights into water body distribution, it only classified water bodies as either seasonal or stable without determining specific coverage frequency, limiting its applicability for dynamic water monitoring. Similarly, water quality assessments primarily relied on transparency inversion, offering only a partial perspective on water conditions. To achieve a more comprehensive understanding, a broader range of water quality parameters—such as turbidity, nutrient levels, and pollutant concentrations—needs to be incorporated into future assessments. Strengthening water resource monitoring through multi-source data integration and advanced modeling techniques will significantly improve the accuracy of hydrological assessments in the region.

Socio-economic integration: This remains a critical aspect that requires further attention. The population density data, sourced from the WorldPop website at a 1 km^2 resolution, did not accurately capture real population distributions, particularly in densely populated urban and agricultural areas. The Kriging interpolation method used to refine these estimates failed to correct these discrepancies, impacting the accuracy of unit labor output assessments and related socio-economic analyses. Furthermore, the study did not fully incorporate socio-economic factors and climate change influences on agricultural practices, both of which are essential for a holistic understanding of regional agricultural dynamics.

7.3 Way Forward

Building on the key findings of this study, future research in the HKH region should focus on enhancing methodologies, addressing identified limitations, and further strengthening the capacity for sustainable development monitoring and decision-making. The challenges highlighted in this study—including data gaps, model optimization needs, water resource assessment limitations, and socio-economic integration—necessitate a comprehensive and coordinated response. To this end, advancing remote sensing techniques, integrating additional data sources, and expanding analytical frameworks will be essential.

(1) Enhancing Vegetation and Agricultural Monitoring for SDG 2 and SDG 15

To improve the assessment of vegetation dynamics and agricultural productivity, future research should expand the scope of existing models by incorporating a broader range of vegetation types and integrating key climatic and soil variables. This will enhance the predictive stability of models across various temporal scales, improving their utility for ecological monitoring and sustainable land management. Additionally, refining neural network structures and incorporating diverse, high-quality training datasets will enhance the generalization capabilities of vegetation monitoring models.

Addressing field data limitations will be crucial for improving crop yield estimation models. Expanding field surveys and utilizing crowdsourced or ground-based observation networks can provide valuable validation data, enhancing the accuracy of agricultural assessments. Similarly, the development of more regionally adaptive land classification methods will help mitigate uniform identification challenges in cultivated land mapping, ensuring better alignment with local environmental conditions.

To further support SDG 15, improving forest and grassland monitoring methodologies will be necessary. While this study leveraged machine learning models to extract vegetation coverage data, challenges such as low classification accuracy and spatiotemporal inconsistencies due to complex terrain remain. Future work should explore the integration of multi-source remote sensing data—including China's high-resolution satellite data—to enhance classification precision. Moreover, harmonizing classification frameworks across different vegetation types will improve cross-temporal and cross-regional comparability, allowing for more reliable trend analyses.

(2) **Strengthening Water Resource Monitoring for SDG 6**

Effective water resource management requires more frequent and accurate monitoring of water body coverage and quality. Future research should move beyond percentile synthesis features to analyze the full time-series variation characteristics of spectral data, capturing seasonal and interannual fluctuations more comprehensively. Integrating high revisit-frequency satellite datasets with high-resolution sensors will improve spatial and temporal coverage, enabling better detection of hydrological changes.

The classification of water body types remains a limitation in this study, as the current methodology only distinguishes between seasonal and stable water bodies. Expanding classification frameworks to include detailed coverage frequency and hydrological cycle characteristics will provide a more nuanced understanding of water dynamics.

In terms of water quality assessment, the reliance on transparency inversion provides only a partial perspective. Future efforts should incorporate additional parameters—such as nutrient levels, suspended sediments, and organic pollutants—by integrating satellite-derived indices with multi-source ground-based water quality data. The development of deep learning-powered water quality inversion models, using satellite-ground collaborative observations, will enhance predictive accuracy and expand the range of measurable indicators.

Additionally, the impact of climate change on water resources, particularly in relation to permafrost melting and glacier retreat, requires further investigation. Future research should conduct scenario-based analyses to predict long-term hydrological trends, assessing their implications for water security and ecosystem stability in the HKH region.

(3) **Advancing Integrated Monitoring Through the Sustainable Development Remote Sensing Assessment Cloud Service Platform**

A key component of future research will be the continued development of the Sustainable Development Remote Sensing Assessment Cloud Service Platform, which serves as a foundation for data sharing, analysis, and decision support. While the platform currently focuses on SDG 2, SDG 6, and SDG 15, expanding its scope to include additional SDGs will enhance its role as a comprehensive sustainable development monitoring system.

(1) Strengthening Data Sharing and Regional Interoperability

Although the platform already provides online data access and analysis tools, it currently operates in isolation from other regional data infrastructures. Future improvements should focus on integrating the platform with other HKH regional monitoring systems, allowing for seamless data exchange and collaborative research. Establishing standardized data-sharing protocols will ensure consistency and facilitate large-scale comparative studies.

(2) Expanding Decision-Support Capabilities

Currently, the platform primarily functions as a data repository and analysis tool. To maximize its impact, future enhancements should introduce decision-support functionalities, enabling policymakers to derive actionable insights from remote sensing data. This could include AI-driven predictive modeling, scenario simulations, and customized policy recommendations, supporting evidence-based decision-making in environmental management, agriculture, and water governance.

(3) Integrating Additional Remote Sensing Data and Models

Incorporating more diverse remote sensing datasets, particularly from China's high-resolution satellites, will significantly improve the platform's analytical capacity. The inclusion of Sentinel, Landsat, Gaofen, and SDGSAT-1 data will provide enhanced spatial and temporal coverage, facilitating more detailed and accurate monitoring. Additionally, expanding the range of machine learning models available within the platform—such as advanced transfer learning techniques—will refine data interpretation and enhance predictive capabilities.

(4) Developing High-Value Public Data Products and Universal Technical Methods

Expanding the availability of publicly accessible, high-value datasets will be critical for supporting research and policy applications. This includes standardized long-term datasets on land cover, water quality, and agricultural productivity, ensuring continuity and comparability across studies. Additionally, developing universal remote sensing methodologies for complex mountainous environments will help harmonize monitoring efforts across different institutions and regions.

(4) **Long-Term Vision and Strategic Goals**

The ultimate objective of this research and platform development is to establish a regional public platform for SDG monitoring, evaluation, and decision support in the HKH region. By continuously integrating new data sources, enhancing analytical capabilities, and improving interoperability with global monitoring frameworks, this initiative aims to provide science-based support for sustainable development efforts. Strengthening international collaboration, particularly in areas such as climate resilience, natural resource management, and agricultural sustainability, will be crucial for ensuring long-term impact.

By addressing the challenges outlined in this study and leveraging advancements in remote sensing, big data analytics, and decision-support technologies, future research can significantly improve environmental monitoring, resource management, and sustainable development planning in the HKH region, ultimately contributing to the successful implementation of the United Nations Sustainable Development Goals.

Glossary

- **Fractional Vegetation Cover (FVC)**

A metric that quantifies the extent of vegetation over a given area. It provides insights into vegetation health and productivity, helping to monitor changes in plant life due to climate change, land use, and environmental factors.

- **Convolutional Neural Network (CNN)**

A deep learning architecture designed to process and analyze visual data, especially remote sensing images. CNNs utilize multiple convolution layers to detect spatial patterns and structures, such as vegetation cover, in satellite imagery.

- **Deep Spatiotemporal Transfer Learning Inversion Model**

A machine learning approach that combines spatial and temporal data to enhance vegetation cover extraction. By integrating seasonal and spatial information, it identifies complex patterns across time and space, particularly useful for regions like the HKH.

- **Global Human Settlement Layer (GHSL)**

A dataset produced by the European Commission that maps global human settlements. It is instrumental in studies that seek to differentiate natural land cover and water bodies from anthropogenic influences by excluding urban areas.

- **Google Earth Engine (GEE)**

A cloud-based platform for environmental analysis, providing access to a large collection of satellite imagery and geospatial datasets. It is widely used for monitoring land cover and environmental changes.

- **Long Short-Term Memory (LSTM)**

A type of recurrent neural network (RNN) designed to capture long-term dependencies in time-series data. LSTMs are particularly useful in predicting vegetation cover trends, accounting for seasonal and long-term variations.

S. Lu et al., *Remote Sensing Assessment of Sustainable Development in the Hindu Kush-Himalayan (HKH) Region*, https://doi.org/10.1007/978-981-96-8675-9

- **PROSAIL (PROspect + SAIL) Model**

A radiative transfer model combining PROspect, which models leaf optical properties, and SAIL, which models canopy reflectance. It is used in remote sensing to estimate vegetation properties, especially in forests and grasslands.

- **Regional Land Cover Monitoring System (RLCMS)**

A system designed for monitoring and mapping annual land cover changes, particularly in regions like the HKH. RLCMS utilizes satellite imagery to track vegetation dynamics and land cover alterations, offering insights into environmental and human impacts.

- **10-m Dynamic Water Body Map of the HKH (DWH10)**

DWH10 is a high-resolution map focused on monitoring the distribution and changes of water bodies within the Hindu Kush Himalayan (HKH) region. The map uses satellite remote sensing data, with a spatial resolution of 10 m, to capture dynamic water bodies such as rivers, lakes, and reservoirs, with a focus on detecting changes in water bodies over time.

- **Vegetation Radiative Transfer Model**

A model that simulates light interactions with vegetation surfaces, helping to estimate vegetation properties like leaf area index (LAI) and canopy reflectance from remote sensing data.

- **Water Transparency Inversion Model**

A model used to estimate water transparency from remote sensing data. It often employs indices like the Forel-Ule Index (FUI) to link water color characteristics to transparency, aiding in water quality assessments.

- **Recursive Feature Elimination (RFE)**

A technique in machine learning for selecting the most important features by iteratively removing the least significant ones and evaluating model performance to refine the dataset.

- **Random Forest Algorithm**

A machine learning algorithm that constructs an ensemble of decision trees to perform classification or regression. It is particularly effective for large datasets and provides robust predictions through aggregation of individual tree results.

- **Grey Level Co-occurrence Matrix (GLCM)**

A statistical method used to analyze image texture by examining the spatial relationships between pixel values. It is frequently applied in remote sensing for classifying land cover types based on texture patterns.

- **SeLI (Surface Energy Balance Algorithm for Land)**

An algorithm used to estimate the Leaf Area Index (LAI) from remote sensing data by integrating ground-based measurements to improve the accuracy of LAI estimates, especially in diverse landscapes.

- **LAI Inversion**

The process of estimating the Leaf Area Index (LAI) from remote sensing data using mathematical models or empirical formulas. LAI inversion helps in assessing vegetation cover and its health.

- **Empirical Formula**

A mathematical expression based on real-world observations that defines the relationship between variables. In remote sensing, empirical formulas are often used to model vegetation parameters such as LAI.

- **XGBoost (Extreme Gradient Boosting)**

A machine learning algorithm that utilizes decision trees to perform classification or regression. XGBoost is highly efficient, accurate, and fast, often employed in remote sensing for tasks like land classification and water body detection.

- **SDGSAT-1**

A satellite that collects thermal infrared (TIR) data, which is especially useful for detecting water bodies at night. Its data is valuable for water body detection in regions with complex terrain, such as the HKH, where mountain shadows may interfere with visibility.

- **Land Surface Water Index (LSWI)**

An index used to differentiate water bodies from other land surfaces by analyzing satellite data in the shortwave infrared (SWIR) and near-infrared (NIR) bands. It is effective in detecting water bodies in areas with intricate terrain.

- **Self-Normalized Water Index (SNWI)**

An index that compares normalized values from specific spectral bands (such as TIR and MNDWI) to differentiate stable water bodies from temporal ones, enhancing water detection in dynamic landscapes.

- **Percentile Synthesis**

A method for aggregating data by using specific percentiles (e.g., 15th and 85th percentiles) from a set of satellite images, helping to distinguish stable water bodies from seasonal or temporary ones.

- **GRU (Gated Recurrent Unit)**

A machine learning algorithm used in time-series prediction tasks. GRUs, a type of recurrent neural network (RNN), are particularly effective at capturing temporal dependencies in data and are used in tasks such as vegetation cover prediction.

- **WOFSOT (Water and Nitrogen Stress Model)**

A crop growth model that simulates the dynamics of crop growth under varying environmental conditions, with a focus on water and nitrogen stress.

- **Joint Research Centre Global Surface Water (JRC-GSW)**

A global dataset that provides information on the distribution of surface water bodies, offering valuable insights for water monitoring studies, though its accuracy may be affected by terrain shadows, especially in mountainous areas like the HKH.

- **Forel-Ule Index (FUI)**

A scale used to classify the color of water bodies, which is linked to water transparency. It is employed in remote sensing to estimate water transparency based on the observed color characteristics of water.

- **CIE Tri-Stimulus Values (X, Y, Z)**

Standardized values that represent the color of light as perceived by the human eye. In remote sensing, these values are calculated from satellite reflectance data to estimate the transparency of water bodies.

- **Stable Water Bodies**

Water bodies that remain consistent in size and location over multiple years, contrasting with temporal water bodies that may fluctuate due to seasonal or episodic events.

- **Temporal Water Bodies**

Water bodies that appear seasonally or periodically, often linked to temporary water sources such as floodplains, seasonal lakes, or other short-term water features.

MIX
Papier aus verantwortungsvollen Quellen
Paper from responsible sources
FSC® C105338

If you have any concerns about our products,
you can contact us on
ProductSafety@springernature.com

In case Publisher is established outside the EU,
the EU authorized representative is:
Springer Nature Customer Service Center GmbH
Europaplatz 3, 69115 Heidelberg, Germany

Printed by Libri Plureos GmbH
in Hamburg, Germany